Quantum Biocomputing in Quantum Biology
Volume I

Hafiz Md. Hasan Babu

Quantum Biocomputing in Quantum Biology Volume I

Arithmetic and Combinational Circuits

 Springer

Hafiz Md. Hasan Babu
Department of Computer Science
and Engineering
University of Dhaka
Dhaka, Bangladesh

ISBN 978-981-97-7153-0   ISBN 978-981-97-7154-7   (eBook)
https://doi.org/10.1007/978-981-97-7154-7

This Springer imprint is published by the registered company Springer Nature Singapore Pte Ltd.
The registered company address is: 152 Beach Road, #21-01/04 Gateway East, Singapore 189721,
Singapore

If disposing of this product, please recycle the paper.

To my respected great parents and also to my lovely wife, daughter, and son who made me possible to write this book

Preface

Quantum computing is a cutting-edge method of computing that is based on quantum mechanics and its incredible phenomena. It combines physics, mathematics, computer science, and information theory beautifully. It achieves tremendous processing power, low energy consumption, and exponential speed above traditional computers by regulating the behavior of minuscule physical things such as atoms, electrons, photons, and other microscopic particles. With the advent of nanotechnology, quantum computing vibrates an incredibly immense role in developing more compact and less power-consuming computers. The quantum computer is an entirely new notion from regular computing, and it does not employ binary logic. Quantum computers may evaluate several probabilities at the same time because of a phenomenon known as quantum entanglement.

Biocomputing or DNA computing or biological computing performs computations using biological molecules using cutting-edge technologies rather than traditional silicon chips. It is a modern area of science that recognizes biomolecules as fundamental elements of electronic devices and it is connected to chemistry, software engineering, cell genomics, physics, and mathematics. Biocomputing carries the promise of cheap, huge, accessible data storage and an exponential increase in computing power and speed. Biological computing, the amalgamation of biology and computer science, is reshaping the landscape of scientific inquiry. Imagine the intricacies of living organisms harnessed to process information, solve problems, and revolutionise technology. In this intricate dance between the natural world and computational algorithms, biological computing is unearthing new possibilities that extend far beyond the boundaries of conventional science.

Delving into the historical background, we uncover the roots of this interdisciplinary marvel, tracing its evolution from theoretical concepts to real-world applications. You can imagine a future where diseases are diagnosed swiftly, environmental issues are tackled with precision, and artificial intelligence learns and adapts like never before – all thanks to the power of biological computing.

Biological computing, or biocomputing or DNA computing, refers to a form of computing that uses biological organisms or processes as a primary component or mechanism for storage, processing, and transferring information. It encompasses a broad spectrum of technologies, including DNA computing, molecular computing, and cellular computing.

Yet, amidst the awe-inspiring potential lies a labyrinth of challenges and ethical dilemmas. As we journey deeper, we confront these complexities, exploring the delicate balance between innovation and responsibility. The ethical considerations surrounding genetic manipulation, data privacy, and environmental impact force us to question the boundaries of scientific exploration.

The combination of quantum computing and biocomputing (DNA computing or biological computing) is the first revolutionary computing process in the modern world. This combination is called "quantum biocomputing" where "computing in quantum biology" is described. Quantum computing and biocomputing have caught the attention of researchers working with computer science, physics, biochemistry, molecular biology, and quantum biology. The high-speed computation capability has already proved that quantum computing is the fastest system for solving the computational problems. The different approaches to bimolecular coding of DNA computing have made the possible revolutions in the modern world of computation. The imaginary impacts of these technologies are trying to be predicted by scientists; they are working on the limitations of quantum and DNA computing to get the best from it. Though these computing techniques are completely different in nature and there are no similarities between them, the new concepts can be evolved from these two. An innovative idea of combining these two computing technologies to get the maximum benefit of both at a time to utilize all advantages collectively to perform calculations in the best way ever. This is a new platform of the computing system, which consists of the combination of quantum computing and Biocomputing—it can be called the cross-platform system of quantum and DNA computing which can be done in two ways namely "quantum-DNA computing or quantum biocomputing or quantum biological computing" and "DNA-quantum computing or bioquantum computing or biological quantum computing".

The book *Quantum Biocomputing in Quantum Biology Volume I* starts with the basics of quantum computing, biocomputing, quantum biology, quantum-DNA computing, and DNA-quantum computing. It also discusses the fundamental operations in quantum computing and biocomputing. Different types of quantum arithmetic circuits, quantum-DNA arithmetic circuits, and DNA-quantum arithmetic circuits such as basic and universal gate operations, half adder, full adder, half subtractor, full subtractor, N-qubit adders, multipliers, dividers, etc. are explained clearly. Nuclear magnetic resonance (NMR), NMR relaxation, quantum cache memory, heat conduction circuit, and trap ion are also discussed. The readers will get a clear idea about different types of quantum, DNA, quantum-DNA, and DNA-quantum circuits (arithmetic, combinational, and sequential circuits) and will be able to design their own circuits. As a whole, this book is a great resource for quantum computing, DNA computing or biocomputing or biological computing, quantum-DNA computing or quantum biocomputing or quantum biological computing, and DNA-quantum computing or bio-quantum computing or biological quantum computing; it is the book where computing in quantum biology is introduced to quantum biology researchers, students, and academicians. It is a novel approach to writing a book in this field. This book will quench the thirst of beginners to advanced-level readers.

Dhaka, Bangladesh   Hafiz Md. Hasan Babu

Acknowledgements

I would like to express my sincerest gratitude and special appreciation to the various researchers in the field of quantum-DNA computing or quantum biocomputing or quantum biological computing. The contents of the quantum-DNA computing or quantum biocomputing or quantum biological computing and DNA-quantum computing or bio-quantum computing or biological quantum computing book have been compiled from various research works, in which the researchers are pioneers in their respective fields. All the research articles related to the contents are listed at the end of each chapter.

I am grateful to my great parents and dear family members for their endless supports. Most of all, I want to thank my lovely wife Mrs. Sitara Roshan, sweet daughter Ms. Fariha Tasnim, and sweet son Md. Tahsin Hasan for their invaluable cooperation to complete this book.

Finally, I am also thankful to all of those, specially to my beloved students Nitish Biswas, Md. Tareq Hasan, and Rownak Borhan Himel who have provided their immense support and important time to finish this book.

Introduction

Quantum computing combines quantum physics and conceptual computer science together that enables it more unique and quicker than existing silicon-based computing systems. Quantum computing solves many difficult traditional computer issues, making it an intriguing concept for computing approaches. Shor's factorization algorithm and Grover's search algorithm consider complex problems in traditional computing that quantum computing solves in polynomial time. In a real quantum computer, qubits can be represented by various physical systems, such as electrons with spin, photons with polarization, trapped ions, and semiconducting circuits. With the ability to perform complex operations exponentially faster, quantum computers have the potential to revolutionize many industries and solve problems that were previously thought impossible.

Besides, biological molecules are utilized to store information in biomolecular programming, and operations are carried out through various processes. Deoxyribose nucleic acid (DNA) is utilized to make molecular computers because of its predictable molecular activity. Biological and biochemical approaches are used in DNA computing or biocomputing. This computer technology has been used to solve several combinatorial search problems. DNA computing or biological computing or biocomputing (biological computing) from other traditional computing approaches in that it uses parallel processing and performs nano-level computations. All biological organisms have the ability to self-replicate and self-assemble into functional components. The economical benefit of biocomputers lies in this potential of all biologically derived systems to self-replicate and self-assemble given appropriate conditions. Additionally, researchers explore biomolecular logic circuits composed of engineered protein-protein and protein-DNA interactions designed to sense environmental signals and respond with computational outputs. The field of synthetic biology also aims to construct custom biological networks within living cells that can perform computation through precisely programmed cell-cell communication. Nature has optimized massively distributed and parallel information processing within multicellular organisms, providing inspiration for biocomputing's architectural models. The main advantage of this technology over other like technologies is the fact that through it, a doctor can focus on or find and treat only damaged or diseased cells. Selective cell treatment is made possible. The biological computer can also perform simple mathematical calculations.

After combining quantum computing and DNA computing or biological computing or biocomputing, quantum-DNA computing or quantum biological computing or quantum biocomputing and DNA-quantum computing or biological quantum computing or bio-quantum computing can be obtained that can perform the parallel operation very quickly which is the best feature that is different from traditional computing. Quantum computing is faster than supercomputers and the generated data of quantum operations are needed to be stored. Besides, DNA has the capability to store a large amount of information.

This book introduces an innovative combination of quantum physics and molecular biology named "quantum-DNA computing or quantum biological computing or quantum biocomputing" and "DNA-quantum computing or biological quantum computing or bio-quantum computing." Part I named "Quantum Computing and Biocomputing" has five chapters in which the first chapter contains the quantum computing definition, merits, demerits, challenges, and motivations. The second chapter explains the fundamental operations (AND, OR, NOT, NAND, XOR, XNOR operations) of quantum computing. The third chapter contains the definition of biocomputing, its merits, demerits, challenges, and motivations. The fourth chapter discusses the fundamentals of biocomputing. The fifth chapter describes the details of quantum biology, quantum-DNA computing, and DNA-quantum computing. In Part II, arithmetic circuits in quantum biocomputing such as half adder, full adder, half subtractor, full subtractor, N-qubit adders, multipliers, dividers, etc. in quantum and DNA computing are described. Cross-platforms (quantum biocomputing) such as quantum-DNA computing or quantum biocomputing or quantum biological computing and DNA-quantum computing or bio-quantum computing or biological quantum computing are also introduced in this part. Part III considers the quantum and DNA combinational circuits such as quantum, quantum-DNA, and DNA-quantum encoder, decoder, multiplexer, and demultiplexer. Part IV discusses the sequential circuits such as flip-flop, register, and counters.

Contents

About the Author

Dr. Hafiz Md. Hasan Babu is currently working as a Professor in the Department of Computer Science and Engineering, University of Dhaka, Bangladesh, as well as the Dean of the Faculty of Engineering and Technology of the University of Dhaka, Bangladesh. In addition, at present, he is a member (part-time) of Bangladesh Accreditation Council, Ministry of Education of the Government of the People's Republic of Bangladesh. Dr. Hasan Babu was the Chairman of the Department of Computer Science and Engineering of the University of Dhaka from 19-02-2003 to 18-02-2006 and pro-vice-chancellor of National University of Bangladesh from 12-07-2016 to 12-07-2020. He was also a Professor and the Founding Chairman of the Department of Robotics and Mechatronics Engineering, University of Dhaka, Bangladesh. He served as a World Bank Senior Consultant and General Manager of the Information Technology and Management Information System Departments of Janata Bank Limited, Bangladesh. Dr. Hasan Babu was the World Bank Resident Information Technology Expert of the Supreme Court Project Implementation Committee, Supreme Court of Bangladesh. He was also the Information Technology Consultant of Health Economics Unit and Ministry of Health and Family Welfare in the project "SSK (Shasthyo Shurokhsha Karmasuchi) and Social Health Protection Scheme" under the direct supervision and funding of German Financial Cooperation through KfW. Professor Dr. Hafiz Md. Hasan Babu received his M.Sc. degree in Computer Science and Engineering from the Brno University of Technology, Czech

Republic, in 1992 under the Czech Government Scholarship. He obtained the Japanese Government Scholarship to pursue his Ph.D. from the Kyushu Institute of Technology, Japan, in 2000. He also received DAAD (Deutscher Akademischer Austauschdienst) Fellowship from the Federal Republic of Germany.

Professor Dr. Hasan is a very eminent researcher. He was awarded the best paper awards in three reputed international conferences. In recognition of his valuable contributions in the field of Computer Science and Engineering, he received the Bangladesh Academy of Sciences Dr. M. O. Ghani Memorial Gold Medal Award for the year 2015, which is one of the most prestigious research awards in Bangladesh. He was also awarded the University Grants Commission of Bangladesh (UGC) Gold Medal Award-2017 for his outstanding research contributions in Computer Science and Engineering. He has written more than 100 research articles published in reputed international journals (*IET Computers and Digital Techniques, IET Circuits and Systems, IEEE Transactions on Instrumentation and Measurement, IEEE Transactions on VLSI Systems, IEEE Transactions on Computers, Elsevier Journal of Microelectronics, Elsevier Journal of Systems Architecture, Springer Journal of Quantum Information Processing*, etc.) and joined international conferences. According to Google Scholar Prof. Hasan has already received around 1949 citations with h-index 23 and i10-index 50.

He is a regular reviewer of reputed international journals and international conferences. He presented invited talks and chaired scientific sessions or worked as a member of the organizing committee or international advisory board in many international conferences held in different countries. For his excellent research record, he has also been appointed as the Associate Editor of *IET Computers and Digital Techniques*, published by the Institution of Engineering and Technology of the UK. Professor Dr. Hasan was appointed as a member of the prime minister's ICT Task Force Committee, Government of the People's Republic of Bangladesh in recognition of his National and International level contributions in Engineering Sciences. Dr. Hasan Babu was also the President of Bangladesh Computer Society for the session 2017–2020. At present, he is the President of International Internet Society, Bangladesh Chapter. In addition, he has published the following Textbooks by famous publishers of the UK, Singapore, and USA for the graduate and postgraduate students:

1. Hafiz Md. Hasan Babu,"Quantum Computing: A Pathway to Quantum Logic Design", IOP (Institute of Physics) Publishing, 2020, Bristol, UK.
2. Hafiz Md. Hasan Babu, "Reversible and DNA Computing", Wiley Publishers, 2021, UK.
3. Hafiz Md. Hasan Babu, "VLSI Circuits and Embedded Systems", CRC Press, July 2022, USA.
4. Md. Jahangir Alam, Guoqing Hu, Hafiz Md. Hasan Babu and Huazhong Xu, "Control Engineering Theory and Applications", CRC Press, September 2022, USA.
5. Hafiz Md. Hasan Babu, "Multiple-Valued Computing in Quantum Molecular Biology", Volume I, CRC Press, 2023, USA.
6. Hafiz Md. Hasan Babu, "Multiple-Valued Computing in Quantum Molecular Biology", Volume II, CRC Press, 2023, USA.
7. Hafiz Md. Hasan Babu, "DNA Logic Design: Computing with DNA", World Scientific Publishing Company, May 2024, Singapore.

Acronyms

BCD	Binary coded decimal
CLB	Configurable logic block
DNA	Deoxyribonucleic acid
EMR	Electron magnetic resonance
LUT	Look-up table
MUX	Multiplexer
NMR	Nuclear magnetic resonance
NTI	Negative ternary inverter
PCR	Polymerase chain reaction
PTI	Positive ternary inverter
QB	Quantum biology
RAM	Random access memory
RF	Radiofrequency
SIPO	Serial-in parallel-out
SISO	Serial-in serial-out
SPLD	Simple programmable logic devices
STI	Standard ternary inverter
XNOR	Exclusive NOR
XOR	Exclusive OR

List of Figures

List of Tables

Part I
Quantum Computing and Biocomputing

Overview

Quantum computing and biocomputing, both are emerging and exciting fields nowadays. Recently, these two fields are getting big attention from researchers because they can perform advanced level calculations in computer systems. Quantum computing is the combination of physics, mathematics, and computer science. It harnesses the collective properties of quantum states. The properties are superposition, interference, and entanglement. The machine that performs quantum computations is known as a quantum computer. These computer systems are built by using the principle of quantum mechanics. Quantum mechanics is a fundamental theory in physics applied to the elements in the quantum realm. The quantum realm is an area where atoms and particles are the smallest sizes. The size is so small around 100 nanometers or less. On the other hand, biocomputing (DNA computing) uses biological molecules to solve computational problems. Biocomputing, which combines biology and computing, has a significant impact on our daily lives, influencing fields like medicine, agriculture, and even materials science. It enables the development of new drugs, improves crop yields, and leads to innovations in areas like self-healing materials. It also drives innovations in bio-sensing devices that detect pollutants or pathogens in water and soil, ensuring real-time environmental monitoring. The knowledge of computer science and molecular biology is needed to work with DNA computing. Researchers have made a big step forward in efforts to store information in DNA as molecules, which are more secure and long-lasting than any other option. The magnetic hard drives that mass people use to store information can take up lots of space, but in a single gram of DNA, it is possible to store 215 petabytes (215 million gigabytes) of data. So, the impressive impact of quantum computing and DNA computing is amazing undoubtedly. What about combining these two challenging and exciting technologies? Combining these two computing processes, two new computing processes will be formed to get better performance compared with these two individuals and those are quantum-DNA computing or quantum biological computing (quantum biocomputing) and DNA-quantum computing or biological quantum computing (bioquantum computing), for example, quantum biocomputing, the fusion of quantum computing and bioinformatics, promises to revolutionize modern technologies by enabling faster and more accurate analysis of biological data, leading to advancements in drug discovery, precision medicine, and understanding complex biological systems. This field leverages quantum algorithms to tackle problems that are currently intractable for classical computers, offering potential breakthroughs in areas like protein folding prediction, genomic data analysis, and disease modeling. Quantum biocomputing, which harnesses biological and quantum systems together for computation, holds promising applications in various aspects of daily life, including healthcare, environmental monitoring, and even military applications. It could revolutionize personalized medicine, enable real-time environmental sensing, and optimize biomanufacturing processes, while Bioquantum computing is a fascinating field of research is emerging at the intersection of biology and quantum physics, known as bioquantum computing. This revolutionary approach harnesses the principles of quantum mechanics within biological systems, promising unprecedented computational capabilities. In this book, we will explore the concept of bioquantum computing, its potential applications, and the challenges it faces. Bioquantum computing also represents the convergence of quantum computing and biology, offering unprecedented computational power to tackle complex problems in the life sciences. Despite some obstacles, the progress made in bioquantum computing is remarkable, and the possibilities it presents are truly exciting. As researchers delve deeper into the potential of quantum systems within biological frameworks, we can expect groundbreaking advancements that will shape the future of computing, biology, and numerous other scientific domains. With continued research, innovation, and collaboration, we are poised to unlock the extraordinary computational capabilities offered by bioquantum computing, revolutionizing the way we solve complex problems and understand the world around us. The benefits will be more and different. This part will describe quantum computing, fundamentals of quantum computing, biocomputing, fundamentals of biocomputing and quantum biology (quantum-DNA computing and DNA-quantum computing). From now on, quantum biocomputing will be called as quantum-DNA computing or DNA-quantum computing.

Chapter 1
Quantum Computing

1.1 Introduction

The word quantum is a Latin word that means "how much" in modern understanding, the quantum means the smallest possible discrete unit of any physical property, such as energy or matter. The word "Quantum" was first introduced by Max Planck and Niels Bohr, who are the inventors of quantum theory. For this work, they received a Nobel prize in physics. After that, Einstein also described the quantum theory in his theory of photoelectric effect, where the light was considered quanta, and in 1921, he also won a Nobel Prize for his great work.

Quantum computing is the renewed invention of modern science and technology. It mainly focuses on the principle of quantum theory or quantum mechanics. It is the study of atomic structure and functions. Quantum computing was first introduced by Richard Philips Feynman and Yuri Ivanovich Manin (Fig. 1.1) in 1980. They said that the quantum computer can simulate things in such a way that is not possible by a traditional computer. After that, in 1981, Richard Feynman asked a question at the physics and computation conference at the Massachusetts Institute of Technology and the question was "Is it possible to combine physics with the computer?" This combination was quite challenging because Quantum mechanics is a branch of physics that focuses on the laws of nature of individual atoms and particles. So, if it is tried to simulate that the elementary problem may be taken place. That is why the question which was asked by Feynman was not supported by other scientists who were present at that conference. Later, he thought that if quantum mechanics could not be simulated on a computer, then perhaps a quantum mechanical computer might be built. His plan eventually came to fruition.

There are two main principles in quantum mechanics: Superposition and Entanglement. A quantum computer is built by using the knowledge of these principles. Quantum computing has three collective properties including these two. Another one is interference.

© The Author(s), under exclusive license to Springer Nature Singapore Pte Ltd. 2025   3
H. M. Hasan Babu, *Quantum Biocomputing in Quantum Biology Volume I*,
https://doi.org/10.1007/978-981-97-7154-7_1

Fig. 1.1 Richard Philips Feynman and Yuri Ivanovich Manin

Superposition : It combines all possible states of quantum particles where a particle can be in a single or mixed state. Unlike classical computers, quantum computers are made up of quantum bits, or "qubits". The basic unit of quantum computing is a qubit that encodes the data. A quantum particle can be |0> or |1> or the superposition of both where they can remain both 1 and 0 simultaneously.

A prevalent example is a spinning coin which can be the proper way to explain the superposition principles. A coin has two possible states—Head and Tail. If a coin is thrown over the air, the coin will start spinning. When it spins, it gets a situation where it is head and tail at the same time until it falls to the ground. This state is called the superposition state for the coin. Figure 1.2 shows the superposition of any particle.

Another example can be the electron, where an electron has two quantum states; one is spin up, and another is spin down. So when an electron goes to the superposition state, it can be either spin up or spin down state or a complex combination of both states which is the Superposition state.

Entanglement: At first, Albert Einstein, Boris Podolsky, and Nathan Rosen gave the idea of entanglement in 1935. They explored that quantum states would strongly interact with each other. Albert Einstein said that two objects of quantum mechanics could impact each other's behavior even over a very long distance. This behavior means entanglement where two or more particles can be linked. They can exchange the properties of the particles with the linked ones. That means that changing the state of an entangled qubit in quantum computing will immediately change the state of the paired qubit.

Therefore, this entanglement property improves the processing and computation speed of quantum computers. Quantum entanglement is shown in Fig. 1.3.

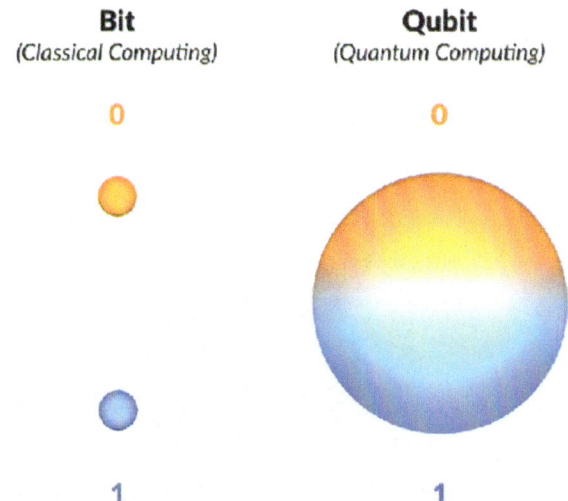

Fig. 1.2 Quantum superposition

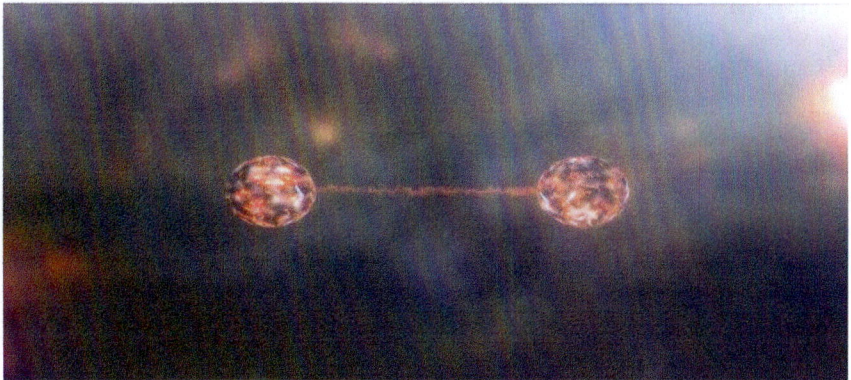

Fig. 1.3 Quantum entanglement

The superposition and entanglement properties of quantum computers have made this different from classical computers. Interference is another essential property of quantum computing.

Interference: This concept states that elementary particles cannot be in more than one place at a time (by superposition), but a single particle like a photon which is a particle of light, can traverse its path and interfere with the direction of its path. When the wave function of quantum particles overlap with each other, the wave functions of particles can either reinforce or diminish each other. This phenomenon is called interference. Most of the time it is not beneficial to the system. Quantum interference is depicted in Fig. 1.4. In other words, quantum interference, a

Fig. 1.4 Quantum
interference

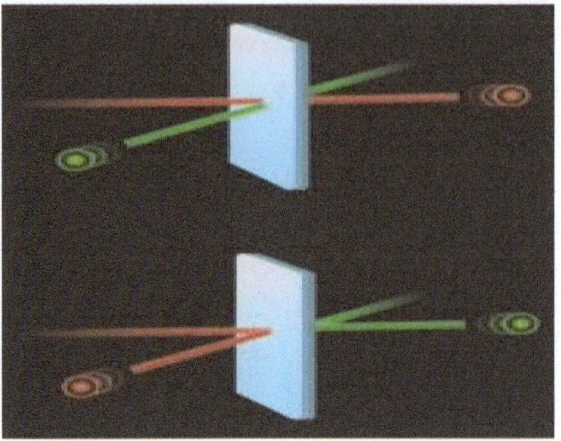

fundamental phenomenon in quantum mechanics, occurs when the probability amplitudes of quantum states interact, either constructively or destructively, affecting the likelihood of different outcomes. This phenomenon is crucial for quantum computing, allowing for the manipulation of probability amplitudes in ways that classical computers cannot. Quantum algorithms like Shor's algorithm and Grover's algorithm rely on quantum interference to achieve speedups in computation. Quantum interference plays a crucial role in quantum information technologies, such as quantum communication and quantum cryptography. Quantum interference helps us understand the fundamental nature of quantum systems and the behavior of matter at the atomic and subatomic level. The principles of quantum interference are being explored for various potential applications, including quantum sensors, quantum imaging, and quantum materials.

1.2   Merits of Quantum Computing

The invention of quantum computing has made daily life easier, simpler and compelled us to believe that every impossible thing can be possible with modern science. It has the following advantages:

High-speed computations: The main advantage of quantum computing is fast computing. According to Professor Catherine McGeoch at Amherst University, quantum computers are 1000 times faster than any classical computer. These computers perform some calculations within a few seconds, whereas a classical computer may need 1000 years to complete the same task. So, it has a high-performance rate of computations.

High storage capacity: Qubits remain in a superposition state in quantum computing. So, if qubits are added to the registers, they will store more data than classical computers.

Data simulations: These computers are very effective for data simulations. Anyone can simulate many algorithms with chemical simulations, traffic optimization, and weather forecasting. For example, Volkswagen has already shown the traffic flow for 10,000 taxis in Beijing.

Data security: Quantum computers ensure very high privacy because they are good at cryptography. To break the security of quantum computers quite impossible. China launched satellites using quantum computers, and they claimed that their satellites would never be hacked.

Less power: Quantum computers can reduce power consumption from 100 to 1000 times because they use quantum tunneling. As it requires less energy than other computers, it will also reduce costs.

Problem optimization: A quantum computer will be beneficial for optimizing any problem and determining the best delivery routes for the FedEx truck to the airport.

Drug development: It is evident that inventing a medicine and bringing it to the market takes many years. Billions of dollars are needed to discover a new drug. However, the use of quantum computing improves the front end of the process, reducing cost and time to the market.

Google search: Google uses quantum computers to refine searches. As a result, this computer allows Google to speed-up every search.

Self-driving cars: Using quantum computers Google is developing batteries, self-driving cars, and transportation. These self-driving cars will remove pollution, congestion and ensure safe journeys.

Artificial intelligence: Quantum computers may be used in artificial intelligence because they can make decisions more precisely than the classical computer.

Machine learning: Quantum computers are used in machine learning techniques to increase the efficiency of machine learning and reduce code lines.

Computational chemistry: One of the most promising applications of quantum computing can be the computational chemistry, which can solve critical problems such as improving the nitrogen-fixation process for creating ammonia-based fertilizer.

1.3 Demerits of Quantum Computing

Every innovation carries some disadvantages also and quantum computers are also the same. It has the following drawbacks:

Complex design and large size: The main disadvantages of this technology are that its format is so complex that it is difficult to understand, and the scope of this computer is so enormous. It is not easy to carry this computer from somewhere to anywhere like a classical computer because of its large size.

High cost: This computer is more expensive than any other computer, and nobody can afford it easily. The maintenance cost is also very high.

Internet security: Scientists believe that if a quantum computer is implemented optimally, all the Internet security will be breached. This is because these computers can decode all the code on the Internet.

High error rate: Quantum computers have a high error rate. As it uses qubits, it cannot detect errors like conventional computers.

Fragile: Atoms are needed to make a quantum computer. Maintaining particle is challenging because any vibration can affect them, and also cause decoherence.

Temperature problem: Quantum computers need to be cooled so that atoms cannot be collapsed with each other changing their energy state. That is why it is required to provide nearly hostile 460-degree Fahrenheit which is very difficult to maintain.

Heating problem: The CPU of a quantum computer generates heat from itself, which is the biggest problem because this heating problem can create anarchy in the whole system.

Algorithm creation: Each time, a quantum computer needs a new algorithm for computation. It is unlike classical computers; they must perform unique algorithms in their environment. So, it is another big problem that is difficult to solve.

1.4   Challenges of Quantum Computing

The revolutionary invention of quantum computing can solve any task other than regular computers. Complicated problems are performed using quantum algorithms based on fundamental quantum phenomena such as superposition and entanglement.

Some scientists say that if a quantum computer is made perfectly, the quantum computer will change the whole world altogether. However, scientists also say that it is never possible to make it accessible to the public and easier for people because of the overall complexity of these machines.

There are some challenges that need to be overcome. If these problems are solved, modern science may create a computer commercially that might change the whole world. The critical challenges of this are described below:

Quality of qubit: Several challenges exist in building a large-scale quantum computer—fabrication, verification, and architecture. The power of quantum computing comes from storing a complex state in a single bit. This complex state makes quantum systems challenging to build, verify, and design. The first challenge is to make qubits effectively that will generate valuable instructions. The famous quantum computing algorithm is Shor's algorithm which can quickly factor a large number. He discovered a quantum error-correcting code that can store the information of one qubit onto a highly entangled state of nine qubits.

Qubit maintenance: Need to control multiple qubits, including error detection schemes applying complex algorithms. That control must have low quiescence on the order of 10's, and it has to come from CMOS-grounded adaptive feedback control circuits.

Numerous cables/wires: Another challenge is to scale up the number of qubits within a quantum chip, which requires multiple control wires or cables to create each qubit. These too many wires create anarchy within a computer system, which should be organized properly.

Moreover, there are more challenges such as algorithm creation and Internet security problems in quantum computing. Each time it needs to write a new algorithm. They cannot work as a classical computer. They need unique algorithms to perform tasks in their environment, which is tough to implement each time.

Another challenge is that they need a negative temperature of 460 degrees which is a shallow temperature and difficult to maintain. If a quantum computer is implemented in the best way, then complete internet security breaks because these computers are good at decrypting all the codes on the internet.

1.5 Motivation for Quantum Computing

Quantum computing has several advantages, an excellent achievement for very large-scale integrated circuit design that makes people more attractive in this field. It can work faster than any digital circuit. These quantum circuits are used to build a quantum computer. The qubits are used to store information and have the property of superposition which means that the value of qubits can stay more than one position at a time. Qubits can work in a parallel way that makes the quantum computer more effective and faster. Researchers are building efficient algorithms so that quantum computers can deliver enormous speed for solving a specific problem. The rate of quantum computer computation improves many technologies such as machine learning, cyber security, bullet trains, and many more. The most significant advantage of quantum computing logic is that it can hold multiple ternaries. Multivalued quantum circuits consume less power than other logical systems, and circuit simulation is also more straightforward. The multivalued logic aids quantum circuits which is becoming more efficient and powerful. This system can give high privacy because it has high encryption, and can consume less power. Even it can be applied in machine learning fetch for boosting outcomes. These things are possibly done by a quantum computing system that is never possible in classical computers. The properties mentioned above motivate us to work in this exciting field.

1.6   Summary

In this chapter, some basic information about quantum computing is presented. This chapter only focuses on the fundamental concept, some basic information and the history of quantum computing. Quantum computation promises the ability to compute solutions to any problem. Though it has potential advantages, it is still underdeveloped. Therefore, researchers should emphasize the quantum field and work to reduce disadvantages. This chapter covers the advantages, disadvantages, challenges, and motivations. Quantum computing has a wide range of potential applications across various fields, including simulating complex molecular structures for drug discovery and materials science, optimizing logistics and financial models, and developing new cryptographic methods. It also holds promise for advancements in artificial intelligence and machine learning. After reading this chapter, the reader must be able to achieve the fundamental concept and the background of this field.

Bibliography

1. S. Aaronson, *Quantum Computing and Hidden Variables II: The Complexity of Sampling Histories* (2004). arXiv preprint quant-ph/0408119
2. D.S. Abrams, S. Lloyd, Simulation of many-body fermi systems on a universal quantum computer. Phys. Rev. Lett. **79**(13), 2586 (1997)
3. C.G. Almudever, L. Lao, X. Fu, N. Khammassi, I. Ashraf, D. Iorga, S. Varsamopoulos, C. Eichler, A. Wallraff, L. Geck, et al., in *Design, Automation & Test in Europe Conference & Exhibition (DATE), 2017*. The engineering challenges in quantum computing (IEEE, 2017), pp. 836–845
4. A. Barenco, A universal two-bit gate for quantum computation. Proc. R. Soc. Lond. Ser. A Math. Phys. Sci. **449**(1937), 679–683 (1995)
5. S.S. Clarke, *Quantum Computing: A Mathematical Analysis of Shor's Algorithm* (2020)
6. A. Datta, A. Shaji, Quantum discord and quantum computing—an appraisal. Int. J. Quant. Inf. **9**(07n08), 1787–1805 (2011)
7. Hafiz Md. Hasan Babu, "Quantum Computing: A Pathway to Quantum Logic Design", IOP (Institute of Physics) Publishing, Bristol, UK (2020)
8. Hafiz Md. Hasan Babu, "Multiple-Valued Computing in Quantum Molecular Biology", Volume I, CRC Press, USA (2023)
9. Hafiz Md. Hasan Babu, "Multiple-Valued Computing in Quantum Molecular Biology", Volume II, CRC Press, USA (2023)
10. M.V. Gotarane, S.S.M. Gandhi, Quantum computing: future computing. Int. Res. J. Eng. Technol. **3**(2), 1424–1427 (2016)
11. A.D. Córcoles, A. Kandala, A. Javadi-Abhari, D.T. Mcclure, A.W. Cross, K. Temme, P.D. Nation, M. Steffen, J.M. Gambetta, Challenges and opportunities of near-term quantum computing systems. Proc. IEEE **10** (2019)
12. N.P. Landsman, When champions meet: rethinking the Bohr–Einstein debate. Stud. Hist. Philos. Sci. Part B Stud. Hist. Philos. Modern Phys. **37**(1), 212–242 (2006)
13. F. Mahfoud. PhD thesis summary
14. D.B. Malament, On the time reversal invariance of classical electromagnetic theory. Stud. Hist. Philos. Sci. Part B Stud. Hist. Philos Modern Phys. **35**(2), 295–315 (2004)
15. M.J. Nene, G. Upadhyay, in *Advanced Computing and Communication Technologies*. Shor's algorithm for quantum factoring (Springer, 2016), pp. 325–331

16. M.A. Nielsen, I. Chuang, *Quantum Computation and Quantum Information* (2002)
17. S.K. Shukla, R. Iris Bahar, *Nano, Quantum and Molecular Computing* (Springer, 2004)
18. Andrew M Steane. Efficient fault-tolerant quantum computing. *Nature*, 399(6732):124–126, 1999

Chapter 2
Fundamentals of Quantum Computing

2.1 Introduction

Compared with today's classical CMOS-based systems, quantum computers guarantee an exponential increase in power. As a result, it is difficult for the human mind to perceive this increasing magnitude. So there is real excitement that quantum computers will deliver benefits that are not possible with today's systems. The design of quantum computing-based systems begins with reversible logic circuit synthesis, low-power CMOS circuits, and nanotechnology-based systems. Quantum mechanical processes have been proved to be a good choice for constructing reversible gates, and these gates are known as quantum gates because quantum mechanics is basically reversible.

Reversible computing is a form of unconventional computing, where reversible logic is a computing paradigm that allows for the recovery of input information from outputs, minimizing energy dissipation and enabling reversibility of computations. Its key properties include information preservation, one-to-one mapping, reduced energy dissipation, and the ability to reverse operations. These properties make reversible logic particularly relevant in areas like quantum computing and low-power design. Because of the unity of quantum mechanics, quantum circuits are relatively reversible unless they simply "collapse" the quantum states in which they operate.

Now some questions may arise in mind. How many fundamental quantum gates are there in quantum computing? Is it the same as the conventional Boolean logic where OR, AND, and NOT gates are the fundamental gate? How to construct the basic binary logic circuit in quantum computing? How do the basic operations work in quantum computing? Is it too complex to design the circuit diagram of quantum operations? All of these questions will be explored throughout this chapter.

H. M. Hasan Babu, *Quantum Biocomputing in Quantum Biology Volume I*,
https://doi.org/10.1007/978-981-97-7154-7_2

2.2   Basic Quantum Gate Operations

If the input and output assignments of a function or circuit are one-to-one, the function or circuit is said to be reversible. That is, the outputs of a reversible circuit can be calculated uniquely from the inputs, and the inputs may be recovered from the outputs.

A quantum logic gate, also known as a quantum gate, is a fundamental quantum circuit that works with a minimal number of qubits. They are equivalent to classical logic gates in conventional digital computers for quantum computers. Unlike many classical logic gates, quantum gates are reversible. Unitary matrices are used to represent them. The most common quantum gates work with one or two-qubit spaces. This means that quantum gates can be characterized by orthonormal 2×2 or 4×4 matrices.

There are only three fundamental gates in quantum computing based on the synthesization of reversible logic circuits are given as follows:

1. Controlled NOT (CNOT) Gate
2. Controlled V Gate
3. Controlled V+ Gate

2.2.1   Quantum Controlled NOT Gate

The quantum controlled-NOT gate, or simply CNOT gate is a two-qubit operation in which the first qubit is known as the control qubit and the second qubit is known as the target qubit. Using a combination of CNOT gates and single-qubit rotations, any quantum circuit can be simulated to an arbitrary degree of accuracy.

The CNOT gate utilizes a quantum register with two qubits. The CNOT gate flips the second qubit (the target qubit) if and only if the first qubit (the control qubit) is $|1>$. That means it performs a classical NOT on the target whenever the control is in the state of $|1>$. It is also known as the controlled-x or CX gate. Table 2.1 shows the function of a quantum controlled-NOT gate.

Here, the inputs $|A_0>$ and $|A_1>$ are the control qubit and the target qubit. Quantum CNOT gate produces two outputs where the controlled output is $|Q_1>$ which gets

Table 2.1 Operations in quantum CNOT gate

Input		Output	
$|A_0>$	$|A_1>$	$|Q_0>$	$|Q_1>$
$|0>$	$|0>$	$|0>$	$|0>$
$|0>$	$|1>$	$|0>$	$|1>$
$|1>$	$|0>$	$|1>$	$|1>$
$|1>$	$|1>$	$|1>$	$|0>$

Fig. 2.1 Quantum CNOT
gate

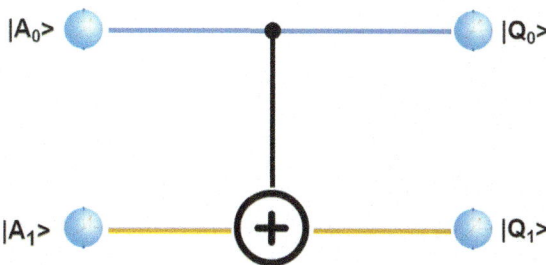

flipped only if the controlled input $|A_0>$ is $|1>$. During this time the output $|Q_0>$
always remains the same as per the input $|A_0>$.

The next question is, how to draw a circuit diagram of a quantum CNOT gate?
This is pretty simple. In Table 2.1, CNOT is the quantum NOT operation which
is controlled by a qubit. Therefore, Pauli X-gate with a controlled qubit is used to
design the quantum CNOT gate. Figure 2.1 shows the logic circuit diagram of the
quantum CNOT gate.

2.2.1.1 Design Procedure of Quantum CNOT Gate

In quantum logic circuits, this gate has the simplest architectural design which is
easy to construct only in three steps

1. At first, two superposition-mode qubits are drawn as $|A_0>$ and $|A_1>$, where $|A_0>$
   is the control qubit, and $|A_1>$ is the target qubit.
2. Then two lines are drawn from these inputs to the target outputs $|Q_0>$ and $|Q_1>$.
3. Finally, on the $|A_1>$ line, a NOT gate is drawn which is connected to the $|A_0>$
   line, because $|A_0>$ controls the NOT gate of the target output $|Q_1>$.

2.2.1.2 Working Principle of Quantum CNOT Gate

The only input to the NOT gate is $|A_1>$, and the gate control input is $|A_0>$, which
are termed as CNOT gate. The block diagram of the quantum CNOT gate is given in
Fig. 2.2.

The working procedure of the quantum CNOT gate is given below:

1. The output value $|Q_0>$ is always the same as the value of input $|A_0>$ because
   there is no operation through this line;
2. In the case of the output value of $|Q_1>$

   (a) When the control qubit $|A_0>$ is $|0>$, the gate will close and will work as a
       buffer. That means the output value of $|Q_1>$ is the same as the input value
       $|A_1>$. According to CNOT table, when $|A_1> = |0>$ or $|1>$ with control qubit
       $|A_0> = |0>$, the outputs ($|Q_1>$) are $|0>$ and $|1>$ respectively.

Fig. 2.2 Block diagram of
quantum CNOT gate

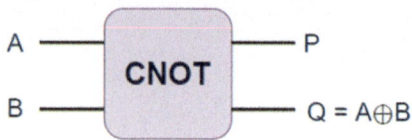

Fig. 2.2 Block diagram of
quantum CNOT gate

(b) When the control qubit $|A_0>$ is $|1>$ the gate will open and works as an
   inverter. That means the output value of $|Q_1>$ is opposite to the input value
   $|A_1>$. According to CNOT table, when the $|A_1> = |0>$ **or** $|1>$ with control
   qubit $|A_0> = |1>$, it flips the $|A_1>$ qubit and produces $|1>$ and $|0>$ **as** outputs
   in $|Q_1>$ respectively.

From the input and output relation, it is observed that this gate works exactly like
a classical XOR operation. Therefore, the CNOT gate can be used for the quantum
XOR operation.

So, the CNOT gate can be described as the gate that maps the basic states of
inputs ($|A_0>$, $|A_1>$) to the outputs ($|Q_0> = |A_0>$, and $|Q_1> = (|A_0>$ **XOR** $|Q_1>)$).
Figure 2.2 shows the block diagram of the quantum CNOT gate.

The first experimental realization of a CNOT gate was accomplished, in 1995.
A single Beryllium ion in a trap was employed in this experiment. The two qubits
were encoded into the optical state and the vibrational state of the trapped ion. The
CNOT-operation reliability was estimated to be on the order of 90% at the time of
the trial.

2.2.2 Quantum Controlled-V Gate

The 2nd fundamental gate in quantum computing is the controlled-V gate (or V gate)
which is shown in Fig. 2.3.

When the control signal $|A_0> = |0>$, the qubit $|A_1>$ will pass through the con-
trolled part and remain unchanged, i.e., $|Q_1> = |A_1>$. When $|A_0> = |1>$, the unitary
operation

$V = \frac{i+1}{2} \begin{pmatrix} 1 & -i \\ -i & 1 \end{pmatrix}$ is applied to the input $|A_1>$, that is, $|Q_1> = V(|A_1>)$.

Fig. 2.3 Quantum
controlled V gate

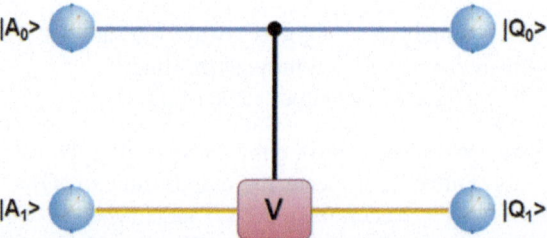

Table 2.2 Operations in the quantum controlled-V gate

Input		Output	
$|A_0>$	$|A_1>$	$|Q_0>$	$|Q_1>$
$|0>$	$|X>$	$|0>$	$|X>$
$|1>$	$|0>$	$|1>$	$|v>$
$|1>$	$|1>$	$|1>$	$|V>$
$|1>$	$|v>$	$|1>$	$|1>$
$|1>$	$|V>$	$|1>$	$|0>$
$|1>$	$|w>$	$|1>$	$|0>$
$|1>$	$|W>$	$|1>$	$|1>$

2.2.2.1 Design Procedure of Quantum Controlled-V Gate

Figure 2.3 shows the circuit diagram of the quantum controlled-V gate. This gate is also a two-input gate, and the number of output is the same here as well.

Three steps to construct quantum controlled V gate are given below:

1. Two superposition-mode qubits are drawn, $|A_0>$ and $|A_1>$. In this gate, the control qubit is $|A_0>$, and the target qubit is $|A_1>$.
2. Two lines are drawn from these inputs to the outputs $|Q_0>$ and $|Q_1>$.
3. At last, the **V** gate is drawn on the $|A_1>$ line and connect it to the $|A_0>$ line because $|A_0>$ is the control qubit that controls the V gate of output $|Q_1>$.

2.2.2.2 Working Principle of Quantum Controlled V Gate

The quantum controlled-V gate's working principle is not as straightforward as the quantum CNOT gate. Let's have a look at the operational table of the controlled V gate (Table 2.2).

It is called quantum controlled-V gate, where Table 2.2 shows that the outputs of the gate depend on the control qubit $|A_0>$.

The working procedure of the quantum controlled-V gate is listed below:

1. Consider the controlled-V gate, where the $|A_1>$ values are $|0>$, $|1>$, $|v> = |0>$, $|V> = |1>$, $|w> = |0>$ and $|W> = |1>$ and the control input $|A_0>$ values are $|0>$ and $|1>$.
2. The output value $|Q_0>$ is always the same as the input value $|A_0>$ due to the fact that this line does not have any operations.
3. For the output value of $|Q_1>$, the working principle of the controlled-V gate is as follow:

   (a) The controlled V gate will not open if the control qubit $|A_0>$ is $|0>$, eventually it will work as a buffer. That means the output value of $|Q_1>$ will be the same as the input value of $|A_1>$ when the input value of $|A_0>$ is $|0>$.

(b) When the control qubit $|A_0>$ is $|1>$, the quantum controlled-V gate will open and it will work with the input values of $|A_1>$ to produce output $|Q_1>$ according to the operational table (Table 2.2) of the controlled V gate. Two intermediate inputs $|v>$ and $|V>$ are used for controlled V gate output values. Here the $|v>$ value is assigned to the V gate output if its input $|A_1>$ is $|0>$, otherwise $|V>$ value if its input $|A_1>$ is $|1>$. Now, if the input of the V gate is $|v>$ then the output will be $|1>$ (flipped) because the $|v>$ value is the result of the input $|0>$ to the V gate. In the same way, applying $|V>$ input to a V gate results in an output of $|0>$. If the input of a V gate is $|w>$ then its output will be $|0>$, and if the input of a V gate is $|W>$ then the target output will be $|1>$. The inputs $|w>$ and $|W>$ maintain the properties of quantum controlled-V^+ gate.

2.2.3   Quantum Controlled V+ Gate

The last of three fundamental quantum gates is the quantum-controlled V^+ gate (or simply, V+ gate). In the controlled-V+ gate when the control signal $|A_0> = |0>$ the qubit $|A_1>$ will pass through the controlled part which will be unchanged, that is, $|Q_1> = |A_1>$. When $|A_0> = |1>$ the unitary operation $V+ = V^{-1}$ is applied to the input $|A_1>$, that is, $|Q_1> = V^+ (|A_1>)$. The controlled-V^+ gate is shown in Fig. 2.4.

2.2.3.1   Design Procedure of Quantum Controlled-V+ Gate

The design architecture of the quantum controlled-V+ gate is shown in Fig. 2.4. The construction is similar to the quantum controlled-V gate, but the difference is at the operational part.

The construction of this fundamental gate is easy which has three steps like the other two.

1. At first, two superposition-mode qubits are drawn, $|A_0>$ and $|A_1>$, where the control qubit is $|A_0>$, and the target qubit is $|A_1>$.
2. Then two lines from these inputs produce outputs $|Q_0>$ and $|Q_1>$.

Fig. 2.4 Quantum controlled V+ gate

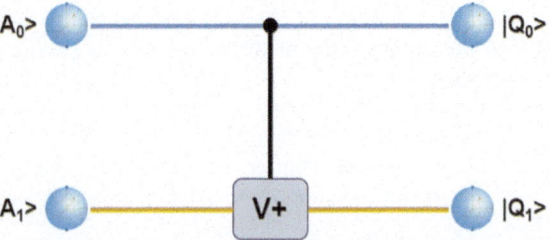

3. Subsequently, **V+** gate is drawn on the $|A_1\rangle$ line and connect it to the $|A_0\rangle$ line because $|A_0\rangle$ controls the **V+** gate of the output $|Q_1\rangle$ depending on the input $|A_1\rangle$.

2.2.3.2 Working Principle of Quantum Controlled V+ Gate

The Controlled-V and the Controlled V+ gates are the two types of square-root-of-NOT gates. If two Controlled-V or two Controlled-V+ gates are triggered in series, they act as an inverter.

It is similar to the controlled-V gate in which the control input is $|1\rangle$, and the corresponding unitary operator is propagated to the second output, where the unitary operation for the Controlled-V+ is $V+ = \frac{1}{i+1} \begin{pmatrix} 1 & \frac{-1}{i} \\ i & 1 \end{pmatrix}$. The operation on this gate is shown in Table 2.3.

The operations of the quantum controlled-V+ gate are described below:

1. The input qubits $|A_0\rangle$ and $|A_1\rangle$ are the target qubits.
2. The input qubit can be one from $|A_1\rangle$ values which are $|0\rangle$, $|1\rangle$, $|w\rangle = |0\rangle$, $|W\rangle = |1\rangle$, $|v\rangle = |0\rangle$, and $|V\rangle = |1\rangle$ and for control input $|A_0\rangle$ values are $|0\rangle$ and $|1\rangle$.
3. Again, the output value $|Q_0\rangle$ is always the same as the input value $|A_0\rangle$ because there is no operation across this line.
4. For the value of $|Q_1\rangle$'s output, operations are as follows:

   (a) When the control qubit $|A_0\rangle$ is $|0\rangle$ the controlled-V+ gate will not open. Consequently, the input $|A_1\rangle$ will work as a buffer. That means the output value of $|Q_1\rangle$ is the same as the input value of $|A_1\rangle$.
   (b) When the control qubit $|A_0\rangle$ is $|1\rangle$ the controlled-V+ gate will work as per Table 2.3. So with the input qubits of $|A_1\rangle$, the output $|Q_1\rangle$ will be produced according to Table 2.3. The intermediate inputs variables $|w\rangle$ and $|W\rangle$ produce output from the controlled-V+ gate. Here the $|w\rangle$ value is assigned to the controlled-V+ gate output if the input $|A_1\rangle$ is $|0\rangle$; and $|W\rangle$ value is assigned if the input $|A_1\rangle$ is $|1\rangle$. Now, if the input of the controlled-V+ gate is $|w\rangle$ then the output will be $|1\rangle$ because the $|w\rangle$ value is the result by applying input $|0\rangle$ to a controlled-V+ gate. Correspondingly, applying

Table 2.3 Operations in the quantum controlled V+ gate

Input		Output	
$|A_0\rangle$	$|A_1\rangle$	$|Q_0\rangle$	$|Q_1\rangle$
$|0\rangle$	$|X\rangle$	$|0\rangle$	$|X\rangle$
$|1\rangle$	$|0\rangle$	$|1\rangle$	$|w\rangle$
$|1\rangle$	$|1\rangle$	$|1\rangle$	$|W\rangle$
$|1\rangle$	$|v\rangle$	$|1\rangle$	$|0\rangle$
$|1\rangle$	$|V\rangle$	$|1\rangle$	$|1\rangle$
$|1\rangle$	$|w\rangle$	$|1\rangle$	$|1\rangle$
$|1\rangle$	$|W\rangle$	$|1\rangle$	$|0\rangle$

|W> input to controlled-V+ gate results in an output of |0>. Finally, if the input of a controlled-V+ gate is |v> (might come from a controlled-V gate as output) then |0> is obtained as output, and if the input is |V> then the target output will be generated as |1>.

2.3 Basic Operations in Quantum Logic

To perform quantum logic operations, the quantum computing has three fundamental gates, namely OR gate, AND gate, and NOT gate. Also, four additional quantum gates are made by the fundamental quantum gates such as NOR gate, NAND gate, XOR gate, and XNOR gate. In a circuit, the quantum gates produce decisions based on a combination of signals coming from its inputs. All quantum computations are is performed at the hardware level using those seven quantum gates.

Specifically, it is possible to construct the quantum fundamental gates from the classical gates which behave exactly as quantum computing nature and with them, it is easy to perform all necessary quantum operations in the quantum realm. This section will present how to construct the quantum fundamental gates for quantum computing to perform quantum operations. On the other way, in quantum logic, basic operations involve manipulating the state of qubits using unitary transformations, typically represented as matrices. Key operations include manipulating superposition states, creating entanglement, and performing measurements. These operations are fundamental to building quantum algorithms. Main features of basic operations in quantum logic can be explained as follows:

1. **Qubits and Superposition:**

   i) Qubits, unlike classical bits, can exist in a superposition of states, meaning they can be both 0 and 1 simultaneously.
   ii) Basic operations like the Hadamard gate (H) can put a qubit into a superposition, giving it equal probabilities of being measured as 0 or 1.

2. **Unitary Transformations (Quantum Gates):**

   a) Quantum gates are unitary operators that manipulate the state of qubits.
   b) Examples of basic quantum gates include:
      i) Pauli operators (X, Y, Z): These operators perform rotations on the Bloch sphere, manipulating the qubit's state.
      ii) CNOT (Controlled-NOT): This gate entangles two qubits, flipping the target qubit if the control qubit is 1.
      iii) Hadamard gate (H): Creates superposition.
      iv) S gate: Performs a phase rotation.

3. **Entanglement:**

   a) Entanglement is a quantum phenomenon where two or more qubits become correlated in such a way that they share the same fate, regardless of the distance between them.

   b) The CNOT gate can be used to create entanglement between qubits.

4. **Measurement:**

   i) Measurement is the process of collapsing the superposition of a qubit into a definite state (0 or 1).

   ii) Measurements are crucial for extracting information from quantum systems.

5. **Quantum Circuits:**

   i) Quantum circuits are sequences of quantum gates that operate on qubits.

   ii) By combining different quantum gates, complex quantum algorithms can be constructed.

2.3.1 Quantum OR Operation

A circuit that performs quantum OR (QOR) operation in quantum logic can be described as $|Y> = |A_0> \text{ QOR } |A_1>$, or $|Y> = |A_0> + |A_1>$. And if the output value of the OR is for more than two inputs, then merge them as chain-OR which will give the output as $|Y> = |A_0> + |A_1> + \cdots + |A_n>$.

Figure 2.5 shows the circuit diagram of quantum OR operations which includes three quantum controlled-V gates that are connected linearly, one quantum CNOT gate, two input qubits $|A_0>$ and $|A_1>$, and a constant qubit (also known as Ancilla qubit) $|0>$ which are the actual inputs of the 1st controlled-V gate. Ancilla bits are extra bits that are used in computation to achieve certain goals. All these are configured in the way depicted in Fig. 2.5. And $|Q>$ is the resulting qubit which considers the quantum value of the quantum OR operations of $|A_0>$ and $|A_1>$.

2.3.1.1 Working Principle of Quantum OR Operation

Table 2.4 displays the functionality of the quantum OR operation. It will always produce output $|1>$ if either one of two input qubits or both of the input qubits is $|1>$. The output is $|0>$ only when both input qubits are $|0>$.

Fig. 2.5 Circuit diagram of quantum OR operation

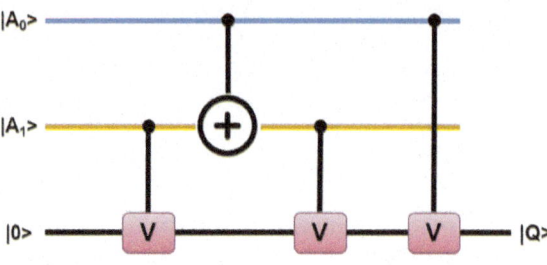

Table 2.4 Truth table of quantum OR operation

$|A_0>$	$|A_1>$	$|Q>$
$|0>$	$|0>$	$|0>$
$|0>$	$|1>$	$|1>$
$|1>$	$|0>$	$|1>$
$|1>$	$|1>$	$|1>$

Assume this for each pattern of input qubits of $|A_0>$ and $|A_1>$. Remember that, each gate is controlled by an input qubit and it will open only if the control qubit is $|1>$.

1. When both inputs $|A_0>$ and $|A_1>$ are $|0>$,

    (a) The first V gate will not be open as the control input $|A_1>$ is $|0>$.
    (b) The 2nd gate is CNOT which is controlled by $|A_0>$. Thus, this gate will pass the unchanged value of $|A_1>$ (i.e., $|0>$).
    (c) Therefore, the 2nd V gate will also remain closed (as the control qubit is $|0>$, from Step b) and it will work as a buffer. The output from this gate will be $|0>$ as well which will be the input for the 3rd V gate.
    (d) Finally, the last V gate which will produce the final output of the quantum OR operation, it gets input qubit $|0>$ and control qubit also $|0>$ (value of $|A_0>$). Therefore, no operational change will occur. And it will generate the final output as $|0>$.

1. When inputs $|A_0> = |0>$, and $|A_1> = |1>$,

1. (a) The first V gate will be open as the control input $|A_1>$ is $|1>$. The 1st V gate gets input qubit $|0>$ and the control qubit $|1>$. Therefore, it will produce $|v>$ as output. This output will work as input for the 2nd V gate.
    (b) Sequentially, the 2nd gate is CNOT which is controlled by $|A_0>$ and its value is $|0>$. Thus, this gate will pass the unchanged value of $|A_1>$ (i.e., $|1>$).
    (c) As a result, the 2nd V gate will also be open (as the control qubit is $|1>$, from Step b) and it will perform an operation based on the input. The input of the 2nd V gate is $|v>$ (from Step a), so this gate will produce $|1>$ as output, which will be the input of the 3rd V gate.

(d) In the last V gate, it gets input qubit $|1>$ and the control qubit $|0>$ (value of $|A_0>$). Therefore, no operational change will be occurred and this gate will work just like a buffer. And the final output will be $|1>$.

1. When inputs $|A_0> = |1>$, and $|A_1> = |0>$,

1. (a) The first V gate will not be open as the control input $|A_1>$ is $|0>$. Thus no operational change will be occurred and it will pass $|0>$ as output, which will be the input for the 2nd V gate.
   (b) It is seen that the CNOT gate is controlled by $|A_0>$ and the value of $|A_0>$ is $|1>$ now. Thus, this gate will invert the value of $|A_1>$ and give $|1>$ as output. This output will work as the control qubit for the 2nd V gate.
   (c) Now, the 2nd V gate will be able to carry out the operation (as the control qubit is $|1>$, from Step b). The input of the 2nd V gate is $|0>$ (from Step a). So, this gate will produce $|v>$ as output, which will be the input of the 3rd V gate.
   (d) In the last V gate, it gets input qubit $|v>$ and control qubit $|1>$ (value of $|A_0>$). Therefore, the last V gate will produce the output $|1>$ (from Table 2.5). And the final output is $|1>$.

1. When inputs $|A_0> = |1>$, and $|A_1> = |1>$,

   (a) The first V gate will be operated because the control input $|A_1>$ is $|1>$. It will produce the output $|v>$ which will be delivered to the 2nd V gate.
   (b) The CNOT gate will also work as the control qubit $|A_0>$ which is $|1>$. Thus, this gate will invert the value of $|A_1>$ which will produce $|0>$ as output. As usual, this output will work as the control qubit for the 2nd V gate.
   (c) The 2nd V gate will remain closed, as the control qubit is $|0>$ (from Step b). Thus $|v>$ is obtained from this gate, which will be the input of the last V gate.
   (d) The last V gate gets input qubit $|v>$ and the control qubit $|1>$ (value of $|A_0>$). Hence, the last V gate will produce the output $|1>$. So, the expected final output is $|1>$.

Table 2.5 is further designed to understand the working procedure of quantum OR operation thoroughly. Table 2.5 points out the characters of the outputs from each gate in the quantum OR circuit with respect to each input pattern.

2.3.2 Quantum NOR Operation

The quantum NOR operation is the inverted operation of quantum OR operation as shown in Table 2.6. This implies that it will give the output $|1>$ only when both of the

Table 2.5 Operations on each gate in the quantum OR operational circuit

$|A_0>$	$|A_1>$	1st V gate	CNOT gate	2nd V gate	Final output (3rd V gate)
$|0>$	$|0>$	$|0>$	$|0>$	$|0>$	$|0>$
$|0>$	$|1>$	$|v>$	$|1>$	$|1>$	$|1>$
$|1>$	$|0>$	$|0>$	$|1>$	$|v>$	$|1>$
$|1>$	$|1>$	$|v>$	$|0>$	$|v>$	$|1>$

Fig. 2.6 Circuit diagram of a quantum NOR operation

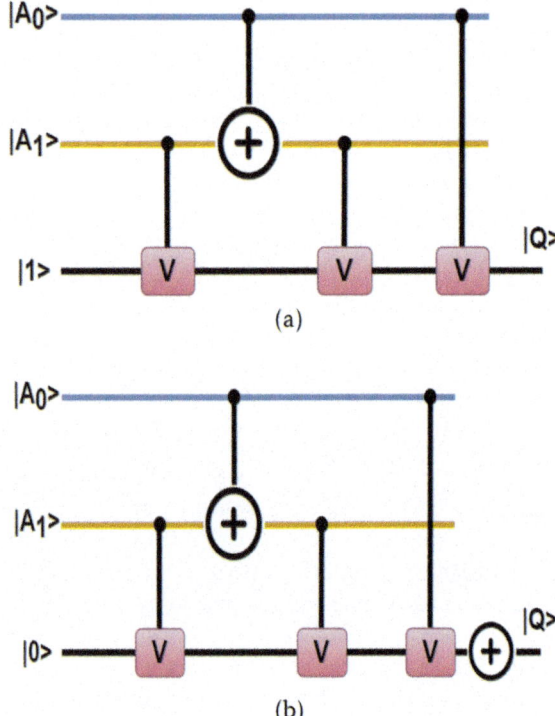

input qubits are $|0>$. Otherwise, the output is always $|0>$. Therefore, an extra quantum NOT operation is needed along with the quantum OR circuit to construct the quantum NOR operation circuit diagram. The circuit diagram of quantum NOR operation with the above idea is shown in Fig. 2.6(a). But have a close look at Fig. 2.6(b). Why these two circuits? Any of them can be used to conduct the quantum NOR operation. In Fig. 2.6(b), $|0>$ is used as the constant qubit instead, which makes the differences in the two circuits. But the result is always the same (Fig. 2.6).

Table 2.6 Truth table of quantum NOR operation

$|A_0>$	$|A_1>$	$|Q>$
$|0>$	$|0>$	$|1>$
$|0>$	$|1>$	$|0>$
$|1>$	$|0>$	$|0>$
$|1>$	$|1>$	$|0>$

2.3.2.1 Working Principle of Quantum NOR Operation

As mentioned earlier the operation in Fig. 2.6a is exactly the same as the quantum OR operation but it includes a quantum NOT operation that inverts the outputs of the quantum OR operation and produces the output of the quantum NOR operation (shown in Table 2.6).

This section will describe the working procedure of Fig. 2.6b and it will show how the quantum NOR results are formed with that circuit for each pattern of input qubits of $|A_0>$ and $|A_1>$.

When both inputs $|A_0>$ and $|A_1>$ are $|0>$,

1. The 1st V gate will not be open as the control input $|A_1>$ is $|0>$. It leads $|1>$ (the constant qubit) as an input for the 2nd V gate.
2. The 2nd gate is CNOT which is controlled by $|A_0>$, which will pass the value of $|A_1>$ unchanged (i.e., $|0>$) because the value of control qubit is $|0>$.
3. Now, the 2nd V gate will also remain closed (as the control qubit is $|0>$, from Step b) which will work as a buffer. The output from this gate will be $|1>$ as well, which will be the input of the 3rd V gate.
4. Finally, the last V gate will produce the final output of the quantum NOR operation which gets input qubit $|1>$ and control qubit $|0>$ (value of $|A_0>$). Therefore, no operational change will occur and the final output will be $|1>$.

1. When inputs $|A_0> = |0>$, and $|A_1> = |1>$,

   (a) The 1st V gate will be open as the control input $|A_1>$ is $|1>$. The 1st V gate gets input $|1>$ and the control qubit $|1>$. Therefore, it will produce $|V>$ as output (Table 2.6). This output will work as the input for the 2nd V gate.
   (b) Chronologically, the 2nd gate is CNOT which is controlled by $|A_0>$ that is $|0>$. Thus, this gate will pass the value of $|A_1>$ unchanged (i.e., $|1>$).
   (c) Now, the 2nd V gate will also be open (as the control qubit is $|1>$, from Step b) and it will be operated based on the input. The input of the 2nd V gate is $|V>$ (from Step a), so this gate will produce $|0>$ as output, which will be the input of the 3rd V gate.
   (d) Finally, the last V gate gets input qubit $|0>$ and control qubit $|0>$ (value of $|A_0>$). Therefore, no operational change will be occurred and the gate will work just like a buffer. So, the final output will be $|0>$.

Table 2.7 Operations on each gate in the quantum NOR operational circuit

$|A_0>$	$|A_1>$	1st V gate	CNOT gate	2nd V gate	Final output (3rd V gate)
$|0>$	$|0>$	$|1>$	$|0>$	$|1>$	$|1>$
$|0>$	$|1>$	$|V>$	$|1>$	$|0>$	$|0>$
$|1>$	$|0>$	$|1>$	$|1>$	$|V>$	$|0>$
$|1>$	$|1>$	$|V>$	$|0>$	$|V>$	$|0>$

1. When inputs $|A_0> = |1>$ and $|A_1> = |0>$,

1. (a) The 1st V gate will not be open as the control input $|A_1>$ is $|0>$. Thus no operational change will be occurred and it will pass $|1>$ as output, which will be the input for the 2nd V gate.
   (b) It is seen that the CNOT gate is controlled by $|A_0>$ and the value of $|A_0>$ is $|1>$ now. Thus this gate will invert the value of $|A_1>$ and give $|1>$ as output. This output will work as the control qubit for the 2nd V gate.
   (c) Now, the 2nd V gate will be able to carry out the operation (as the control qubit is $|1>$, from Step b). The input of the 2nd V gate is $|1>$ (from Step a). So, this gate will produce $|V>$ as output, which will be the input of the 3rd V gate.
   (d) Finally, the last V gate gets input qubit $|V>$ and control qubit $|1>$ (value of $|A_0>$). Therefore, the last V gate will produce output $|0>$ (from Table 2.7). And the final output will be $|0>$.

1. When inputs $|A_0> = |1>$ and $|A_1> = |1>$,

1. (a) The 1st V gate will perform operation because the control input $|A_1>$ is $|1>$. It will produce output $|V>$ which will be delivered to the 2nd V gate.
   (b) Now, the CNOT gate will also work as the control qubit $|A_0>$ is $|1>$. Thus, this gate will invert the value of $|A_1>$ which produces $|0>$ as output. Ordinarily, this output will work as the control qubit for the 2nd V gate.
   (c) The 2nd V gate will remain closed, as the control qubit is $|0>$ (from Step b). Thus $|V>$ will be produced from this gate which will be the input of the last V gate.
   (d) Finally, the last V gate gets input qubit $|V>$ and control qubit $|1>$ (value of $|A_0>$). Hence, the last V gate will produce the output $|0>$ (from Table 2.7). And as expected, the final output will be $|0>$.

Table 2.7 is further designed to understand the working procedure of quantum NOR operation thoroughly as shown in Fig. 2.6(b).

Fig. 2.7 Circuit diagram of
a quantum AND operation

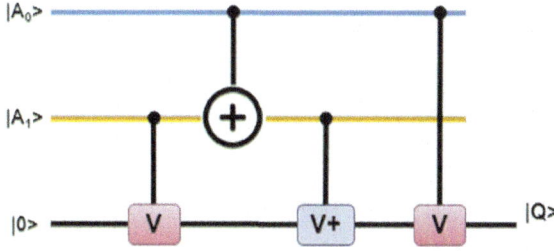

2.3.3 Quantum AND Operation

A circuit that performs quantum AND or QAND operation in quantum logic can be described as $|Y> = |A_0>$ QAND $|A_1>$, or $|Y> = |A_0>.|A_1>$ and to do AND operation for more than two input values, it needs to combine them as chain-AND to produce the output as

$|Y> = |A_0>.|A_1>. … .|A_n>$.

Figure 2.7 shows the circuit diagram of the quantum AND operation, which includes two controlled-V gates, one CNOT gate, and one controlled-V+ gate. And it requires three input qubits $|A_o>$, $|A_1>$, and a constant qubit of $|0>$, respectively, where the 1st V gate is controlled by the value of $|A_1>$ and the CNOT gate; and the 2nd controlled-V (which produces the final output) is controlled by the value of $|A_0>$, and the controlled-V+ gate is controlled by the output value of the CNOT gate. On the other way, in the context of quantum mechanics, "quantum AND operation" refers to a specific type of quantum operation that acts on two qubits (quantum bits) and has a similar effect to the classical AND operation, but operates on the quantum states of the qubits. It is a fundamental building block in quantum computing and is used in various quantum algorithms. The operation typically involves using a CNOT gate where one qubit acts as a control and the other as a target. The CNOT gate flips the target qubit's state only if the control qubit is in the $|1>$ state. By carefully applying CNOT gates and other operations, a quantum AND operation can be performed. Quantum AND operations are essential for implementing various quantum algorithms, such as quantum search and quantum factorization.

2.3.3.1 Working Procedure of Quantum AND Operation

The quantum AND operation's functionality is shown in the truth table (Table 2.8). It will always produce the output $|0>$ if either one of two input qubits or both of the input qubits is $|0>$. The output is $|1>$ only when both input qubits are $|1>$.

Consider each pattern of input qubits of $|A_0>$ and $|A_1>$. Keeping in mind that, each gate is controlled by an input qubit and it will be open only if the control qubit is $|1>$. Besides, always closely examine the shift between V and V+ gates.

Table 2.8 Truth table of a quantum AND operation

$|A_0>$	$|A_1>$	$|Q>$
$|0>$	$|0>$	$|0>$
$|0>$	$|1>$	$|0>$
$|1>$	$|0>$	$|0>$
$|1>$	$|1>$	$|1>$

1. When both inputs $|A_0>$ and $|A_1>$ are $|0>$,

   (a) The 1st V gate will not be open as the control input $|A_1>$ is $|0>$. And it will pass the constant bit $|0>$ to the next V+ gate as input.
   (b) The 2nd gate is CNOT which is controlled by $|A_0>$. Thus, this gate will pass the value of $|A_1>$ unchanged (i.e., $|0>$).
   (c) Now, the controlled-V+ gate will also remain closed (as the control qubit is $|0>$, from Step b) and it will work as a buffer. The output from this gate will be $|0>$ as well, which will be the input of the 2nd V gate.
   (d) Finally, the last V gate (2nd) which will produce the final output of the quantum AND operation, and it will get the input qubit $|0>$ and control qubit also $|0>$ (value of $|A_0>$). As a result, there will be no operational changes. And the final output will be $|0>$.

1. When inputs $|A_0> = |0>$ and $|A_1> = |1>$,

1. (a) The 1st V gate will be open as the control input $|A_1>$ is $|1>$. The 1st V gate gets input $|0>$ and the control qubit $|1>$. Therefore, it will produce $|v>$ as output (Table 2.2). This output will work as input for the V+ gate.
   (b) Sequentially, the 2nd gate is CNOT which is controlled by $|A_0>$ and it produces $|0>$. Thus this gate will pass the value of $|A_1>$ unchanged (i.e., $|1>$).
   (c) Now, the V+ gate will also be open (as the control qubit is $|1>$, from Step b) and it will be operated based on the input. The input of the V+ gate is $|v>$ (from Step a). So, this gate will produce $|0>$ as output (Table 2.3), which will be the input of the 2nd V gate.
   (d) Finally, the last V gate, it gets input qubit $|0>$ and control qubit $|0>$ (value of $|A_0>$). Therefore, no operational change will be occurred and the gate will work just like a buffer. So, the final output will be $|0>$.

1. When inputs $|A_0> = |1>$ and $|A_1> = |0>$,

1. (a) The 1st V gate will not be open as the control input $|A_1>$ is $|0>$. Thus no operational change will be occurred and it will pass $|0>$ as output, which will be the input for the V+ gate.
   (b) It is seen that the CNOT gate is controlled by $|A_0>$, where the value of $|A_0>$ is $|1>$ now. Thus this gate will invert the value of $|A_1>$ and give $|1>$ as output. This output will work as the control qubit for the V+ gate.

Table 2.9 Operations on each gate in the quantum AND operational circuit

$|A_0>$	$|A_1>$	1st V gate	CNOT gate	V+ gate	Final output (2nd V gate)
$|0>$	$|0>$	$|0>$	$|0>$	$|0>$	$|0>$
$|0>$	$|1>$	$|v>$	$|1>$	$|0>$	$|0>$
$|1>$	$|0>$	$|0>$	$|1>$	$|w>$	$|0>$
$|1>$	$|1>$	$|v>$	$|0>$	$|v>$	$|1>$

(c) Now, the V+ gate will be able to carry out the operation (as the control qubit is $|1>$, from Step b). If the input of the V+ gate is $|0>$ (from Step a), then this gate will produce $|w>$ as output (from Table 2.3), which will be the input of the 2nd V gate.

(d) Finally, the last V gate gets input qubit $|w>$ and control qubit $|1>$ (value of $|A_0>$). Therefore, the last V gate will produce output $|0>$ (from Table 2.2), and the final output is $|0>$.

1. When inputs $|A_0> = |1>$ and $|A_1> = |1>$,

1. (a) The 1st V gate will perform the operation because the control input $|A_1>$ is $|1>$. It will produce output $|v>$ which will be delivered to the V+ gate.

(b) Then, the CNOT gate will also work as the control qubit $|A_0>$ is $|1>$. Thus, this gate will invert the value of $|A_1>$ which will produce $|0>$ as output. As usual, this output will work as the control qubit for the V+ gate.

(c) Now, the controlled-V+ gate will remain closed, as the control qubit is $|0>$ (from Step b). Thus $|v>$ is obtained from this gate which will be the input of the last V gate.

(d) Finally, the last controlled-V gate gets input qubit $|v>$ and control qubit $|1>$ (value of $|A_0>$). Hence, the last V gate will produce the output $|1>$ (from Table 2.2), and the expected final output is $|1>$.

To understand comfortably the working mechanism of the quantum AND operation, consider the Table 2.9. For each input pattern, Table 2.9 presents the outputs from each gate in the quantum AND logic circuit.

2.3.4 Quantum NAND Operation

The quantum NAND operation is the opposite of the quantum AND operation. This implies that it will give the output $|0>$ only when both of the input qubits are $|1>$. Otherwise, the output is always $|1>$. Therefore, to construct the quantum NAND operation circuit diagram, it is needed to apply an extra quantum NOT operation

Fig. 2.8 Circuit diagram of
a quantum NAND operation

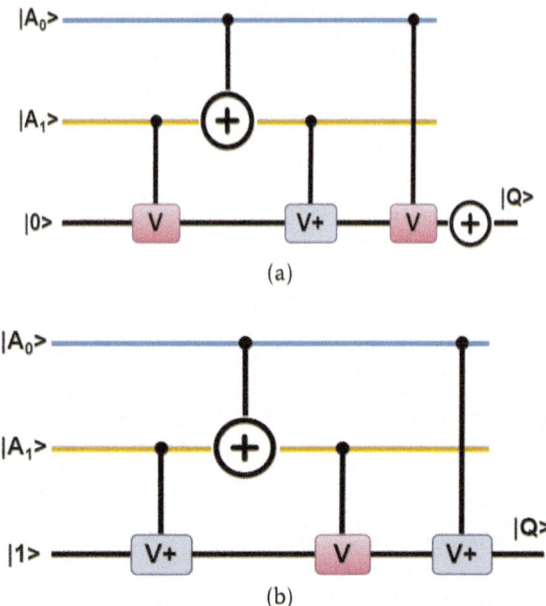

in addition to the quantum AND circuit. The circuit diagram of quantum NAND operation using the concept mentioned above is shown in Fig. 2.8a.

Let's have a close look at how to construct quantum NAND in a different way.

In (a) and (b) of Fig. 2.8, both are operational circuit diagrams of quantum NAND operation. The quantum NAND operation can be done by using any of them. So far, the most familiar idea is in Fig. 2.8a. But have a perceptive look at Fig. 2.8b. In Fig. 2.8b, the circuit was changed completely. Here two V+ gates, one CNOT gate, and one V gate are required, where $|1>$ is used as the ancilla qubit instead of $|0>$.

2.3.4.1 Working Principle of Quantum NAND Operation

As mentioned earlier the operation in Fig. 2.8a is exactly the same as the quantum AND operation, where it includes a quantum NOT operation that inverts the outputs of the quantum AND operation and produces the output of the quantum NAND (QNAND) operation (shown in Table 2.10). On the other way, a quantum NAND operation, also known as a NOT-AND operation, is a quantum logic operation that performs a NAND operation on two or more qubits. It outputs $|1>$ only if not all input qubits are $|1>$, and outputs $|0>$ if all input qubits are $|1>$. Quantum NAND operations are reversible and can be implemented using a sequence of other quantum operations, like the Toffoli gate followed by a Pauli X gate on the output

Table 2.10 Truth table of a quantum NAND operation

$|A_0>$	$|A_1>$	$|Q>$
$|0>$	$|0>$	$|1>$
$|0>$	$|1>$	$|1>$
$|1>$	$|0>$	$|1>$
$|1>$	$|1>$	$|0>$

qubit. Quantum operations, including NAND, are reversible, meaning you can undo the operation and return to the initial state. The quantum NAND operation, like its classical counterpart, is considered a universal operation, meaning any quantum circuit can be implemented using only NAND operations. In essence, a quantum NAND operation functions identically to its classical counterpart, performing the logical NAND operation on qubits, ensuring reversibility and allowing for universal quantum computation.

On that account, in this section, the working procedure of Fig. 2.8b is described which will show how the QNAND results are formed with that circuit for each pattern of input qubits of $|A_0>$ and $|A_1>$. Just a reminder, carefully notice the shift from V gate to V+ gate and vice versa.

1. When both inputs $|A_0>$ and $|A_1>$ are $|0>$,

1. (a) The 1st V+ gate will not be open as the control input $|A_1>$ is $|0>$. It leads $|1>$ (the ancilla qubit) as an input for the V gate.
   (b) The 2nd gate is CNOT which is controlled by $|A_0>$. Thus this gate will pass the value of $|A_1>$ unchanged (i.e., $|0>$) because the value of control qubit is $|0>$.
   (c) Now, the V gate will also remain be closed (as the control qubit is $|0>$, from Step b) and it will work as a buffer. The output from this gate will be $|1>$ as well, which will be the input of the 2nd V+ gate.
   (d) Finally, last V+ gate will produce the final output of the quantum NAND operation. It will get the input qubit $|1>$ and control qubit also $|0>$ (value of $|A_0>$). Therefore, no operational change will be occurred, and the final output is $|1>$.

1. When inputs $|A_0> = |0>$ and $|A_1> = |1>$,

1. (a) The 1st V+ gate will be open as the control input $|A_1>$ is $|1>$. The 1st V+ gate gets input $|1>$ and the control qubit $|1>$. Therefore, it will produce $|W>$ as output (Table 2.3). This output will work as input for the V gate.
   (b) Chronologically, the 2nd gate is CNOT which is controlled by $|A_0>$ and the value of $|A_0>$ is $|0>$. Thus, this gate will pass the value of $|A_1>$ unchanged (i.e., $|1>$).

(c) Now, the V gate will be open (as the control qubit is |1>, from Step b) and it will be operated based on the input. The input of the V gate is |W> (from Step a). So, this gate will produce |1> as output (Table 2.2), which will be the input of the next V+ gate.

(d) Finally, the last V+ gate gets input qubit |1> and the control qubit |0> (value of |A₀>). Therefore, no operational change will be occurred and this gate will work just like a buffer. So, the final output is |1>.

1. When inputs |A₀> = |1> and |A₁> = |0>,

1. (a) The 1st V+ gate will not be open as the control input |A₁> is |0>. Thus no operational change will be occurred, and it will pass |1> as output, which will be the input for the V gate.

(b) It is seen that the CNOT gate is controlled by |A₀> and the value of |A₀> is |1> now. Thus, this gate will invert the value of |A₁> and give |1> as output. This output will work as the control qubit for the V gate.

(c) Now, the V gate will be able to carry out the operation (as the control qubit is |1>, from Step b). The input to the V gate is |1> (from Step a). So, this gate will produce |V> as output, which will be the input of the next V+ gate.

(d) Finally, the last V+ gate gets the input qubit |V> and the control qubit |1> (value of |A₀>). Therefore, the last V gate will produce output |1> (from Table 2.2). So, the final output will be |1>.

1. When inputs |A₀> = I1> and |A₁> = |1>,

1. The 1st V+ gate will perform its operation because the control input |A₁> is |1>. It will produce output |W> which will be delivered to the V gate.

2. Then, the CNOT gate will also work as the control qubit |A₀> which is |1>. Thus, this gate will invert the value of |A₁> and it will produce |0> as output. Ordinarily, this output will work as the control qubit for the V gate.

3. Now, the V gate will remain closed, as the control qubit is |0> (from Step b). Thus, |W> is obtained from this gate which will be the input of the last V+ gate.

4. Finally, the last V+ gate gets input qubit |W> and control qubit |1> (value of |A₀>). Hence, the last V+ gate will produce output |0> (from Table 2.3). And as expected, the final output will be |0>.

Table 2.11 shows the working procedure of quantum NAND operation thoroughly for Fig. 2.8b.

2.3.5 *Quantum XOR Operation*

The logic behind Quantum XOR or simply QXOR for a two-valued system (binary system) is very simple; if the qubits are the same, the result is |0> and if the qubits are

Table 2.11 Operations on each gate in the quantum NAND operational circuit

$|A_0>$	$|A_1>$	1st V+ gate	CNOT gate	V gate	Final output (2nd V+ gate)
$|0>$	$|0>$	$|1>$	$|0>$	$|1>$	$|1>$
$|0>$	$|1>$	$|W>$	$|1>$	$|1>$	$|1>$
$|1>$	$|0>$	$|1>$	$|1>$	$|V>$	$|1>$
$|1>$	$|1>$	$|W>$	$|0>$	$|W>$	$|0>$

Fig. 2.9 Circuit diagram of a quantum XOR operation

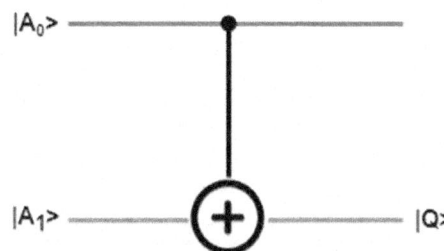

Table 2.12 Truth table of a quantum XOR operation

$|A_0>$	$|A_1>$	$|Q>$
$|0>$	$|0>$	$|0>$
$|0>$	$|1>$	$|1>$
$|1>$	$|0>$	$|1>$
$|1>$	$|1>$	$|0>$

different, the result is $|1>$. This operational circuit has already been designed because the quantum XOR is exactly the same as the quantum CNOT gate. Figure 2.9 shows the circuit diagram of quantum XOR operation, and Table 2.12 shows the truth table for quantum XOR operation. On the other way, in quantum computing, the XOR operation or exclusive-or, is often implemented using a CNOT (Controlled-NOT) gate. This gate's action on a two-qubit computational basis state is $|a, b>$ $\rightarrow |a, a \oplus b>$, where a and b are qubits. Essentially, the CNOT gate leaves the first qubit (the control qubit) unchanged and performs an XOR on the second qubit, effectively flipping it if the control qubit is 1. The CNOT gate achieves this in the quantum realm. It's a unitary operation (reversible), which is crucial for preserving coherence in quantum circuits. In essence, the CNOT gate allows quantum computers to perform the equivalent of an XOR operation on qubits, while maintaining the necessary quantum coherence and reversibility for complex quantum algorithms.

As it is equivalent to the CNOT gate, its working procedure is already shown.

In short, $|A_0>$ is the control input and the operation will occur only if the value of $|A_0>$ is $|1>$. Therefore, for four input patterns, in two cases the operation will be held and for the other two the output will be the same as the input value of $|A_1>$. As a result, for input values $|A_0> = |1>$, $|A_1> = |0>$ it will produce $|1>$ as output, and for input values $|A_0> = |1>$ and $|A_1> = |1>$ it will produce $|0>$ as output accordingly.

Fig. 2.10 Circuit diagram of
a quantum XNOR operation

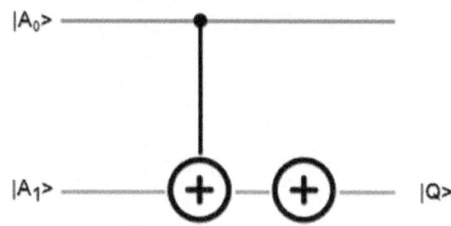

Table 2.13 Truth table of a
quantum XNOR operation

$|A_0>$	$|A_1>$	$|Q>$
$|0>$	$|0>$	$|1>$
$|0>$	$|1>$	$|0>$
$|1>$	$|0>$	$|0>$
$|1>$	$|1>$	$|1>$

2.3.6 Quantum XNOR Operation

Quantum Exclusive NOR or quantum XNOR or simply QXNOR is the inverted operation of quantum XOR operation. It inverts the output of the QXOR operation. Therefore, adding a quantum NOT gate to the output of the QXOR can construct the QXNOR operational circuit. On the other way, in quantum computing, the exclusive-NOR (XNOR) operation, a logical operator where the output is true only when both inputs are the same, is not directly implemented by a single quantum gate. Instead, it's obtained by combining other quantum gates, primarily the XOR gate (CNOT) and a NOT gate (X gate). To obtain the XNOR operation, we can follow these steps:

1. Apply CNOT: Use a CNOT gate with two qubits as inputs.
2. Apply X (NOT) gate: Apply an X gate (NOT gate) on the target qubit of the CNOT.
3. Result: This combination effectively mirrors the behavior of the XNOR gate.

Alternatively, we can use a CNOT gate and apply an X gate on the second qubit to implement XNOR operation in a quantum circuit. Figure 2.10 shows the circuit diagram of the quantum XNOR operation and Table 2.13 shows the truth table for quantum XNOR operation.

From Table 2.13, it is observed that the output will be $|1>$ only if both the input qubits are the same. Otherwise the output will be $|0>$.

2.4 Summary

Quantum logic has several fundamental quantum gates such as Pauli gate, Pauli-X gate, Hadamard gate, Toffoli gate, Fredkin gate, Deutsch gate, and Swap gate. But the synthesized fundamental quantum gates are the Controlled-NOT gate, the

Controlled-V gate, and the Controlled-V+ gate. The quantum logical expression and the classical logical expression are the same but the working procedure is different as the bit operations and qubit operations are not the same. The quantum logic gates and their operations using quantum circuits are presented in this chapter with their truth tables and working principles.

Bibliography

1. G.A. Barbosa, Quantum half-adder. Phys. Rev. A **73**(5), 052321 (2006)
2. L. Diósi, *A Short Course in Quantum Information Theory: An Approach from Theoretical Physics*, vol 827 (Springer, 2011)
3. N. Isailovic, Y. Patel, M. Whitney, J. Kubiatowicz, in *33rd International Symposium on Computer Architecture (ISCA'06)*. Interconnection Networks for Scalable Quantum Computers (IEEE, 2006), pp. 366–377
4. L.B. Levitin, T. Toffoli, Z. Walton, *Operation Time of Quantum Gates* (2002). arXiv preprint quant-ph/0210076
5. S.T. Marella, H.S.K. Parisa, in *Quantum Computing and Communications*. Introduction to Quantum Computing (IntechOpen, 2020)
6. T.S. Metodi, D.D. Thaker, A.W. Cross, F.T. Chong, I.L. Chuang, in *38th Annual IEEE/ACM International Symposium on Microarchitecture (MICRO'05)*. A Quantum Logic Array Microarchitecture: Scalable Quantum Data Movement and Computation (IEEE, 2005), 12 pp
7. M. Mohammadi, M. Eshghi, in *2008 5th International Multi-Conference on Systems, Signals and Devices*. Behavioral Description of Quantum v and v+ Gates to Design Quantum Logic Circuits (IEEE, 2008), pp. 1–5
8. M. Morrison, *Design of a Reversible ALU based on novel reversible logic structures* (University of South Florida, 2012)
9. A. Muthukrishnan, *Classical and Quantum Logic Gates: An Introduction to Quantum Computing Quantum Information Seminar* (1999)
10. D.D. Thaker, T.S. Metodi, A.W. Cross, I.L. Chuang, F.T. Chong, in *33rd International Symposium on Computer Architecture (ISCA'06)*. Quantum Memory Hierarchies: Efficient Designs to Match Available Parallelism in Quantum Computing (IEEE, 2006), pp. 378–390
11. H. Thapliyal, N. Ranganathan, Design of reversible sequential circuits optimizing quantum cost, delay, and garbage outputs. ACM J. Emerg. Technol. Comput. Syst. **6**(4), 1–31 (2010)

Chapter 3
Biocomputing

3.1 Introduction

Biocomputing or DNA (Deoxyribonucleic Acid) computing or biological computing, also known as molecular computing, is a new approach to massively parallel computation. So, biocomputing or DNA computing is an emerging interdisciplinary field that harnesses the information processing capabilities of biological
substrates like DNA, proteins and cells to perform computational tasks. Rather
than relying solely on conventional silicon-based computers, bio computing leverages the innate computational properties of biomolecules to encode, store, process
and transmit information in unconventional ways. Core approaches include DNA
computing, which uses DNA biochemistry to solve problems in a massively parallel fashion. Protein computing utilizes protein conformational dynamics to implement logic gates and communication modules for molecular information processing. Cellular computing focuses on engineering gene circuits and synthetic biology tools to program computational behaviours in living cells. Neural computing builds artificial neural networks inspired by biological brains. Key application areas include biomedicine, smart drug delivery systems, biosensing, hybrid
organic-inorganic electronics, and biomolecular manufacturing. While still facing
challenges around biocompatibility, programming complexity and ethical concerns,
bio computing has achieved major technical milestones demonstrating its promise.
Continued progress at the interface of biology and computing could enable future
technologies like bio processors, in-vivo biocomputers, living materials and bio-
intelligent systems. With responsible development, bio-inspired computation may
catalyse the next revolution in human technological capabilities. This emerging
field thus warrants enthusiastic attention as computation further converges with
the living world. DNA computing was introduced as the means of solving a class
of intractable computational problems in which the computing time can grow
exponentially with problem size (the 'NP-complete' or the non-deterministic polynomial time complete problems). A DNA computer is basically a collection of

specially selected DNA strands whose combinations will result in the solution to some problems depending on the problem at hand. Technology is currently available both to select the initial strands and to filter the final solution. DNA computing is a new computational paradigm that employs molecular manipulation to solve computational problems at the same time exploring natural processes as computational models. Biocomputing or DNA computing, which uses biological components for computation, offers significant potential for breakthroughs in healthcare, environmental monitoring, and bio-manufacturing. It promises natural interfaces, cognitive enhancements, and potentially even consciousness transfer. However, it also faces challenges like scalability, ethical concerns, and the need for precise control of biological systems. Some examples of biocomputing include

1. DNA computing: Using DNA and molecular biology tools to solve mathematical problems. DNA molecules can encode information and molecular operations on DNA like annealing can perform parallel computations.
2. Protein computing: Using protein interactions and conformational changes to perform logic operations and calculations. Proteins can switch between different confirmations in response to inputs like other molecules binding, allowing them to mimic logic gates and circuits.
3. Cellular computing: Programming gene circuits and networks within living cells to carry out sensing, information processing and actuation tasks. Synthetic biology allows engineering cells with toggle switches, oscillators, logic gates etc.
4. Neural computing: Building artificial neural networks that are inspired by information processing in biological brains. The connections between neural network nodes mimic the synaptic signaling between neurons in the brain.
5. Molecular computing: Designing and synthesizing molecules with specific structures so they can implement algorithmic functions and calculations when reacting with each other. The molecules effectively act as tiny programmable computers.

   The DNA found in living cells is composed of four bases, viz. Adenine (A), Guanine (G), Thiamine (T), and Cytosine (C). Order of these bases is unique in each individual and determines the unique characteristics of that particular individual. Each base is attached to its neighboring base in the sequence via phosphate bonding. The base, sugar, and phosphate are together called a nucleotide. Two DNA sequences bond with each other via hydrogen bonding between each Watson-Crick complementary base pairs (A with T, and C with G) form DNA double helix. Each DNA strand has two ends: 5′-end and 3′-end that determine the polarity of the DNA strand. During the formation of DNA double-strand two complementary single strands bond with each other in an anti-parallel fashion. From now on, biocomputing will be called as DNA computing.

   DNA computing accomplishes the computations using genetic molecules rather than traditional silicon chips. A computation may be assumed as the execution of an algorithm, which itself may be defined as a step-by-step list of well-defined instructions that take some inputs, process it, and yield a result. In DNA computing, information is represented using the four-character genetic alphabet (A, G, T and C), rather than the binary alphabet (1 and 0) used by traditional computers. This is feasible because short DNA molecules of any arbitrary sequence may be produced to order.

DNA computers use strands of DNA to perform computing operations. The computer consists of two types of strands - the instruction strands and the input data strands. The instruction strands splice together the input data strands to generate the desired output data strand. DNA computing holds the promise of important and significant connections between computers and living systems, as well as promising parallel computations massively.

The DNA is the major information storage molecule in living cells, and billions of years of evolution have tested and refined both this wonderful informational molecule and highly specific enzymes that can either duplicate the information in DNA molecules or transmit this information to other DNA molecules. Instead of using electrical impulses to represent bits of information, the DNA computer uses the chemical properties of these molecules by examining the patterns of combination or growth of the molecules or strings. DNA can do this through the manufacture of enzymes, which are biological catalysts that could be called the 'software', used to execute the desired calculation.

Problems are solved in a DNA computer by encoding a problem using A, G, T and C. However, a compact DNA computer is constructed by synthesizing corresponding DNA strands which performs several operations on those strands by various methods. Main powers of DNA computing over conventional ones are as follows:

1. Massively parallel operation: A single test tube of DNA can contain trillions of DNA strands, and all strands respond to the biological operations in parallel.
2. Huge information density: Information density of DNA is huge over silicon. Estimated storage capacity of 490 Exa-bytes per gram of DNA.

DNA computing has drawn the attention of many a researcher in recent years for its applicability to solve computationally hard problems. It can perform faster than conventional computers with its inherent massively parallelism nature. Proper synthesis of reversible circuit is a well-researched computing problem in recent days for its particularly low power consumption (ideally zero) and inherent reversible nature of reversible logic. The optimal synthesis of a reversible truth table means finding the reversible circuit made up of reversible gates satisfying the given truth table with the optimum cost. Nowadays, reversible logic has emerged as a promising computing paradigm having its solicitations in low power computing, quantum computing, nanotechnology, optical computing, and DNA computing. The classical set of gates such as AND, OR, and EX-OR are not reversible. Recently, it was shown how to encode information in DNA and use DNA amplification to implement some reversible gates. Furthermore, in the past reversible gates such as Fredkin gates have been constructed using DNA, whose outputs are used as inputs for other reversible gates. Thus, it can be concluded that arbitrary circuits of reversible gates can be constructed using DNA. This has been the driving force leading to the design of reversible adder and multipliers using Fredkin gate.

DNA stands for deoxyribonucleic acid, is the hereditary material in humans and almost all other organisms. Nearly, every cell in a person's body has the same DNA. Most DNA is located in the cell nucleus (where it is called nuclear DNA), but a small amount of DNA can also be found in the mitochondria. Challenges and limitations currently facing the field of biocomputing or DNA computing or biological computing are as follows:

1. One major challenge is creating biocompatible systems that can integrate and function effectively within biological environments and subjects. Biological tissues present a complex milieu of molecules, cell types and interactions that engineered systems must adapt to. Immunogenicity issues can arise whereby implanted bio computing devices trigger unwanted immune reactions. Approaches to improve biocompatibility include biomimetic designs using natural biological materials, bio-inert surface coatings, and localized release of immunosuppressant drugs.

2. Programming and encoding complexity is another hurdle. Engineering robust gene circuits or neural networks requires sophisticated design tools, modeling frameworks, and debugging cycles.82 Synthetic biology is working to create modular, well-characterized genetic "parts" that can be predictably assembled. Abstraction layers and computer-aided design software also help hide low-level complexity. DNA sequence optimization algorithms assist in filling design specifications.

3. Wet lab experimentation remains time-consuming and laborious. Standardizing protocols, automation technology like liquid handling robotics, and foundries for fabrication can relieve workflow bottlenecks. Microfluidics miniaturizes experiments onto chips and allows precise environmental control over reactions. High-throughput screening tools test libraries of design variants in parallel.

4. Analysing and characterizing the dynamics of engineered networks is non-trivial. Researchers are devising mathematical models and multi-scale computational simulations to predict system behaviours before costly lab work. Advanced microscopy and "omics" tools facilitate quantitatively tracking molecular mechanisms.

5. Maintaining the viability of engineered organisms and cells is an issue, as synthetic gene circuits add metabolic load. Strategies like genome streamlining, component optimization for low toxicity, and nutritional feedback controls help improve durability. Decoupling designs into separate survival and task-based modules also helps.

6. Interfacing engineered systems with the complexity of real-world environments remains challenging. Bio-hybrid interfaces that connect synthetic biology with traditional electronics and hardware are still maturing. Onboard power sources or wireless power delivery are active research areas. Orthogonal communication schemes isolate synthetic systems from natural biological crosstalk.

7. Safety and ethical concerns exist around bio computing applications like human augmentation or environmental release. Robust safeguards against unintended effects, molecular containment, and reversible engineering are important areas

of investigation. Policy groups also advocate early awareness, monitoring and regulation around such engineering.

8. While significant hurdles exist, researchers are making steady progress through foundational engineering principles like modularity, model-based design, optimization, and characterization. Continued technology innovation and inter-disciplinary collaboration will aid in systematically addressing the challenges on the path ahead.

Biocomputing finds applications in various fields, including medicine, diagnostics, and even cryptography.

3.2   Structure and Function of DNA

The information in DNA is stored as a code made up of four chemical bases: Adenine *(A)*, Guanine *(G)*, Cytosine *(C)*, and Thymine *(T)*. Human DNA consists of about 3 billion bases, in which more than 99 percent of those bases are the same in all people. The order or sequence, of these bases, determines the information available for building and maintaining an organism, similar to how letters of the alphabet appear in a certain order to form words and sentences. DNA bases pair up with each other, *A* with *T* and *C* with *G*, to form units called base pairs. Each base is also attached to a sugar molecule and a phosphate molecule. Together, a base, sugar, and phosphate are called a nucleotide. Nucleotides are arranged in two long strands that form a spiral called a double helix. Figure 3.1 shows the Hydrogen Bonds of the Interior DNA and Fig. 3.2 shows A Double Helix Structure; Fig. 3.3 shows Phosphodiester Bonds; and Fig. 3.4 shows DNA structure, Major and Minor Grooves are binding sites for DNA binding proteins during processes such as transcription (the copying of RNA from DNA) and replication (Figs. 3.1, 3.2, 3.3, and 3.4).

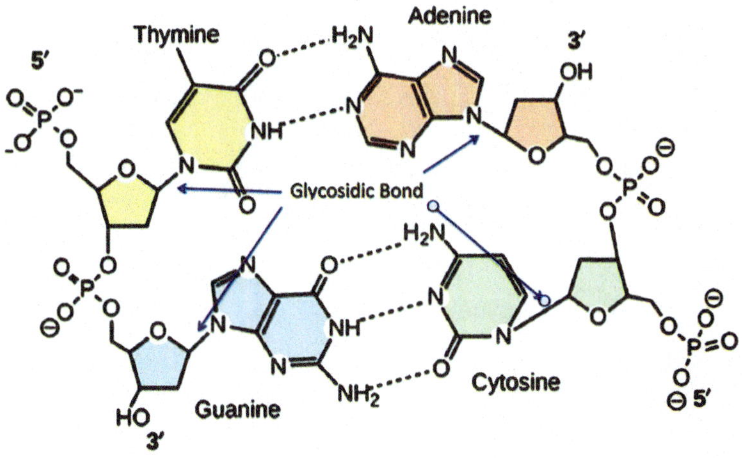

Fig. 3.1 Hydrogen bonds of the interior DNA

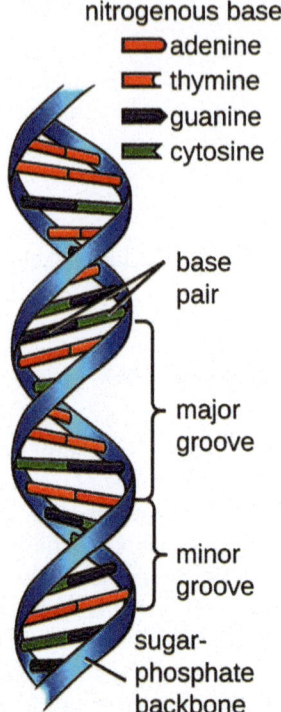

Fig. 3.2   A double helix structure

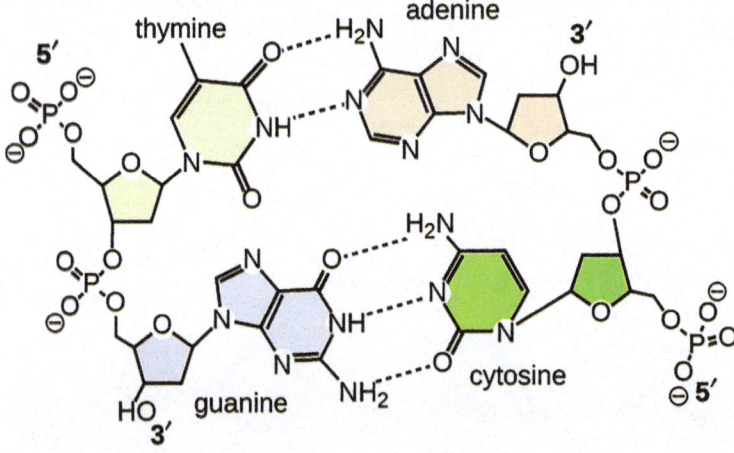

Fig. 3.3   Phosphodiester bonds

Fig. 3.4 DNA structure

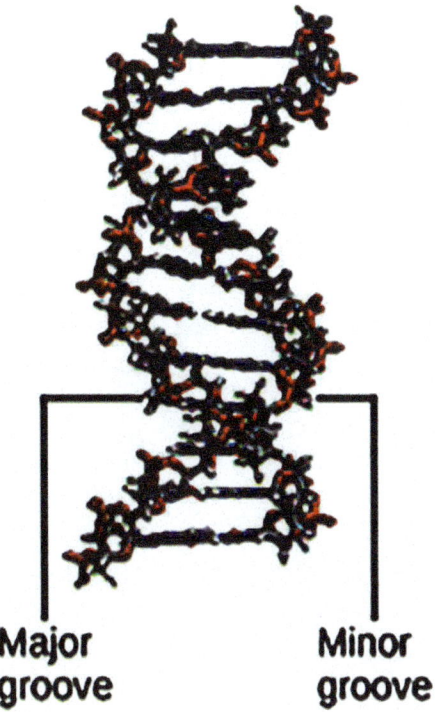

**Major
groove**

**Minor
groove**

The structure of the double helix is somewhat like a ladder, with the base pairs forming the ladder rungs and the sugar and phosphate molecules forming the vertical side pieces of the ladder. An important property of DNA is that it can replicate, or make copies of itself. Each strand of DNA in the double helix can serve as a pattern for duplicating the sequence of bases. This is critical when cells divide because each cell needs to have an exact copy of the DNA present in the old cell (Fig. 3.5). The DNA structure is shown in Fig. 3.5.

DNA denaturation is the process of conversion of double-stranded DNA into single strands due to heat or alkali-induced breakage of hydrogen bonds and van der Walls force between every two nitrogen bases from two different strands.

By slow cooling, the denatured single strands can reunite according to the complementarily principle ($A = T$; G^G) allowing DNA to regain its native double helix; this phenomenon is called denaturation.

Ligation is the chemical process of joining DNA molecules together by forming phosphodiester bond between two deoxyribose sugars of different terminal nucleotides of those molecules with the activity of ligase enzyme (Fig. 3.6). The ligation process is shown in Fig. 3.6.

The polymerase chain reaction (PCR) is a biochemical technology to amplify a single or a few copies of a piece of DNA across several orders of magnitude, generating thousands to millions of copies of a particular DNA sequence. It relies on thermal

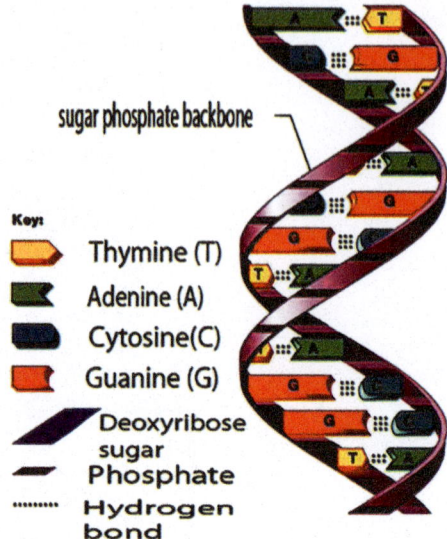

Fig. 3.5 Structure of DNA

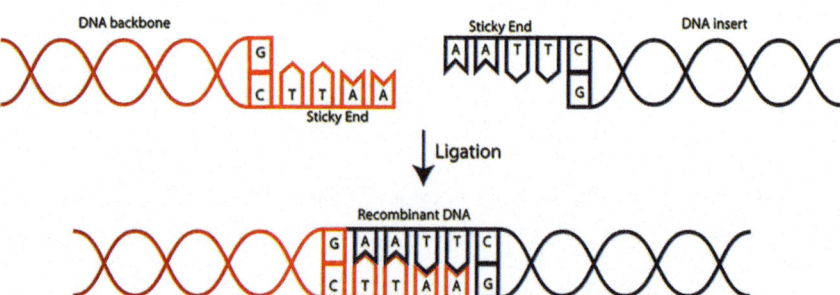

Fig. 3.6 Ligation process of DNA

cycling, consisting of cycles of repeated heating and cooling of the reaction for DNA melting and enzymatic replication of the DNA. Primers containing sequences complementary to the target region along with a DNA polymerase are key components to enable selective and repeated amplification. As PCR progresses, the DNA generated is itself used as a template for replication, setting in motion a chain reaction in which the DNA template is exponentially amplified, PCR can be extensively modified to perform a wide array of genetic manipulations.

Capillary electrophoresis, also known as capillary zone electrophoresis, can be used to separate ionic species by their charge and frictional forces and hydrodynamic radius. It is the most efficient separation technique available for the analysis of both large and small molecules of DNA.

3.3 DNA Computing

DNA computing or biocomputing or biological computing is a form of computing which uses DNA, biochemistry, and molecular biology, instead of the traditional silicon-based computer technologies. DNA computing, or, more generally, biomolecular computing, is a fast developing interdisciplinary area. Research and development in this area concerns theory, experiments, and applications of DNA computing. Future outlook and emerging trends in biocomputing are as follows:

1. Looking ahead, bio computing is poised to integrate more deeply with fields like artificial intelligence, robotics, and the Internet of Things. One avenue is developing bio-hybrid AI systems, combining biological computing substrates and learning algorithms for perception and inference tasks. Engineered organisms that synthesize their sensors and logic could enable fully autonomous, adaptable biocomputers.
2. Within the body, networks of engineered cells may one day run physiological regulation and repair routines like biological robots. In-vivo biocomputers could monitor organ health and coordinate therapeutic responses, forming a distributed treatment system. Nano-bioelectronics will miniaturize bio/organic interfaces for seamless integration. Implanted neural lace devices could allow direct brain-computer communication.
3. DNA digital data storage is emerging as an ultra-dense, stable alternative to silicon memory. Entire datasets, books and videos have been encoded as DNA sequences. Integrating DNA memory with biological processors will enable storing and accessing massive information troves for AI. DNA could also allow on-chip training of nanoscale neural nets.
4. Cell-free synthetic biology promises to expand bio manufacturing capabilities beyond living organisms. Printing hybrid bio-electronic materials containing engineered proteins and nucleic acids may support wearable, self-repairing soft robotics for human augmentation. Bio computing could thus distribute "enhanced intelligence" ubiquitously via engineered biomaterials.
5. Self-organizing cellular systems that reshape and reconfigure on command will lead to programmable, morphing biohybrid materials for drug delivery or tissue engineering. Viral engineering for nanofabrication will create manufacturing platforms integrating top-down and bottom-up processes. Bio computing could thereby revolutionize digital fabrication, smart materials, and sustainable manufacturing.
6. Protecting privacy and security of biometric data will be crucial as human-machine biointerfaces become intimate and pervasive. Ethical guidelines must shape applications for human improvement versus entrenching inequity. Overall, bio computing could fundamentally reshape our information infrastructure— while navigating immense opportunities and challenges along the way. Interdisciplinary collaboration, public awareness and appropriate regulations will help guide responsible progress.

3.3.1   Watson-Crick Complementary

By the natural of DNA molecule, Watson-Crick complementary plays the most important role in the DNA computing. DNA consists of four bases of nucleic acid, such as Adenine *(A)*, Guanine *(G)*, Cytosine *(C)*, and Thymine *(T)*. Adenine can only connect with Thymine, and Cytosine that can only connect with Guanine (not *A* always *T* and not *C* always equals *G*).

3.3.2   Adleman's Breakthrough

The first ever wet experiment that proved DNA and biochemical processes could be used as a computing tools to solve complex computational problem has been done by Prof. L.M. Adleman in 1994. In seven days of lab experiment, Adleman became successful to solve Hamiltonian Path Problem (HPP) of seven cities. HPP is a special case of the traveling salesman problem, obtained by setting the distance between two cities to a finite constants if they are adjacent and infinity otherwise. In HPP, it can assume G consisting of vertices $v_1, v_2, \ldots, v_n$, and v_{start} and v_{end}. One of the main requirements that should be met by HPP is, the directed graph if and only if there exists a sequences of compatible one-way edges. $v_1, v_2, \ldots, v_n$, begin with v_{start} and end with v_{end}, enter another vertex exactly only one time. Figure 3.7 shows the HPP problem that was solved by Adleman in his first wet experiment. To find out the unique Hamiltonian path from the directed graph, Adleman has followed a non-deterministic algorithm as presented below:

Algorithm 3.3.1 Non-deterministic algorithm

1: Begin
2: Generate random paths through the graph. Keep only those paths that begin with v_{start} and end with v_{end}.
3: If the graph has n vertices, keep only those paths that enter exactly n vertices.
4: Keep only those paths that enter all of the vertices of the graphs at least once.
5: If any path remains, say Yes, otherwise say No.
6: End

In Algorithm 3.3.1, Adleman has generated randomly 20-*mer* of DNA sequences to represent each city denoted as O. On the other hand, to represent a vertex that connected between two different cities, Adleman suggested the DNA designed a combination of O_i and O_{i-1}. After all DNA has been synthesize to represent all available vertices, 50 *mol* of 0 and 50 mol of O (represent WC complementary of all vertices) added together in one test tube for mixing together in single ligation process. This process will produce randomly all possible combinatorial solution for the graph. Only vertex that passes all cities will be considered as a feasible solution. Therefore, Adleman employed PCR technique to check whether the strand pass all

Fig. 3.7 HPP on seven
vertices

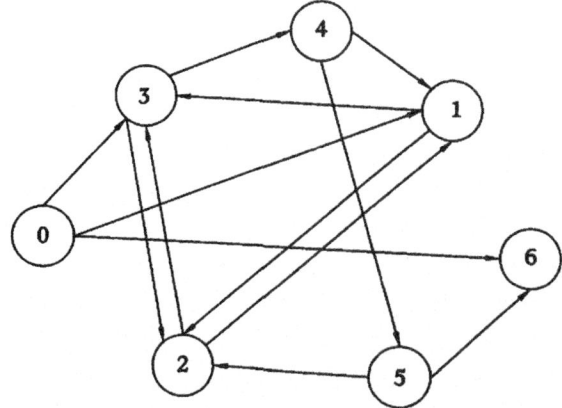

cities or not. By using and as a primer, biochemical technique allowed selecting only
related strands and discharging unrelated strands that do not fulfill the requirement.
At the end of this process, Adleman can only have thus strands that start with city
and end with city as earlier requirement in his hand. This process is explained in
Line 3 in his algorithm.

The output of Line 3 will undergo electrophoresis process. For this purpose,
Adleman chose to run the electrophoresis process. This process will sort the strand
according to their size. Because, here the number of cities is seven, so only strands
have 140-bp (base pair) band (corresponding to double-stranded strand) was excited
and soaked in double distilled $H_2O\,(dd\,H_2O)$. So, only strands that pass seven cities
will be extracted in this process to pursue the next process.

This process is explained in Line 4 in Adleman algorithm. In order to realize Line
4 in Adleman's algorithm, he employed a magnetic beads separation process. It is
most labor-intensive work. In this process, each complementary of city was used to
check either all the cities are existed in the strands. The procedure will be iterated
until the result is obtained. Even though the experiment took almost seven days'
work bench experiment, and labor-intensive work, but the result is very acceptable
arid, it gives a new approach to solve most complicated calculation especially when
deal with huge amount of variables.

3.4 Relationship Between Binary Logic and DNA Logic

Several DNA-based designs describe the relationships between 0/1 logic and DNA
logic. However, the DNA-based design shows a cleaver approach to encode 0/1 logic
using two types of single DNA strands, one for input bit and another for operand
bit. Also in the design, a reversible Toffoli gate is realized using different DNA
strands for control inputs and target input of Toffoli gate where control input signals

are constructed using the dinucleotide $5'$-AG and $5'$-CT representing bit 1 and 0 respectively. The target input is constructed with $5'$-ψA and $5'$-UA as bit 1 and 0 respectively. Similarly, the final output is constructed using mixer of $3'$-AU, $3'$-AD, $3'$-AT, and $3'$-AX representing bit 1, 1, 0, and 0 respectively.

3.5 Relationship Between DNA Logic and Quantum Logic

DNA logic and quantum logic are distinct concepts, though both involve manipulating information. DNA logic uses the physical properties of DNA molecules (like binding and shape) to perform logical operations, while quantum logic utilizes the principles of quantum mechanics (superposition, entanglement) to achieve computational tasks.

DNA Logic: DNA logic relies on the natural interactions of DNA strands. For example, specific DNA sequences can bind to each other, creating a logical "AND" gate where the output (binding) occurs only if both input sequences are present. This approach leverages the chemical and physical properties of DNA for computation.

Quantum Logic: Quantum logic operates on the principles of quantum mechanics. A key concept is the use of qubits, which can exist in a superposition of states (both 0 and 1 simultaneously), unlike classical bits. Quantum logic gates manipulate these qubits to perform calculations, often offering potential advantages in speed and complexity for certain types of problems.

Similarities: Both DNA logic and quantum logic aim to process information and perform computations.

Differences:

Physical Basis: DNA logic relies on the molecular structures and interactions of DNA, while quantum logic uses quantum mechanical principles.

Qubit vs. Bits: Quantum logic uses qubits, which can be in a superposition of states, whereas DNA logic typically works with classical bits (0 or 1) in a way that can be described using classical logic.

Potential for Quantum Advantage: Quantum logic has the potential to offer computational advantages, such as faster computations for certain types of problems, due to quantum phenomena like entanglement and superposition.

Complexity: DNA logic can be complex due to the many factors influencing DNA interactions (like temperature, pH, and the presence of other molecules), while quantum logic's complexity stems from controlling and manipulating entangled qubits.

Connection: While distinct, there's research exploring how DNA might be used in quantum computing, potentially to create qubits or implement quantum logic gates. For example, certain DNA sequences have been shown to behave as qubits or to act as quantum logic gates.

DNA as a potential quantum computing platform: Some research suggests that DNA could be a potential medium for quantum computing, leveraging the properties of DNA molecules (like base pairing) to implement qubits and logic gates. However, the challenges of scaling up and maintaining the coherence of these systems are significant.

3.6 Advantages of DNA Computing

The main advantages of DNA-based circuits over the silicon chips are as follows:

- In double-strand DNA the data density will be one base per square nanometer and the data density will be over one million Gbit per square inch, where typical high-performance hard drive is used and the data density is about 7-Gbit per square inch.
- Base pair complement gives a unique error correction mechanism, which works as like RAID 1 array.
- As many copies of the enzyme can work on many DNA molecules simultaneously. This is the power of DNA computing, that it can work in a massively parallel fashion.
- In DNA replication, enzymes start on the second replicated strand of DNA even before they are finished copying the first one. So data rate jumps to 2 times of initial speed (initially it is 1000 bits/sec). After each replication is finished, the number of DNA strands increases exponentially. For example: After 30 iterations, it increases to *1000-Gbit/sec*. This is beyond the sustained data rates of the fastest hard drives.
- DNA is a stable molecule that never suffers any changes (mutation) unless it faces harsh (very high temperature, corrosive agents, etc.) environment.
- DNA logic gates can be preserved for a very long time (more than a decade) by maintaining and varying the temperature.
- A tiny energy is required to break the bound when operating DNAs. For example, the energy required to break the bond between A and T is $\equiv 21$ *KJ/mol.* where A denotes Adenine and T denotes Thymine. The same 21 *KJ/mol* will be gained if a bond between them is formed again. That means, energy is reserved in DNA-based logic circuits (Fig. 3.8).

Figure 3.8 presents the DNA replication process. Some potential benefits of bio-computers or DNA computers can be summarized as follows:

1. Increased speed and efficiency: Biocomputers have the potential to drastically increase the speed and efficiency of many processes.
2. Greater accuracy: Biocomputers may be able to perform complex calculations with greater accuracy than traditional computers.
3. Lower power consumption: Biocomputers can potentially operate using less energy than traditional computers.
4. Adaptability: Biocomputers can potentially be reprogrammed for different tasks by using different sequences of DNA.
5. Parallel processing: Biocomputers or DNA computers can potentially perform many operations in parallel, which could lead to further increases in processing speed.

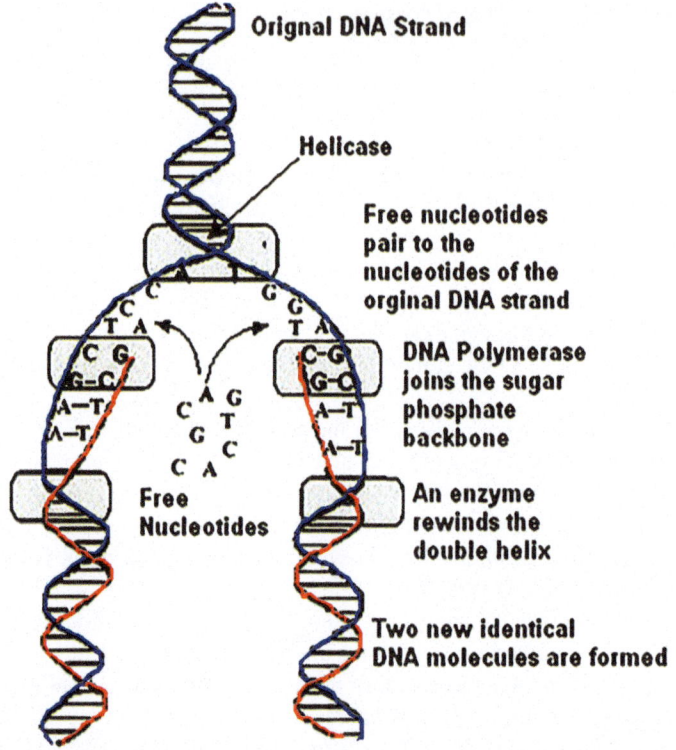

Orignal DNA Strand

Helicase

Free nucleotides pair to the nucleotides of the orginal DNA strand

DNA Polymerase joins the sugar phosphate backbone

Free Nucleotides

An enzyme rewinds the double helix

Two new identical DNA molecules are formed

Fig. 3.8 DNA replication process

3.7 Summary

This chapter introduces the concept of DNA computing, main properties of DNA computing, and the process how the computing is done using DNA. Some graphical representations corresponding to the structures of DNA are also illustrated. In addition, the relationship of binary logic with DNA is also described. Moreover, the DNA replication process is also presented in this chapter.

Bibliography

1. T.A. Brown, How ancient dna may help in understanding the origin and spread of agriculture. Philos. Trans. R. Soc. Lond. Ser. B Biol. Sci. **354**(1379), 89–98 (1999)
2. G. Cui, C. Li, H. Li, X. Li, in *2009 Fifth International Conference on Natural Computation*. DNA Computing and Its Application to Information Security Field, vol 6 (IEEE, 2009), pp. 148–152

3. S.A. El-Seoud, R. Mohamed, S. Ghoneimy, DNA computing: challenges and application. Int. J. Interact. Mobile Technol. **11**(2) (2017)
4. Z. Ezziane, DNA computing: applications and challenges. Nanotechnology **17**(2), R27 (2005)
5. Hafiz Md. Hasan Babu, "Reversible and DNA Computing", Wiley Publishers, UK (2021)
6. Hafiz Md. Hasan Babu, "DNA Logic Design: Computing with DNA", World Scientific Publishing Company, May 2024, Singapore
7. E. Katz, in *DNA-and RNA-Based Computing Systems*. DNA Computing: Origination, Motivation, and Goals—Illustrated Introduction (2021), pp. 1–14
8. M. Kaur, DNA computing: its advantages and future. J. Teach. Educ. **1**(7), 51–59 (2012)
9. X. Liang, W. Zhu, Z. Lv, Q. Zou, *Molecular Computing and Bioinformatics* (2019)
10. D. Sharma, M. Ramteke, in *DNA-and RNA-Based Computing Systems* (Methodologies and Challenges, DNA Computing, 2021), pp.15–29
11. A.A. Sori, A.A. Sori, DNA computer; present and future. Int. J. Eng. Res. Appl. **4**(6), 228–232 (2014)
12. J. Watada, in *Computational Intelligence: A Compendium*. DNA Computing and Its Application (Springer, 2008), pp. 1065–1089
13. J.G. Wetmur, DNA probes: applications of the principles of nucleic acid hybridization. Crit. Rev. Biochem. Mol. Biol. **26**(3–4), 227–259 (1991)

Chapter 4
Fundamentals of Biocomputing

4.1 Introduction

Biocomputing or DNA computing or biological computing is a field that uses DNA, biochemistry, and molecular biology rather than traditional silicon chips for computations. Recently researchers have emphasized exploring this field because of several advantages of DNA computing. For example, Boolean logic uses two input or output states, and binary logic faces challenges from non-Boolean logic when processing uncertain or imprecise information. Biocomputing is an emerging branch of computing that replaces traditional electronic computing with deoxyribonucleic acid (DNA), biochemistry, and molecular biology hardware. It may seem strange that computation may be done in a test tube using biological molecules rather than semiconductor chips.

Millions of natural supercomputers exist inside living organisms. DNA (deoxyribonucleic acid) molecules, which make up human genes, have the ability to execute calculations hundreds of times quicker than even the most powerful human-built computers. DNA could one day be incorporated into a computer chip to produce a "biochip" that will allow computers to run even quicker. Complex mathematical problems have already been solved using DNA molecules.

This chapter will present how operations can be performed in DNA computing and the detail of the DNA logic operational gate.

4.2 DNA Computing Operations

In classical computing, data storage device stores data by converting them into binary digits. But in DNA computing, instead of binary digits, information or data will now be kept in the form of the nitrogen bases **A**, **T**, **G**, and **C**. These bases will make sequences to store data, and encoding and decoding are required to perform the operation with that DNA sequences so that the outcomes become meaningful.

The capacity to generate short DNA sequences artificially allows these sequences to be used as inputs for algorithms. DNA has properties that will allow it to be used to simulate classical logic processes. Single-stranded DNA naturally migrates toward complementary sequences to form double-stranded complexes, whereas double-stranded DNA wants to be in double-stranded form.

A program on a DNA computer is executed as a series of synthesizing, extracting, modifying, and cloning the DNA strands. Instead of using electrical impulses to represent bits of information, the DNA computer uses the chemical properties of DNA molecules by examining the patterns of combination or growth of the molecules or strings. DNA can do this through the manufacture of enzymes, which are biological catalysts that could be called the "software", used to execute the desired calculation. Enzymes do not function sequentially, working on one DNA at a time. Rather, numerous copies of the enzyme can act massively parallel on many DNA molecules concurrently. DNA computers work by encoding the problem to be solved in DNA's language: the base-four number system, which includes the base-four values **A**, **T**, **C**, and **G**, which is more than enough when compared to an electronic computer, which only requires two numbers, 0 and 1.

DNA has cut, copying, pasting, repairing, and many other operations, just like a CPU has addition, bit-shifting, logical operators, and so on, that allow it to accomplish even the most complex computations. The right sequences are sorted out using genetic engineering methods in a DNA computer, which computes in test tubes or on a glass slide coated in 24K gold.

4.3 Performing Fundamental Operations in DNA Computing

As mentioned earlier, the sequence of base patterns will store pieces of information. In two-valued (binary) DNA operations, consider

ACCTAG = true, which is equivalent to binary "1", and
TGGATC = false, which is equivalent to binary "0".

These base sequences represent data. With these base sequences, all fundamental gates and operations will be shown in the following sections.

4.3.1 DNA NOT Operation

The basic binary logical operation is fully conversant to us. As a result, the logical function of binary logic gates does not need to be explained again. Only how to

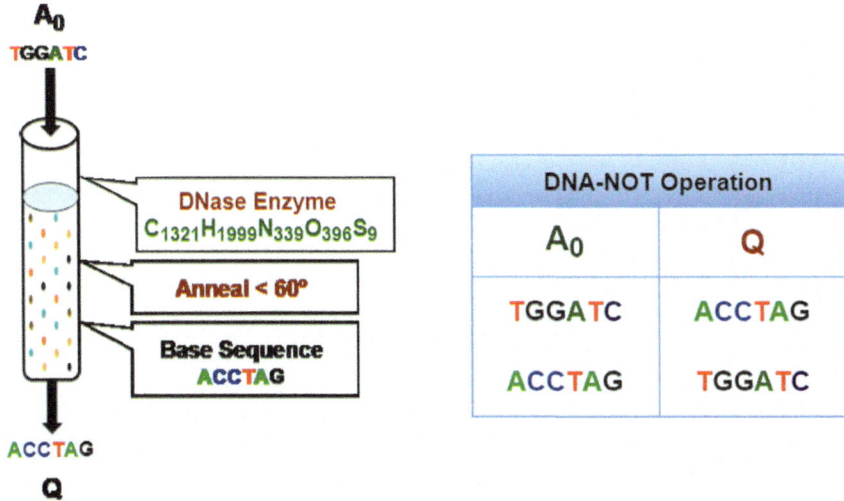

Fig. 4.1 Circuit architecture of DNA NOT operation

Table 4.1 Truth table of DNA NOT operation

A0	Q
TGGATC	ACCTAG
ACCTAG	TGGATC

Fig. 4.2 Pair matching between the DNA base sequences

construct them in the DNA computer to perform DNA computing will be discussed. Figure 4.1 shows the operational diagram of the DNA NOT operation. To design this, a test tube is needed with the DNA mixture, the annealing temperature is less than 60 °C. To perform the operation DNase Enzyme is needed and the base sequence ACCTAG will be used.

The DNA NOT operation inverts the input base sequence. The operational table shows the input-output mapping in Table 4.1. Remember, the base sequences TGGATC, and ACCTAG represents the Boolean false and true respectively.

The operational logic is pretty straightforward. The DNA bases make pairs only if the sequences meet the conditions A-T or C-G. Here, the sequence ACCTAG is treated as the base sequence. Now if the input sequence makes a pair with the base sequence in the test tube (as Fig. 4.2), then they will return the output as true. And if they do not make a pair, then the output will be false.

So, how to detect whether the input sequence and the base sequence make pairs or not? The answer will be detected by the DNase enzyme.

The functional operations are as follows:

1. If the input base sequence is ACCTAG, then it will mix to the test tube where the other base sequence is also ACCTAG. Therefore, DNase enzyme detects no base pair matching. Therefore the result will be false, which represents the sequence TGGATC (same as Table 4.1).
2. If the input base sequence is TGGATC, then it will mix to the test tube where the other base sequence is ACCTAG. Therefore, DNase enzyme will detect a base pair matching between them. Therefore the result will be true, which represents the sequence ACCTAG (same as Table 4.1).

This is how DNA NOT operation will invert the input.

4.3.2   DNA OR Operation

Figure 4.3 shows the operational diagram of DNA OR operation. The logic is as same as the binary OR operation in classical computing. Input–output combinations of DNA OR operation logic are shown in Table 4.2. In DNA, the OR operation refers to a method of combining DNA sequences where the result includes all the sequences of the input sequences, or if any of the input sequences are true. This is similar to the OR logical operation in programming, where the output is true if at least one of the inputs is true. In DNA, this can be used to analyze sequences, identify mutations, or understand genetic relationships. The OR operation can be applied to DNA sequences to find common patterns or regions of overlap between different sequences. For example, if you have two DNA sequences and you want to identify the regions where at least one of the sequences has a specific pattern, the OR operation can help with that. The practical applications of DNA OR operations are as follows:

1. Identifying Mutations: By comparing a patient's DNA sequence with a reference sequence, the OR operation can be used to highlight differences or mutations;
2. Understanding Genetic Relationships: Comparing DNA sequences of individuals can reveal common ancestry and genetic relationships using the OR operation; and
3. Analyzing DNA Microarrays: In DNA microarray experiments, the OR operation can be used to analyze the expression of different genes.

To design DNA OR operation, a test tube with the DNA mixture is needed, when the annealing temperature is 60 °C approximately. To perform the DNA OR operations DNase Enzyme is needed. And here the base sequence TGGATC will be used.

Here the base sequence in the test tube is TCCATC. And the inputs to the DNA OR operation are two DNA sequences. Let the two input sequences be A_0 and A_1. And let the output be Q.

Fig. 4.3 Circuit architecture
of DNA OR operation

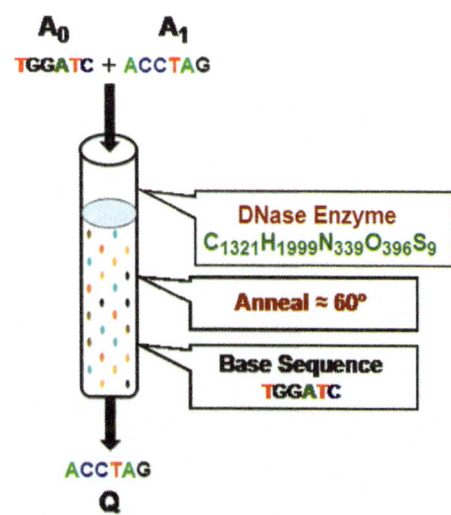

Table 4.2 Truth table of
DNA OR operation

A1	A0	Q
TGGATC	TGGATC	TGGATC
TGGATC	ACCTAG	ACCTAG
ACCTAG	TGGATC	ACCTAG
ACCTAG	ACCTAG	ACCTAG

Therefore, the working principles of DNA OR operation can be described as follows:

1. For A_0 = TGGATC and A_1 = TGGATC, the base sequence in the test tube is also TGGATC. Therefore, no double strands bond will be created. And thus, the DNase enzyme will destroy them all. Therefore the output will be TGGATC (which is equivalent to binary 0).

2. For A_0 = ACCTAG and A_1 = TGGATC, and the base sequence in the test tube is also TGGATC. Therefore, a double strands bond will be created between ACC-TAG and TGGATC. And one base sequence TGGATC will remain in the mixture which will be destroyed by the DNase enzyme. Therefore the output will be ACC-TAG (which is equivalent to binary 1, because a bond has been created and the DNase enzyme will have no effect on them). So, the output result is Q = ACCTAG for the given input.

3. For A_0 = TGGATC and A_1 = ACCTAG, and the base sequence in the test tube is TGGATC. Therefore, a double strands bond will be created between ACCTAG and TGGATC. And one base sequence TGGATC will remain in the mixture which will be destroyed by the DNase enzyme. Therefore the output will be ACCTAG because a bond has been created and the DNase enzyme will have no effect on them). So, the output result is Q = ACCTAG again for the given input.

4. Now, for A_0 = ACCTAG and A_1 = ACCTAG, and the base sequence in the test tube is TGGATC. Therefore, a double strands bond will be created between ACCTAG and TGGATC. And one base sequence ACCTAG will remain in the mixture which will be destroyed by the DNase enzyme because that sequence didn't create any bond with any other base sequence. Therefore, the output will be ACCTAG because a bond has been created and the DNase enzyme will have no effect on them. So, the output result is Q = ACCTAG again for the given input.

So, the design of the operational system of DNA OR operation produces the expected output for the given set of input sequences.

4.3.3 DNA NOR Operation

The NOR operation is the inverted output of the OR operation, where both the DNA OR and DNA NOT operational systems are already designed. Therefore, it is easy to design the operational system for the DNA NOR operation, which is shown in Fig. 4.4. In DNA computing, a NOR operation, often implemented as a logic gate, acts like a "neither-or" operation. It outputs true (or 1) only if both input values are false (or 0). In essence, it negates the result of an OR gate. This operation is crucial in building more complex logical circuits and is used to perform various computational tasks. DNA NOR operations can be implemented using genetic circuits, where the input signals can be, for example, the presence or absence of a specific molecule, and the output can be the expression of a reporter gene. DNA NOR operations are used in synthetic gene circuits and other DNA computing systems to implement logical functions. They can be part of more complex circuits that perform computations within living systems.

From the above circuit, it is easy to understand the architecture of the DNA NOR operation. To perform DNA NOR operation, first DNA OR operation is needed to perform. Then the output of the DNA NOR operation will be inverted by the DNA NOT operation.

The input-output mapping for the DNA NOR operation is shown in Table 4.3. And for example, consider an input-output mapping in Table 4.3, where the inputs are TGGATC and ACCTAG, and they produce TGGATC as output, which is the inverted output from the DNA OR gate.

Table 4.3 Truth table of DNA NOR operation

A1	A0	Q
TGGATC	TGGATC	ACCTAG
TGGATC	ACCTAG	TGGATC
ACCTAG	TGGATC	TGGATC
ACCTAG	ACCTAG	TGGATC

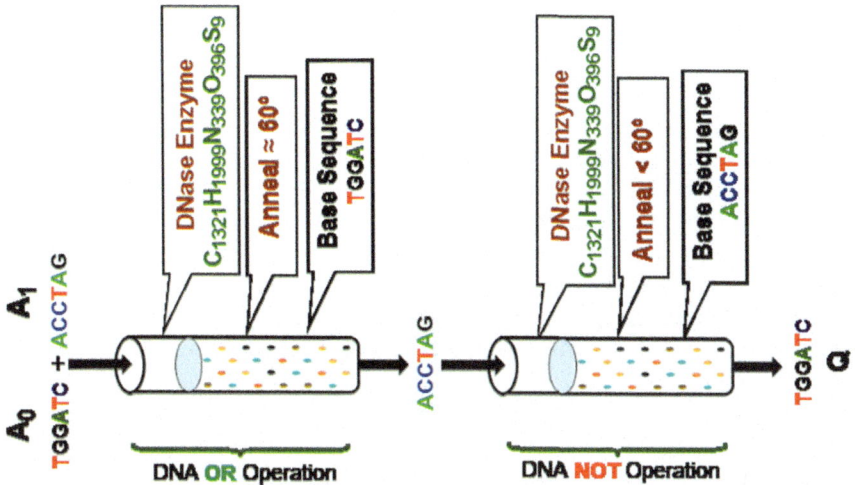

Fig. 4.4 Circuit architecture of DNA NOR operation

As the working procedure of DNA OR is explained before and DNA NOT operations too, therefore it's a doodle for us to understand the working procedure of DNA NOR operation.

The functioning mechanism for the input sequence patterns to perform DNA NOR operation is given below.

1. When the input base sequences both are TGGATC, the DNA OR will generate TGGATC as output. The DNA NOT operation will invert the output of the DNA OR operation. Therefore, the final output will be ACCTAG, which is equivalent to binary 1.
2. When the input base sequences are TGGATC, and ACCTAG the DNA OR will produce ACCTAG as output. And the DNA NOT will generate output TGGATC by inverting the output of the DNA OR operation. And the final output will be obtained as TGGATC.
3. When the input base sequences are ACCTAG, and TGGATC the DNA OR will again produce ACCTAG as output. And the DNA NOT will generate output TGGATC by inverting the output of the DNA OR operation. And the final output will be obtained as TGGATC.
4. And when the input base sequences both are ACCTAG the DNA OR will again produce again ACCTAG as output. And the DNA NOT will generate output TGGATC by inverting the output of the DNA OR operation. And the final output will be obtained as TGGATC.

Therefore the expected output for the given set of input base sequences has obtained (as per Table 4.3).

4.3.4 DNA NAND Operation

In the structure of DNA OR, TGGATC is used as the base sequence in the test tube. The DNA computing system which will perform the operations of DNA NAND operation is containing ACCTAG as the base sequence instead of TGGATC. And the annealing temperature should be more than 60 °C to perform this operation. A DNA-based NAND operation is a logic operation that performs the NOT AND operation using DNA strands. It is a fundamental building block for DNA computing and can be used to implement more complex logic circuits. The NAND operation output is typically detected using a biological process, like DNA translocation through a nanopore, or a chemical reaction like RNA cleaving. Main features of this operation is as follows:

a) How it works:

1. DNA Input Representation: Input values (0 or 1) are represented by specific DNA sequences.
2. Hybridization and Chain Reaction: When a value is inputted, the corresponding DNA strand enters the system and hybridizes with other DNA strands to form a specific path.
3. Output Detection: The output of the NAND operation is detected based on the presence or absence of a particular DNA structure or the occurrence of a chemical reaction, like RNA cleaving or DNA translocation through a nanopore.

b) Applications:

(i) DNA computing: NAND operations are fundamental building blocks for DNA computers, which can perform complex calculations and logic operations;
(ii) Molecular robotics: DNA computing can be used to control the movement and behavior of DNA nanorobots;
(iii) Genetic engineering: DNA logic operations can be used to develop new tools for genetic engineering and diagnostics; and
(iv) Medical diagnosis and treatment: DNA computing has potential applications in medical diagnosis and treatment.

The output for the given input sequences that will be produced by the DNA NAND operation is shown in Table 4.4.

From the above table, it is clear that the DNA NAND will produce an output sequence TGGATC only if the given input sequences both are ACCTAG. Otherwise, it will generate ACCTAG as an output always. Figure 4.5 shows the architecture of the DNA NAND operation.

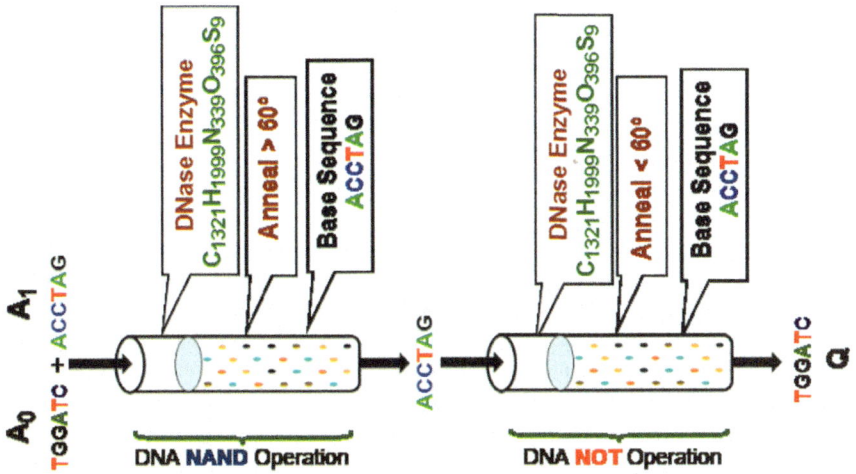

Fig. 4.6 Circuit architecture of DNA AND operation

Table 4.5 Truth table of DNA AND operation

A1	A0	Q
TGGATC	TGGATC	TGGATC
TGGATC	ACCTAG	TGGATC
ACCTAG	TGGATC	TGGATC
ACCTAG	ACCTAG	ACCTAG

DNA AND operation is the inverted value of the DNA NAND operation. So, at first, DNA NAND operation will be performed, and then the DNA NOT operation will be performed to invert the output value of the DNA NAND operation.

The functioning mechanism for the input sequence patterns to perform the DNA AND operation is given below:

1. When the input base sequences both are TGGATC the DNA NAND will generate ACCTAG as output (Table 4.4). And the DNA NOT operation will invert the output of the DNA NAND operation. Therefore the final output will be obtained as TGGATC which is equivalent to binary 1.
2. When the input base sequences are TGGATC, and ACCTAG the DNA NAND will produce ACCTAG as output. And the DNA NOT will generate the output TGGATC by inverting the output of the DNA NAND operation. And the final output will be obtained as TGGATC.
3. When the input base sequences are ACCTAG, and TGGATC the DNA NAND will again produce ACCTAG as output. And the DNA NOT will generate output TGGATC by inverting the output of the DNA NAND operation. And the final output will be obtained as TGGATC.
4. And when the both input base sequences both are ACCTAG the DNA NAND will again produce TGGATC as output. And the DNA NOT will generate output

ACCTAG by inverting the output of the DNA NAND operation. And the final output will be obtained as ACCTAG.

4.3.6   DNA XOR Operation

Table 4.6 shows the truth table of the DNA XOR operation. It will produce ACC-TAG when the given input sequences are not the same sequence pattern. And when both the inputs are the same, then DNA XOR will generate TGGATC as output. To design the architecture of the DNA XOR operation, there is no need for any base mixture as none of the sequences produces the DNA XOR output (which is shown in Table 4.6). The input sequences must be complementary in order to have opposite values, and they will bind together to form a double-stranded sequence.

The logic is very simple. If the input sequences are the same, then they will not be able to make bonds together. If bonds are not created, the DNase enzyme will destroy them. And the output will be false (TGGATC). And when the input sequences are not the same, they will create the DNA double strands, and therefore DNase enzyme will not affect them. As a result, the output will be true (ACCTAG). The circuit architecture of the DNA XOR operation is shown in Fig. 4.7. In other words, DNA-based XOR operations, also known as DNA XOR, utilize the principles of the XOR (exclusive OR) operation in a DNA sequence context. This technique is used in various applications, including data security, storage, and encryption. DNA XOR operations offer unique properties and are used to encrypt data, with DNA sequences serving as key values. DNA sequences are treated as bit strings, and the XOR operation is applied between them. The result is a new DNA sequence that represents the XORed data. This new sequence can be used for data storage, encryption, or other purposes. In essence, DNA XOR provides a unique approach to data security and storage by leveraging the properties of DNA sequences and the XOR operation.

There is no need for a base sequence now to perform DNA XOR operations. The required ideal annealing temperature is more than 60 °C.

Table 4.6 Truth table of a DNA XOR operation

A1	A0	Q
TGGATC	TGGATC	TGGATC
TGGATC	ACCTAG	ACCTAG
ACCTAG	TGGATC	ACCTAG
ACCTAG	ACCTAG	TGGATC

Fig. 4.7 Circuit architecture of a DNA XOR operation

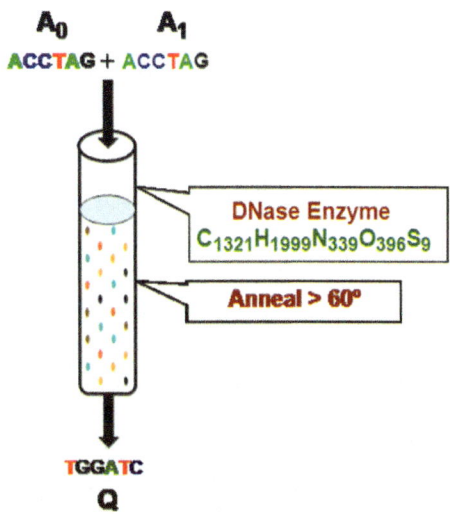

4.3.6.1 Working Principles DNA XOR Operation

The output provided by the designed DNA XOR system for the given set of input sequences is discussed below. Here two input sequences are A_0 and A_1, and the output is Q.

1. When input sequences A_0 and A_1 both are TGGATC, no DNA double strands will form. So, the DNase enzyme will destroy the nitrogen bases in the mixture. Therefore the output sequence will be TGGATC.
2. When input sequences A_0 is $ACCTAG_0$ and A_1 is TGGATC, DNA double strands will form. So, the DNase enzyme will not affect the nitrogen bases in the mixture. Therefore the output sequence will be ACCTAG.
3. When input sequences A_0 is TGGATC and A_1 is ACCTAG, DNA double strands will form. So, the DNase enzyme will not affect the nitrogen bases in the mixture. Therefore the output sequence will be ACCTAG.
4. When input sequences A_0 and A_1 both are ACCTAG, no DNA double strands will form. So, the DNase enzyme will destroy the nitrogen bases in the mixture. Therefore the output sequence will be TGGATC.

So, the designed DNA XOR system provides the expected output results which are shown in the DNA XOR operational Table 4.6.

4.3.7 DNA XNOR Operation

DNA XNOR is the inverted output of the DNA XOR operation. Therefore, a DNA NOT operational system should be added to the output of the DNA XOR operational

Table 4.7 Truth table of a DNA XNOR operation

A1	A0	Q
TGGATC	TGGATC	ACCTAG
TGGATC	ACCTAG	TGGATC
ACCTAG	TGGATC	TGGATC
ACCTAG	ACCTAG	ACCTAG

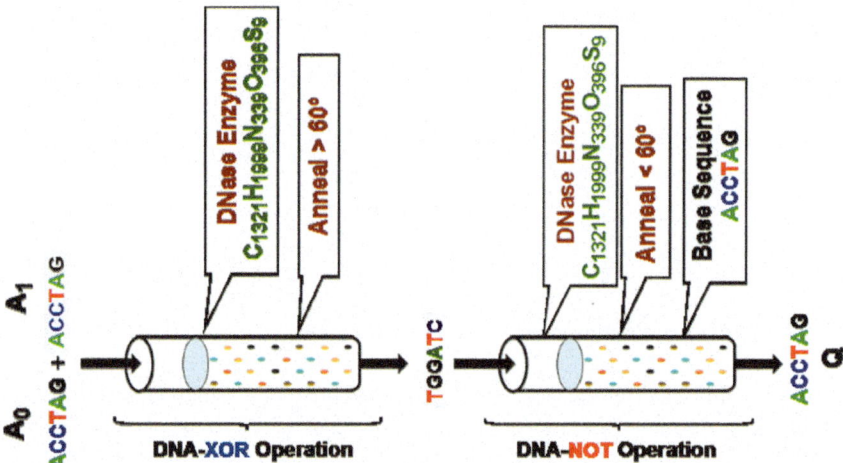

Fig. 4.8 Circuit architecture of a DNA XNOR operation

system to get the output result of the DNA XNOR operation. The DNA exclusive-NOR (XNOR) operation, a fundamental logic operation in DNA computing, can be implemented using biological components like DNA, RNA, or proteins, enabling the creation of bio-inspired computational systems. Here, DNA molecules can be engineered to act as inputs and outputs, and DNA binding can be used to implement logic operations like DNA XNOR. DNA XNOR operations can be used to detect specific biological molecules or conditions, providing diagnostic information. These operations can be incorporated into cellular circuits to control gene expression or cell behavior, leading to therapeutic applications. In addition, DNA XNOR operations can be used to create DNA computing architectures that mimic the functionality of biological neural networks. In essence, the DNA XNOR operation is a versatile logic operation with applications in DNA computing, where it can be implemented using various biological components to create computational systems that can perform a wide range of tasks, from sensing and diagnostics to cellular programming and artificial intelligence. Table 4.7 shows the operations in the DNA XNOR operations. Figure 4.8 shows the circuit architecture of the DNA XNOR operations.

The output provided by the designed DNA XNOR system for the given set of input sequences is given here. Assume that two inputs are A_0 and A_1, and the output is Q.

1. When the input base sequences both are TGGATC, the DNA XOR will generate TGGATC as output (Table 4.6). And the DNA NOT operation will invert the output of the DNA XOR operation. Therefore the final output will be ACCTAG which is equivalent to binary 1.
2. When the input base sequences are TGGATC, and ACCTAG the DNA-XOR will produce ACCTAG as output. And the DNA NOT will generate output TGGATC by inverting the output of the DNA XOR operation. And final output will be TGGATC.
3. When the input base sequences are ACCTAG, and TGGATC the DNA XOR will again produce ACCTAG as output. And the DNA NOT will generate output TGGATC by inverting the output of the DNA XOR operation. And the final output will be TGGATC.
4. And when the input base sequences both are ACCTAG the DNA XOR will again produce TGGATC as output. And the DNA NOT will generate output ACCTAG by inverting the output of the DNA XOR operation. And the final output will be ACCTAG.

So, the designed DNA XOR system provides the expected output results which are shown in the DNA XNOR operational Table 4.7.

4.4 Summary

The design architecture of the DNA computer is not like the classical or quantum computer, it uses chemical reactions performed in a test tube to produce the output. The processes to perform DNA computation are Preparing, Mixing and Annealing, Melting, Amplifying, Separating, Extracting, Cutting, Ligating, Substituting, Marking and Destroying sequences, and Detecting and Reading sequences. The way in which the DNA operations in the truth tables are performed is the same as classical computation. But the design of the circuit architecture is totally different. The total execution time is the required time to perform the largest pipeline in the operation. The heat required to perform a regular operation in DNA computing is specific. The required heat is 284–490 °C. All basic DNA logic operations are presented in this chapter with their designs of the circuit architectures, truth tables, and working principles.

Bibliography

1. K.J. Breslauer, R. Frank, H. Blöcker, L.A. Marky, Predicting DNA duplex stability from the base sequence. Proc. Natl. Acad. Sci. **83**(11), 3746–3750 (1986)
2. S.M. Freier, R. Kierzek, J.A. Jaeger, N. Sugimoto, M.H. Caruthers, T. Neilson, D.H. Turner, Improved free-energy parameters for predictions of RNA duplex stability. Proc. Natl. Acad. Sci. **83**(24), 9373–9377 (1986)
3. Hafiz Md. Hasan Babu, "Multiple-Valued Computing in Quantum Molecular Biology", vol. I (CRC Press, USA, 2023)
4. Hafiz Md. Hasan Babu, "Multiple-Valued Computing in Quantum Molecular Biology", vol. II (CRC Press, USA, 2023)
5. Hafiz Md. Hasan Babu, "DNA Logic Design: Computing with DNA" World Scientific Publishing Company, Singapore (May, 2024)
6. J. Watada, DNA computing and its application, in *Computational Intelligence: A Compendium* (Springer, 2008), pp. 1065–1089
7. T. Yokomori, S. Kobayashi, C. Ferretti, On the power of circular splicing systems and DNA computability, in *Proceedings of 1997 IEEE International Conference on Evolutionary Computation (ICEC'97)* (IEEE, 1997), pp. 219–224
8. Xuedong Zheng, Jing Yang, Changjun Zhou, Cheng Zhang, Qiang Zhang, Xiaopeng Wei, Allosteric DNAzyme-based DNA logic circuit: operations and dynamic analysis. Nucleic Acids Res. **47**(3), 1097–1109 (2019)

Chapter 5
Quantum Biology

5.1 Introduction

In order to maintain the non-equilibrium condition associated with life, biological systems are dynamic, constantly exchanging energy and matter with the environment. We can now investigate biological dynamics at ever-smaller sizes because to advances in observational methods. These investigations have shown evidence of quantum mechanical effects in a variety of biological processes that cannot be explained by traditional physics. These processes are studied by quantum biology.

There is growing evidence that certain mechanisms within living cells use non-trivial aspects of quantum mechanics, such as long-lived quantum coherence, super-position, quantum tunneling, and even quantum entanglement. Previously, these aspects were thought to be relevant primarily at the level of isolated molecular, atomic, and subatomic systems, or at temperatures close to absolute zero, and were therefore not thought to be relevant to the mechanisms responsible for living things.

5.2 What's New in Quantum Biology?

The application of quantum theory to biological concepts that classical physics is unable to adequately describe is known as quantum biology. Despite this straight-forward definition, there is still disagreement within the scientific community over the objectives and place of the area. In this book a new dimension of quantum biology is exposed. Previous all discussions on quantum biology were theoretical, and this book will discuss computing in quantum biology which is completely a new idea. Quantum biology explores the influence of quantum mechanics on biological systems, while biocomputing seeks to harness biological systems for computation. Quantum computing, a related field, utilizes quantum phenomena like superposition and entanglement to perform computations. Quantum biology and biocomputing have a complex relationship, with quantum biology potentially providing insights into the fundamental workings of biological systems that could inform the design of biological computers, and with quantum computing offering tools for analyzing and modeling these biological systems. In essence, quantum biology provides the foundation for understanding how quantum mechanics might be relevant to life, while biocomputing explores the possibility of using biological systems for computation. Quantum computing offers tools for both exploring these quantum effects in biology and for potentially building new types of computers based on them. Note that quantum biology investigates how quantum phenomena, such as electron tunneling and quantum entanglement, play a role in biological processes like photosynthesis, olfaction, and magnetoreception, while biocomputing develops computers based on biological systems, such as using DNA or proteins for computation and quantum computing considers quantum mechanical principles for computation.

5.3   Advantages of Quantum Biology

Recent evidence suggests that a variety of organisms may harness some of the unique features of quantum mechanics to gain a biological advantage.Therefore, quantum biology is necessary if we are to truly comprehend biology and the incredible selectivity of biological processes. In addition, quantum biology has the potential to have a significant impact on a wide range of technologies, including information, sensing, health, and environmental technologies.

5.4   Computing in Quantum Biology

Computations within quantum biology is a hybrid mechanism of quantum physics and molecular biology; the name of this hybrid system is quantum biocomputing. A quantum biocomputer is used to compute different types of mathemetical and logical operations as well as data processing. Quantum biocomputing can be classified into two types. They are quantum-DNA computing and DNA-quantum computing. Though these computing techniques are completely different in nature and there are no similarities between them, the new concepts can be evolved from these two. An innovative idea of combining these two computing technologies to get the maximum benefit of both at a time to utilize all advantages collectively to perform calculations in the best way ever. This is a new platform of the computing system, which consists of the combination of quantum computing and DNA computing-it can be called the cross-platform system of quantum computing and DNA computing. This can be done in two ways and both can be called quantum biocomputing or both can be called as below:

1. Quantum-DNA computing, and
2. DNA-Quantum computing.

Quantum biology, biocomputing, and quantum computing are related because they all explore how quantum mechanics and biology can be harnessed for computational purposes. Quantum biology investigates the role of quantum phenomena in biological systems, while biocomputing utilizes biological materials or structures to perform computations. Quantum computing leverages quantum mechanical principles, like superposition and entanglement, to perform calculations that are difficult or impossible for classical computers. The basic features of quantum biology, quantum computing, and biocomputing are mentioned below.

1. Quantum Biology:

   a) Quantum biology explores how quantum mechanical effects, like superposition and entanglement, might play a role in biological processes, such as protein folding, photosynthesis, and even brain function.
   b) It seeks to understand if quantum phenomena are fundamental to life and how they might be exploited for new technologies.
   c) Examples of quantum biology research include studying how quantum coherence in photosynthesis allows for efficient energy transfer and investigating the potential role of quantum entanglement in neuronal communication.

2. Biocomputing (also known as Biological Computing or DNA Computing):

   a) Biocomputing focuses on using biological systems or molecules, like DNA or proteins, to perform computations.
   b) This can involve using biological systems as the hardware for computation, or designing algorithms that mimic biological processes.
   c) For example, DNA computing uses DNA strands to represent and manipulate data, and protein-based computing explores the potential of using proteins as computational units.

3. Quantum Computing:

   a) Quantum computing utilizes quantum mechanical phenomena, like superposition and entanglement, to perform computations.
   b) Quantum computers use qubits, which can exist in a superposition of states (both 0 and 1 simultaneously), unlike classical bits which can only be 0 or 1.
   c) This allows quantum computers to potentially solve problems that are intractable for classical computers, su1"2ch as drug discovery, materials science, and cryptography.

The connections among these fields are as follows:

1. Quantum biology and quantum computing: Quantum biology provides the scientific foundation for understanding the potential role of quantum phenomena

in biological systems, which can inspire new quantum algorithms and hardware designs.

2. Biocomputing and quantum computing: Biocomputing can provide inspiration for new quantum computing architectures, such as using DNA or proteins as qubits, or designing algorithms that mimic biological systems.

3. Quantum biology and biocomputing: Quantum biology can inform the design of more efficient and powerful biocomputing systems, by understanding the fundamental quantum processes involved in biological systems.

In essence, these three fields are closely intertwined, with advancements in one area often leading to breakthroughs in the others. The convergence of quantum mechanics, biology, and computing promises to unlock new possibilities for scientific discovery and technological innovation.

5.5   Quantum-DNA Computing

The combination of quantum computing and DNA computing is called the quantum biocomputing or quantum-DNA computing or quantum biological computing, the first revolutionary computing process in the modern world that will be described in this chapter. Quantum-DNA computing refers to a field of research exploring the use of DNA molecules, particularly their quantum properties, to perform computations, potentially enabling a new type of quantum computer. This approach aims to leverage the unique structure and properties of DNA to build complex quantum systems, but faces significant challenges in maintaining the delicate quantum states necessary for computation. DNA's double helix structure and the way it interacts with electric fields have led to the suggestion that it could be used to encode and process information using quantum principles. Some researchers theorize that DNA's aromatic bases and hydrogen bonds could function as qubits, the fundamental unit of information in quantum computing. Research in this field includes exploring the use of electric field gradients to manipulate the nuclear spins of nitrogen atoms in DNA, and investigating the possibility of creating a quantum computer that can "grow" instead of being built, potentially transforming computing hardware. DNA-based quantum information processing is a futuristic paradigm that combines biotechnology with quantum technology, aiming to create complex quantum systems by using nature's "construction kit" (DNA). If successful, DNA-based quantum computing could lead to a new generation of quantum computers that are highly scalable and capable of solving complex problems. Quantum computing and DNA computing have drawn the attention of researchers who are working with physics, biochemistry, molecular biology, and computer science. The high-speed computation capability already has proved that quantum computing is the fastest system to solve the computational problem. The many methods to DNA computing with bimolecular coding brought in a revolution in the current world of computer. The imaginary impacts of these technologies are trying to be predicted by the scientists; they are working on the limitations of quantum and DNA computing to get the best from it.

Fig. 5.1 Quantum physics mechanics

5.6 Quantum Computing

Quantum physics is the nature of particles and the study of energy at the most fundamental level and governs the way the universe behaves at the scale of atoms, electrons, and photons. At first, Planck called these individuals unit quanta. A picture of quantum mechanics is depicted in Fig. 5.1.

The word "Quantum" is a Latin word that means "how much". An example of a quantum is- a quantum of light which is known as a photon. Computer and mobile phones are real-life examples of quantum physics that are built-in based on quantum physics principles. An example of quantum physics is shown in Fig. 5.2.

The wave nature of electrons is discussed in quantum physics. It is only possible to manipulate the electrical properties of silicon only because of studying the wave nature of electrons. If the band structure of solid objects is changed, the conductivity will alter as well. Quantum physics knows the answer that how a band structure can be changed.

Fig. 5.2 An example of quantum physics

Fig. 5.3 Quantum AND
operation

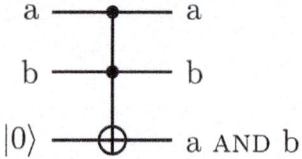

The properties of quantum physics are wave-particle duality, uncertainty principle, entanglement, and superposition. Among these properties, superposition is called when an object has more than one solution or outcome. And this superposition and entanglement is the property of quantum computing. Quantum computing is a new way of computing where the basic unit of data is a quantum bit or qubit. A qubit can be 0 or 1 or the superposition of both, which means 0 and 1 at the same time. Quantum gates are built by quantum bits and quantum circuits are built from quantum gates. Binary logic gates cannot be used directly in the quantum circuits, because the primary property of quantum computing superposition cannot be preserved through this and the actual output will be lost from the circuit. So, it should be quantum gates to build quantum circuits. The quantum gates like Quantum AND gate, Quantum OR gate, and all necessary gates can be implemented in quantum computing to construct quantum circuits. The implementation details of the quantum gates will be discussed later. A simple example of the quantum gate (Quantum AND Operation) is given in Fig. 5.3.

Like this quantum AND gate, it is possible to get all quantum gates that are normally used in a conventional computing system as binary gates. By using these quantum gates, all binary operations can be performed in quantum computing also and get faster performance in less time. So, all circuits in a traditional computing system can be implemented in a quantum computing system also.

5.7 DNA Computing

Molecular biology is the branch of biology at a molecular level that helps to understand molecular synthesis, mechanism, modification. It interacts between the various system of a cell such as DNA, RNA, and protein synthesis. Nowadays molecular biology cell is used for computing which can perform any complex computation very easily. DNA computing is that thing that plays role in computing in this modern world.

DNA computing is a kind of computing where a huge amount of operations can be performed in a high speed and parallel manner. DNA computers perform computations using biological molecules instead of traditional silicon chips. It is another way of computing where DNA strands are used to perform bio-molecular operations. DNA computing can solve the most complex problem very quickly compared to the conventional computer. It works in a parallel manner, and the information storage capacity of a DNA computer is imaginably huge.

5.7.1 DNase Enzyme

Deoxyribonuclease is an enzyme that induces hydrolytic degradation of phosphodiester compounds in the backbone of DNA, leading to DNA destruction. Deoxyribonucleases are a type of nuclease, which is a generic name that includes enzymes that can make phosphodiester compounds bind to nucleotides. There are several types of deoxyribonucleases, each with its own fruit set, chemical structure, and biological activity. The deoxyribonuclease enzyme does not last as long as the DNase enzyme.

DNase is used in many places to clean proteins derived from bacterial organisms. Protein production often involves cell wall damage. Damaged and fragmented cell walls are broken down repeatedly by accident and leaving unwanted DNA in the desired protein. Because the resulting DNA-protein production is so viscous and difficult to purify, DNase is added.

DNase enzymes are crucial in biocomputing for several reasons: they help ensure the quality of RNA samples, particularly for sequencing applications, by removing contaminating DNA. They are also used in protein purification by degrading DNA that can interfere with the process, and they play a role in techniques like DNase-seq that help identify regulatory regions in the genome. Furthermore, DNase enzymes are involved in various biological processes, including apoptosis (programmed cell death) and DNA degradation in certain tissues.

DNase enzymes, particularly DNase I and DNase II, are crucial in quantum biocomputing due to their ability to modify DNA, which serves as the fundamental building block for biological information and potential qubits in this field. DNases are essential for DNA manipulation, including degrading unwanted DNA, cleaning up RNA, and creating fragmented DNA libraries for in vitro reactions. These enzymes play a vital role in research and biotechnology, including DNA sequencing, cloning, and protein purification.

DNase enzymes are crucial in DNA-based quantum computing, particularly for cleaning up DNA samples and ensuring the integrity of the DNA used in quantum calculations. DNase enzymes help by removing unwanted DNA fragments that could interfere with the quantum computations, ensuring a cleaner and more reliable DNA sample for use in DNA-based quantum processors.

5.7.2 Fluorescence Detection

A fluorophore covalently linked to the oligonucleotide primer used in enzymatic DNA sequence analysis is utilized to measure the fluorescence of the DNA fragments. For each of the reactions involving the bases A, C, G, and T, a different fluorophore is utilized. The reaction mixtures are blended and co-electrophoresed down a single polyacrylamide gel tube, the separated fluorescent bands of DNA are identified near the tube's bottom, and the sequence information is obtained directly by the computer.

5.7.3 DNA Logic Operation

DNA computing simulates logic operations and Boolean circuits at the molecular level. A DNA computer consists of DNA-based logic gates, a crucial component of a DNA computer. Many researchers have already shown many approaches to implement DNA basic gates. However, the fundamental DNA logic operations like DNA AND, DNA OR, DNA NOT, DNA XOR, etc. will be implemented in the DNA computing system by using the test tube technique and double DNA strand. If the given input sequences make the bond, this will show the TRUE value or 1. And if they do not make any bond, the DNase enzyme destroys this DNA sequence.

A deoxyribonuclease is known as DNase, such an enzyme that catalyzes the hydrolytic cleavage of phosphodiester linkages in the DNA backbone, thus degrading DNA. This enzyme is responsible for the degradation of the majority of circulating DNA derived from apoptotic and neutrophil extracellular traps.

Implementation details of DNA logic operations are discussed later. An example of a DNA logic operation is given in Fig. 5.4.

The above figure shows that the DNA OR gate will maintain the truth table as the binary logic gate of OR operation but the inputs and output are DNA sequences in DNA OR gate. DNA circuits are built using DNA gates.

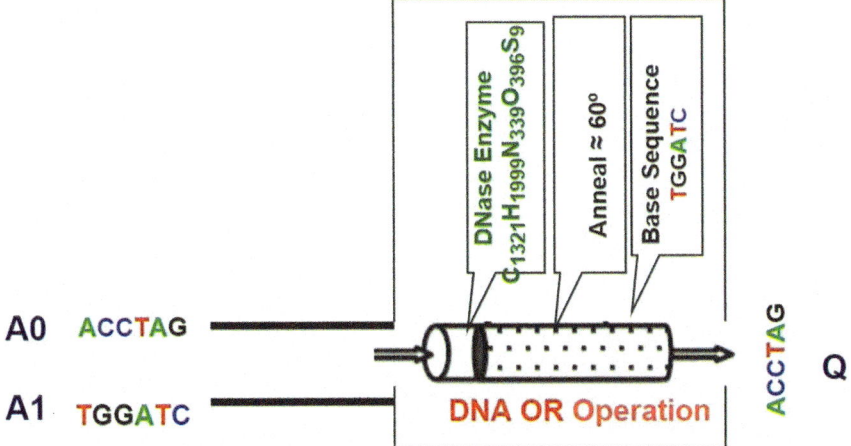

Fig. 5.4 Quantum OR operation

5.8   Quantum Physics and Molecular Biology

Quantum physics is the study of matter and energy at the most fundamental level. It aims to uncover the properties and behaviors of the very building blocks of nature. While many quantum experiments examine very small objects, such as electrons and photons, quantum phenomena are all around us, acting on every scale. However, we may not be able to detect them easily in larger objects. This may give the wrong impression that quantum phenomena are bizarre or otherworldly. In fact, quantum science closes gaps in our knowledge of physics to give us a more complete picture of our everyday lives. The field of quantum physics arose in the late 1800s and early 1900s from a series of experimental observations of atoms that didn't make intuitive sense in the context of classical physics. Among the basic discoveries was the realization that matter and energy can be thought of as discrete packets, or quanta, that have a minimum value associated with them. For example, light of a fixed frequency will deliver energy in quanta called "photons." Each photon at this frequency will have the same amount of energy, and this energy can't be broken down into smaller units. In fact, the word "quantum" has Latin roots and means "how much." Knowledge of quantum principles transformed our conceptualization of the atom, which consists of a nucleus surrounded by electrons. Early models depicted electrons as particles that orbited the nucleus, much like the way satellites orbit Earth. Modern quantum physics instead understands electrons as being distributed within orbitals, mathematical descriptions that represent the probability of the electrons' existence in more than one location within a given range at any given time. Electrons can jump from one orbital to another as they gain or lose energy, but they cannot be found between orbitals.

Quantum biology is a new and evolving field. Quantum biology is the study of applications of quantum mechanics and theoretical chemistry to biological objects and problems. Many biological processes involve the conversion of energy into forms that can be used for chemical transformation and quantum mechanics in nature. Molecular biology is the study of the structure and function of molecules and macromolecular systems associated with biological processes, especially the molecular basis of inheritance and protein synthesis. Quantum biology and molecular biology are related fields, with quantum biology building upon the foundation laid by molecular biology. Molecular biology focuses on understanding biological processes at the molecular level, while quantum biology delves into the role of quantum mechanics in these processes. Quantum effects are increasingly recognized as crucial in various biological phenomena, particularly at the nanoscale. Molecular biology provides the fundamental understanding of biological molecules, including DNA, RNA, and proteins, and how they interact to drive life's processes. It establishes the molecular basis for heredity, gene expression, and cellular functions. Quantum biology explores the potential role of quantum phenomena in biological processes that are traditionally considered to be solely governed by classical physics. On the other hand, quantum computing offers the possibility of simulating complex molecular interactions and biological processes that are computationally intractable with classical computers. Quantum physics and molecular biology are intertwined as molecular biology relies on the fundamental laws of physics, including quantum mechanics. Quantum mechanics describes the behavior of atoms and molecules, which are the building blocks of all living systems. Therefore, quantum principles play a crucial role in understanding biological processes at the molecular level. Atoms and molecules are governed by quantum mechanics, which dictates how electrons are arranged and how they interact with each other and with the nucleus. This knowledge is essential for understanding the structure and properties of biomolecules like proteins and DNA. The intersection of quantum physics and molecular biology has led to the emergence of a new field called quantum biology, which studies the role of quantum phenomena in living organisms.

Scientists are increasingly realizing that life and life processes are strongly associated with the physics of open quantum systems. Without the laws of quantum mechanics, it is not possible to understand the processes of life and existence. Quantum physics has a wide range of practical applications, impacting various fields like technology, medicine, and even everyday life. These applications include lasers, MRI scanners, atomic clocks, solar cells, and quantum computing, whereas quantum biology, the application of quantum mechanics to biological systems, offers potential applications in various fields. These include improving photosynthesis efficiency, enhancing drug discovery through quantum simulations, and understanding how organisms sense magnetic fields. Additionally, quantum biology can inform the development of quantum materials and inspire biologically inspired nanotechnologies for sensing, health, and information technologies. On the other hand, molecular biology finds applications in various fields, including medicine, agriculture, and biotechnology. It enables the development of diagnostic tools, gene therapy, and new drugs. In agriculture, it's used to enhance crop traits, improve yields, and create pest-resistant plants. Biotechnology leverages molecular biology for cloning, protein expression, and the creation of new products.

5.9 Relationship Between Quantum Physics and Molecular Biology

An innovative study has proved that quantum physics plays a role in molecular biology and causes mutations in DNA. Quantum mechanics underpins molecular biology because the interactions between atoms and molecules that drive biological processes are fundamentally quantum mechanical. Quantum effects, such as superposition and entanglement, may play a role in biological phenomena like photosynthesis, olfaction, and enzyme catalysis. Molecular biology is built upon the principles of chemistry, which in turn relies on quantum mechanics to explain the behavior of electrons and atomic nuclei within and between molecules. Quantum effects, like quantum tunneling and entanglement, may be involved in processes like electron transfer in photosynthesis, the sensing of light and odors, and the activity of enzymes. Quantum computing is emerging as a tool to simulate and analyze complex biomolecular systems, offering potential advancements in drug design and other areas of molecular biology. Biological physics uses the principles of physics, including quantum mechanics, to study biological systems at various scales, from the molecular to the macroscopic. Quantum Physics and electrodynamics can shape all molecules and thus it can determine molecular recognition, the working of proteins, and DNA. Many biological processes involve the conversion of energy into forms. These forms are usable for chemical transformation and quantum mechanical in nature. Phenomena such as quantum coherence and quantum tunneling are being involved in performing a very important process for all living cells like enzyme action and energy transfer.

A team from Surrey's Leverhulme Quantum Biology Doctoral Training Center publish a research paper in the journal Physical Chemistry Chemical Physics. They used state-of-the-art computer simulations as well as quantum mechanical methods to determine the role of proton tunneling. A purely quantum phenomenon is observed in spontaneous mutations inside DNA here.

The research team noticed that atoms of hydrogen make the two strands of the DNA's double helix together and behave like spread-out waves. These waves exist in multiple locations at once. Something that can be present in two places at the same time is a known property in the quantum world. This property also exists in molecular atoms that have been already mentioned here. This strongly proves that quantum physics and molecular biology have a compact relationship with each other.

5.10   Quantum-DNA Computing and Quantum Biology

Quantum mechanics plays a role in biological processes and causes mutations in DNA. A nucleotide of DNA may change its form through a process of quantum tunneling which is a major phenomenon of quantum physics. The activities of Cells of every living system are contained within the DNA. DNA is found inside a special area of the cell and organisms have many DNA molecules per cell.

Biology is a natural science concerned with the study of life and living organisms. The cell is the basic unit of life and the basic building block of all organisms. DNA is found inside a cell.

Therefore, it can be said that the quantum-DNA computing or quantum biocomputing or quantum biological computing is quantum biology. Quantum biocomputing offers several potential benefits in modern technology, primarily revolving around its unique computational capabilities and ability to leverage biological systems. These include accelerating complex calculations, revolutionizing fields like drug discovery and materials science, and enhancing artificial intelligence. Quantum biocomputing (quantum-DNA computing) also promises breakthroughs in personalized medicine, environmental monitoring, and biomanufacturing. Key features of quantum biocomputing or quantum-DNA computing or quantum biological computing are as follows:

1. Accelerated Computation: Quantum computers can perform complex calculations much faster than classical computers, potentially solving problems that are currently intractable;
2. Revolutionizing Drug Discovery and Materials Science: By simulating molecular interactions and predicting material properties, quantum biocomputing can accelerate drug discovery and the development of new materials;
3. Enhanced Artificial Intelligence: Quantum computing can improve machine learning algorithms, leading to faster training and more accurate predictions, particularly in areas like anomaly detection;
4. Personalized Medicine: Quantum biocomputers could be used to analyze vast amounts of genomic data, enabling personalized treatments and diagnostics;
5. Biomanufacturing and Environmental Monitoring: Biocomputers can optimize chemical production and sense pollutants in real time, offering new possibilities in these fields;
6. Unbreakable Security: Quantum cryptography, enabled by quantum computing, offers a new level of security for data transmission and storage;
7. Increased Productivity and Efficiency: Faster computational speeds can lead to reduced time-to-market for products and services, and increased productivity across various industries.

5.11 Relationship Between Quantum Computing and DNA Computing with Respect to Quantum Physics and Molecular Biology

Quantum computing is such a branch where phenomena of quantum physics are used to create new ways of computing. It focuses on the principle of quantum theory which deals with modern physics and explains the behavior of matter and energy of an atomic level as well as subatomic level. Quantum computing uses quantum phenomena such as quantum bits, superposition, and entanglement to perform operations.

On the other hand, molecular computing (DNA computing) is a part of molecular biology that uses DNA, biochemistry to perform computations. Quantum computer works with qubits, whereas DNA computer works with base sequences. Researchers designed many operations in both computing systems. So how can we connect them? The simplest thought is to do some operations with the help of quantum computing and other operations with the help of DNA computing.

In other words, quantum computing and DNA computing, while different, are both at the forefront of computational research and share some intriguing parallels. Quantum computing uses quantum mechanics to perform calculations, while DNA computing uses biological molecules for computation, with DNA acting as both a storage and a computation mechanism. Quantum computing leverages quantum phenomena like superposition and entanglement to perform calculations, while DNA computing utilizes the physical and chemical properties of DNA strands. Both fields have the potential to surpass the capabilities of traditional computers, offering the possibility of faster and more efficient computation. Both technologies face significant challenges in terms of scalability, stability, and integration. DNA-based quantum information processing envisions using DNA in various roles within a quantum computer, such as acting as qubits or facilitating quantum interactions. Both quantum computing and DNA computing are groundbreaking technologies with the potential to revolutionize computation. While they differ in their underlying principles and approaches, they share the common goal of pushing the boundaries of what is possible in computing, and DNA might even play a role in the future of quantum computing.

5.12 Establishment of Quantum-DNA Computing Platform

To establish the cross-platform environment of quantum computing and DNA computing, some techniques and procedures need to be introduced. This section will discuss the data conversion circuit, quantum cache memory to control quantum-DNA data flow and heat transfer circuit. Quantum biocomputing or quantum-DNA computing, which combines quantum mechanics with biological systems, offers significant advantages over traditional computing, especially in areas like drug discovery and materials science. It promises faster and more efficient computation, enabling the simulation of complex biological processes and the development of new therapies. The future holds exciting prospects for quantum biocomputing, including break-throughs in personalized medicine and the development of new technologies. The future prospects and key challenges of quantum biocomputing are mentioned below.

Future Prospects:

1. Personalized Medicine: Quantum-powered AI can analyze vast amounts of patient data to develop personalized treatment plans and diagnostics.
2. Development of New Technologies: Quantum biocomputing can enable the development of new materials and devices with unique properties.
3. Solving Complex Biological Problems: Quantum simulations can help researchers understand complex biological processes, like protein folding and disease mechanisms.
4. Advancements in Bioinformatics and Genomics: Quantum-enhanced algorithms can analyze large genomic datasets faster and more accurately, leading to new insights into disease and evolution.
5. Improved Efficiency in Various Industries: Quantum computing can optimize logistics, supply chain management, and other processes, leading to increased efficiency and cost savings.

Key Challenges:

1. Error Correction: Quantum computers are susceptible to errors, and developing effective error correction methods is crucial for their practical use.
2. Hardware Stability: Maintaining the stability of quantum bits (qubits) is a significant challenge.
3. Scalability: Building large-scale, fault-tolerant quantum computers remains a major hurdle.

Despite these challenges, the potential of quantum biocomputing is enormous, and ongoing research is rapidly advancing the field. The future promises a world where quantum computers revolutionize various aspects of our lives, from healthcare to materials science.

5.12.1   The Data Conversion Circuit

The DNA computer doesn't work with qubits. The input of the DNA part is the output of the quantum part, which is definitely in the form of qubits. Therefore, it is needed to develop anything that can convert the qubit into the corresponding DNA base sequence.

Therefore, to work with the quantum-DNA circuit, a data conversion technique is needed to convert the data qubit to DNA base sequence. So the general organization of any quantum-DNA computing can be shown in Fig. 5.5.

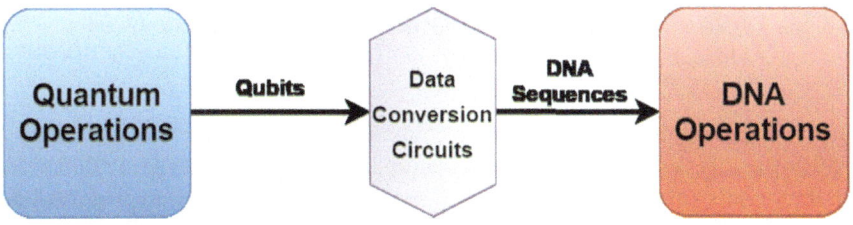

Fig. 5.5 General organization of quantum-DNA computing

Now, the question is how to convert the quantum data (qubit) into DNA base sequences? There are two methods by which qubit will be converted to DNA base sequence:

1. NMR relaxation; and
2. Trap ion.

In Nuclear Magnetic Resonance (NMR), relaxation refers to the process by which excited nuclear spins return to their equilibrium state after being perturbed by a radiofrequency pulse. This process involves the transfer of energy from the spins to the surrounding environment, effectively "cooling" the system. On the other hand an "ion trap" refers to a scientific technique and device that confines charged particles (ions) in a small, isolated region of space using electromagnetic fields. These traps allow for the study, manipulation, and even use of trapped ions in various applications, including quantum computing.

After getting a qubit from a quantum operation, it needs to operate NMR relaxation or trap ion. In NMR relaxation, a room temperature probe or a cryogenic probe can be used at 0 K temperature. By NMR relaxation qubit of superposition state turns into a normal ground state and gets the equivalent DNA sequence. In NMR relaxation at room temperature probes, decoherence problems, polarization, and scaling problems can occur thus cryogenic probes at 0 K temperature can be preferred for quantum-DNA computation systems.

In a DNA-quantum circuit, the inputs are in DNA sequence and provide output in quantum qubits. For providing output, NMR operation or quadrupole trap ion is used for converting DNA sequence to quantum qubits. After getting the DNA sequence from the DNA operation, it needs to be operated by the NMR process to convert it into quantum qubits. In the NMR process, a room temperature probe or a cryogenic probe can be used at 0 K temperature. By applying the NMR process to the DNA sequence, the qubit of the normal ground state turns into a superposition state or ground state according to the input sequence in NMR. The output of the NMR process is a quantum qubit. In NMR at room temperature probes, decoherence problems, polarization, and scaling problems can occur, thus cryogenic probes at 0 K temperature can also be preferred for the DNA-quantum computation system.

5.12.2   Quantum Cache Memory to Control Quantum-DNA Data Flow

Quantum computing is so fast that, according to IBM, the machine performed a mathematically designed complex calculation in 200 s that would take 10,000 years for the world's most powerful supercomputer. This makes quantum computers about 158 million times faster than the world's fastest supercomputer. Therefore, quantum operations provide output instantaneously. But, DNA operations require much time for preparation. Hence, the resulting qubits from the quantum computer cannot enter the DNA system instantaneously. There will be a time delay, qubit properties can be damaged as a result. Eventually, the entire system may become obsolete. Therefore, the necessity of a temporary storage device or system arises where the qubits can be stored for a very short time when working with the quantum-DNA system.

So, what are the options for dealing with this situation? Until the DNA system is ready to receive the qubits, they must be stored somewhere. Therefore, introducing a very fast memory called quantum cache memory, where the qubits can be stored for some while. It can be considered that the quantum cache memory is a transit platform at an airport, where the transit passengers (qubits in this case) wait for their next airplane.

So, another system that needs to be added to the general architecture of quantum-DNA computing and that is quantum cache memory. Quantum Cache Memory (QCM) is an element for preparing quantum data structures for subsequent quantum processing. The QCM has two primary sections: the addressing encoding and information encoding sections. The addressing section, referred to in our development as the Address qubits, comprises a group dedicated to managing the addresses of data points within the QCM. Concurrently, the information section, the information qubits, consists of qubits responsible for storing the data points ready for quantum processing. QCM has dynamic adaptability to fulfill the specific requirements of varying algorithms, with the total count of qubits within the QCM being dynamically determined based on the data size and the algorithmic demands. The block diagram of a quantum cache memory is shown in Fig. 5.6.

After adding the quantum cache memory in the quantum-DNA circuit, the general organization of any quantum-DNA computing is shown in Fig. 5.7.

5.12.3   Heat Transfer Circuit

Two things are added in the designed quantum-DNA computing. A data conversion circuit is added to convert quantum data to the DNA-based sequence, and a quantum cache memory to store quantum data. Another important thing needs to be added with the quantum-DNA computing circuit.

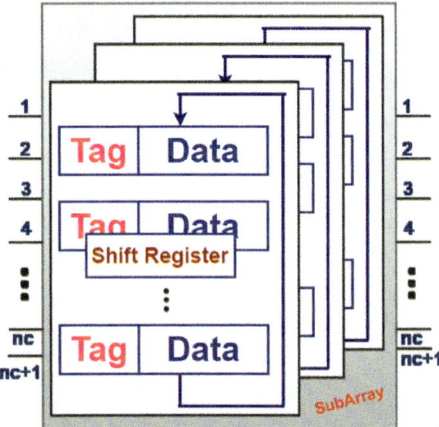

Fig. 5.6   Block diagram of quantum cache memory

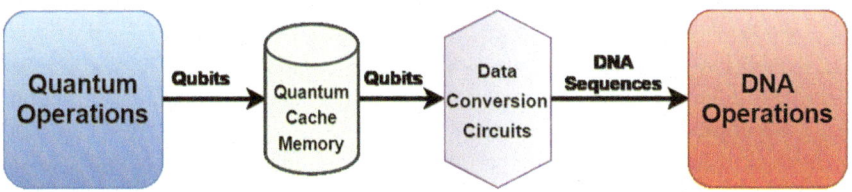

Fig. 5.7   General organization of quantum-DNA computing with quantum cache memory

Quantum computer produces a tremendous amount of heat, on the other hand, DNA computer needs heat. If the quantum computer is connected to the DNA computer, then the extremely excessive amount of heat must be controlled and transferred necessary heat to the DNA computer. This section will discuss how to control this excessive amount of heat.

If a freezer is used, the whole system will be frozen. But it is known that a certain amount of heat is required to perform DNA computations. Therefore, this idea is okay for quantum systems (already used in the quantum computer), but for the quantum-DNA it will not be a great idea.

So, the excessive heat can be transferred so that the proposed system remains at the ideal temperature. A heat conductance circuit will be added to transfer the heat. After adding the heat transfer circuit, The General Organization of Quantum-DNA Computing with Data Conversion Circuits, Quantum Cache Memory, and Heat Conductance Circuit are shown in Fig. 5.8.

A cold storage container will be directly connected to the heat transfer circuit so that the transferred heat can be absorbed by the cooler. Besides, the required heat can be transferred to perform DNA operations from the heat transfer circuit (this is not mandatory, because heat can be provided to the DNA system from the outside as well).

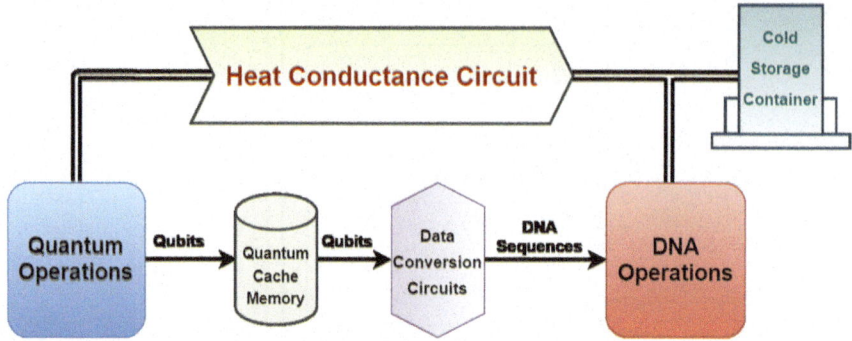

Fig. 5.8 General organization of quantum-DNA computing with data conversion circuits, quantum cache memory, and heat conductance circuit

5.13   Merits of Quantum-DNA Computing

The benefits of quantum computing and DNA computing have been discussed already. In quantum-DNA computing, the individual benefits of quantum computing and DNA computing will be obtained.

It requires less power but the performance of computation will be high. It will be the fastest computing system in the modern computational world. By using the DNA part, the parallelism of computing and the ability to hold massive data and information in a very small space are obtained. The quantum part can provide the capability of solving a very complex problem in a few seconds to save huge time.

By combining these two computing technologies, the whole world could achieve the best success ever for the future world.

5.14   Demerits of Quantum-DNA Computing

Cost is a big obstacle here. This is only for some specific people and organizations. The technologies are rare, so it is not possible to build a quantum-DNA circuit in all aspects of computations.

Heat management is a vital challenge here. The heat emitted from quantum circuits is not easy to manage. Another significant disadvantage is to control it. Without having proper knowledge of quantum computing and DNA computing, it is hard to operate and control.

Quantum biocomputing or quantum-DNA computing, the intersection of quantum computing and biological systems, holds immense potential across various fields like drug discovery, materials science, and bioinformatics. The most important applications of quantum biocomputing are as follows:

1. Drug Discovery and Development:

   a) Accurate Molecular Modeling: Quantum computing allows for precise simulations of molecular interactions, enabling faster and more accurate prediction of drug efficacy and target binding.
   b) Protein Folding and Structure Prediction: Understanding protein folding is crucial for drug development, and quantum algorithms can accelerate the prediction of protein structures, aiding in the design of targeted therapies.
   c) Drug Repurposing: Quantum algorithms can analyze complex metabolic pathways and molecular interactions to identify existing drugs that may have novel therapeutic uses.

2. Materials Science and Engineering:

   a) New Materials Design: Quantum computing can model the behavior of materials at the atomic level, leading to the discovery of new materials with enhanced properties, like stronger, lighter, or more conductive materials.
   b) Battery Technology: Quantum simulations can help optimize battery designs for better energy storage and efficiency.
   c) Quantum Sensors: Quantum sensors offer highly precise measurements, potentially revolutionizing fields like navigation, medicine, and environmental monitoring.

3. Bioinformatics and Genomics:

   a) Genome Analysis: Quantum algorithms can analyze vast genomic datasets more efficiently, leading to a deeper understanding of genetic diseases and personalized medicine.
   b) Data Analysis: Quantum algorithms can accelerate the analysis of large datasets, enabling faster identification of patterns and insights in biological data.
   c) Computational Biology: Quantum computing can be used to model complex biological systems, like gene regulatory networks and metabolic pathways, providing a more comprehensive understanding of biological processes.

4. Artificial Intelligence and Machine Learning:

   a) Quantum Machine Learning: Quantum algorithms can accelerate the training of machine learning models, leading to faster and more accurate predictions.
   b) Quantum Natural Language Processing: Quantum NLP models can process and analyze biological sequences, like DNA, more efficiently.

5. Other Emerging Applications:

   a) Quantum Cryptography: Quantum key distribution offers enhanced security for data transmission and storage.
   b) Financial Modeling: Quantum computing can be used to model complex financial markets and risks more accurately.

c) Combating Climate Change: Quantum computing can contribute to the development of new technologies for carbon capture and energy storage.

5.15   DNA-Quantum Computing

In the world of computing, scientists and researchers constantly seek new ways to enhance computational power and solve complex problems more efficiently. Traditional computing, based on classical bits, has made remarkable progress, but it is reaching its limits. However, a fascinating field of research is emerging at the intersection of biology and quantum physics, known as bioquantum computing or biological quantum computing or DNA-quantum computing. This revolutionary approach harnesses the principles of quantum mechanics within biological systems, promising unprecedented computational capabilities. In this book, we will explore the concept of bioquantum computing, its potential applications, and the challenges it faces. Bioquantum computing or DNA-quantum computing aims to integrate quantum computing techniques with biological systems to tackle complex challenges in the life sciences. It leverages the inherent quantum properties of biological molecules, such as DNA, proteins, and enzymes, to perform computations and simulations that are otherwise intractable for classical computers. The key advantage of bioquantum computing lies in its ability to exploit the massive parallelism and information processing capabilities of biological systems while also benefiting from quantum computational power. DNA-quantum computing or bioquantum computing or biological quantum computing represents the convergence of quantum computing and biology, offering unprecedented computational power to tackle complex problems in the life sciences. Despite some obstacles, the progress made in bioquantum computing is remarkable, and the possibilities it presents are truly exciting. As researchers delve deeper into the potential of quantum systems within biological frameworks, we can expect groundbreaking advancements that will shape the future of computing, biology, and numerous other scientific domains. With continued research, innovation, and collaboration, we are poised to unlock the extraordinary computational capabilities offered by bioquantum computing, revolutionizing the way we solve complex problems and understand the world around us. Quantum computing and DNA computing are both distributed and parallel computing methods. They're useful for solving problems like searching, sorting, merging, pattern recognition, image processing, and encryption that demand high-complexity computations and/or enormous data sets. Because classical computers are incapable of dealing with parallelism, quantum and DNA algorithms cannot be efficiently simulated on them.

It is known that quantum computer's superpower is the ability to solve complex problems at super high speed, and DNA computer's superpower is its high speed along with a massive storage system in a very small amount of DNA molecule. So, a cross-platform can be created that connects quantum computers and DNA computers with the same specific goal (solving complicated problems). In the previous chapter, the first part of the cross-platform which is called quantum-DNA computing is presented. Biological quantum computing, also known as "quantum biology" has applications in understanding and manipulating biological systems at the quantum level. It has the potential to revolutionize fields like drug discovery, personalized medicine, and our understanding of complex biological processes. In essence, biological quantum computing offers the potential to accelerate drug discovery and development, improve our understanding of complex biological processes, lead to breakthroughs in personalized medicine and disease treatment, and open up new avenues for research and discovery in the fields of biology and medicine. In this chapter, the second part of the cross-platform and that is DNA-quantum computing will be discussed.

5.16 Relationship Between Molecular Biology and Quantum Physics with Respect to DNA-Quantum Computing or Bioquantum Computing

Molecular biology and quantum physics have long been regarded as unrelated disciplines. It was never possible thought that one day it would be possible to combine them. Recent research has confirmed that they are correlated. Molecular biology is the study of biological activity, this field mainly focuses on nucleic acids such as DNA, RNA, and proteins that are essential to life processes. And this DNA contains quantum phenomena such as superposition. On the other hand, quantum physics also have quantum properties such as superposition, entanglement, etc. As both have the same property, it is said that they have a compact relationship with each other. Applications of Bioquantum Computing or DNA-quantum computing are as follows:

1. Drug Discovery and Molecular Simulations: Bioquantum computing has the potential to revolutionize the pharmaceutical industry by accelerating drug discovery processes. Quantum algorithms could efficiently analyze vast amounts of molecular data, predict chemical reactions, and simulate the behavior of complex biological systems. This could lead to the discovery of novel drugs and personalized medicine tailored to individual genetic profiles.
2. Neural Networks and Brain-Computer Interfaces: Bioquantum computing can enhance the field of neural networks and brain-computer interfaces. Quantum algorithms could optimize neural network architectures, improve pattern recognition capabilities, and enhance the efficiency of brain-computer interface technologies. This could lead to advancements in neuroprosthetics, neuroimaging, and cognitive computing.
3. Cryptography and Data Security: Quantum computing, including bioquantum computing, has implications for cryptography and data security. Quantum algorithms could provide more secure encryption methods, such as quantum key distribution, protecting sensitive data from potential threats posed by classical computing attacks.
4. Protein Folding: Understanding protein folding, the process by which a protein attains its functional three-dimensional structure, is crucial for advancing

our knowledge of diseases and developing targeted therapies. Bioquantum computing can simulate protein folding dynamics more efficiently than classical methods, leading to breakthroughs in protein structure prediction, drug target identification, and personalized medicine.

5. Quantum Genomics and Bioinformatics: Bioquantum computing can play a vital role in genomics and bioinformatics research. It could assist in DNA sequencing and analyzing large genomic datasets, allowing for faster and more accurate identification of genetic variations and their associations with diseases. Quantum algorithms could also improve the efficiency of genome assembly and gene expression analysis.

6. Optimization and Machine Learning: Quantum algorithms have shown promising results in solving optimization problems and enhancing machine learning algorithms. Bioquantum computing could significantly improve optimization processes in various fields, such as supply chain management, logistics, and financial modeling. Quantum machine learning algorithms could also enhance pattern recognition, data analysis, and predictive modeling.

7. Environmental Modeling and Climate Change Analysis: Bioquantum computing could contribute to environmental modeling and climate change analysis. Quantum algorithms could simulate and optimize complex climate models, helping us understand climate patterns, predict weather phenomena, and develop strategies for mitigating the effects of climate change.

8. Evolutionary Biology and Population Genetics: By harnessing the computational power of bioquantum computing, scientists can delve deeper into understanding the mechanisms of evolution and population genetics. Quantum algorithms could simulate evolutionary processes, analyze genetic diversity, and uncover the underlying factors that drive genetic adaptation and speciation.

Challenges and Future Prospects

While bioquantum computing holds tremendous potential, several challenges need to be addressed for its successful implementation:

1. Hardware Development: Building reliable and scalable bioquantum computing hardware remains a significant challenge. Researchers are exploring various approaches, such as using biomolecules like DNA and proteins as qubits, developing quantum-inspired algorithms for biological systems, and investigating quantum effects in biological molecules.

2. Noise and Error Correction: Quantum systems are highly susceptible to noise and errors caused by interactions with the environment. Developing robust error correction techniques and mitigating decoherence effects are critical for the practical realization of bioquantum computing or DNA-quantum computing.

3. Ethical Considerations: As with any powerful technology, bioquantum computing/DNA-quantum computing raises ethical concerns. Ensuring responsible use and addressing issues like privacy, data security, and potential misuse of advanced computational tools are essential aspects that must be carefully considered.

5.17   Relationship Between DNA Computing and Quantum Computing with Respect to Molecular Biology and Quantum Physics

DNA computing performs millions of operations at the same time using molecular biology rather than traditional silicon chips. American Physicist Richard Feynman presented his idea on nanotechnology. He said individual molecules could be possibly used for computation. And nowadays researchers have confirmed it and made a lot of logical operations. On the contrary, quantum computing is a portion of quantum physics, a type of computation that harnesses the collective properties of quantum states to perform calculations. As both computing have similar properties, it is possible to develop a new way of computing by combining them to get more advantages.

The general organization of DNA-quantum computing with data conversion circuits, DNA cache memory, and heat conductance circuit is depicted in Fig. 5.9

5.18   Establishment of DNA-Quantum Computing Platform

Quantum-DNA computing cross-platform is developed; now DNA-quantum computing needs to be developed to perform computations in a more sophisticated way. Quantum-DNA cross-platform has some challenges that are given as follows:

1. This circuit has to build a data conversion circuit that will convert data from DNA base sequences to quantum qubits.
2. This computing has to build a DNA cache memory where DNA information will be stored and retrieved for further processes.
3. These cross-platforms have to add a quantum heat transfer circuit to diminish the excessive heat.

Therefore, the general organization of a DNA-quantum computing system can be shown in Fig. 5.9 where the block diagram contains data conversion circuits, DNA cache memory, and heat conductance circuit.

The cache memory is developed for the quantum data; therefore, it will not work for the DNA information. So, another data conversion circuit is needed that will convert data from the DNA base sequence to the qubit.

5.18.1   The Data Conversion Circuit

NMR relaxation is used to convert data from qubits to DNA sequences. Now, the inverse process is needed which is NMR process. The structure of NMR and NMR

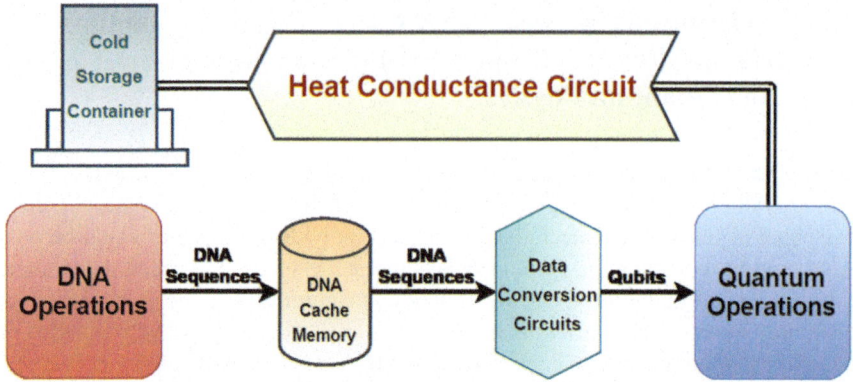

Fig. 5.9 General organization of DNA-quantum computing with data conversion circuits, DNA cache memory, and heat conductance circuit

relaxation is the same because the NMR relaxation is a process of NMR. The only difference is in NMR relaxation it does not emit EMR (Electromagnetic Resonance), whereas in NMR it emits electron magnetic resonance. Another way of converting DNA sequence to qubit is quadruple trap ion.

5.18.2   Cache Memory to Control DNA-Quantum Data Flow

An intermediator system is needed where the output data from the DNA system will be stored and retrieved for further processes. A DNA cache memory is needed to store DNA sequences. The concept of "DNA or biological cache memory' doesn't have a specific, established meaning within the field of biology or neuroscience. However, it's possible to draw an analogy between how computer cache memory works and certain biological systems that prioritize quick access to frequently used information. Just like computer cache, these biological systems can be thought of as "caching" frequently used information, making it quickly available for use. This can involve:

  i) Short-term memory: Information we're actively thinking about is held in a short-term memory "cache," allowing us to quickly access it for tasks;
 ii) Nervous system: Neurons can form connections (synapses) that allow for quick and efficient communication, similar to how cache allows quick access to data.

In summary, the concept of "biological cache memory" is an analogy used to understand how biological systems can quickly access and process frequently used information, similar to how computer cache operates. While it's not a formal biological term, it can be a helpful way to conceptualize the prioritization of information

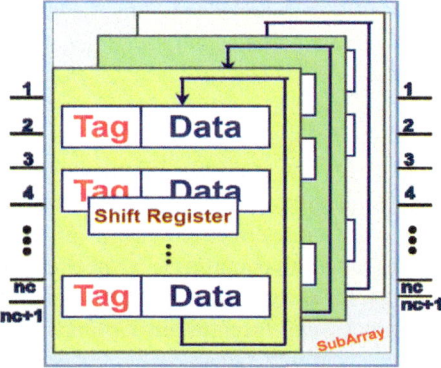

Fig. 5.10 Block diagram of a DNA cache memory

in certain biological processes. It is easy to build a DNA cache memory; it will be discussed later. Figure 5.10 shows the block diagram of DNA cache memory.

5.18.3   Heat Conductance Circuit

The heat conductance circuit is so important here and needs to develop in such a way that the heat of the quantum system should be transferred to some storage or diminished. In DNA-quantum computing, the heat cannot be reused by the DNA system where heat is needed. It may require to provide extra heat to the DNA circuit but cannot take heat from the quantum part. So, the heat emitted by the quantum part should be managed.

If the heat of the quantum part is not managed properly, then the whole circuit could be at risk. The emitted heat of the quantum circuit can destroy the quantum part. As a result, the entire circuit will be useless.

5.19   Quantum Biology and DNA-Quantum Computing

Since quantum-DNA computing is considered quantum biology, DNA-quantum computing should be also called quantum biology. Quantum biology is the combination of quantum computing and DNA computing. It is known that the cross-platform or the combination of quantum computing and DNA computing has two forms. One is quantum-DNA computing, and the another is DNA-quantum computing. So, both forms of computing can be called together quantum biology. Quantum biology investigates the role of quantum phenomena in biological systems, while DNA-quantum computing explores using DNA as a platform for quantum information processing. DNA-quantum computing is a specific application of quantum biology, focusing on harnessing the molecular properties of DNA for quantum computation. Quantum biology aims to understand how quantum mechanics, which governs the behavior of subatomic particles, might influence or be relevant to biological processes at the molecular level. The structure of DNA, with its base pairs and hydrogen bonds, has been suggested as a potential platform for encoding and manipulating quantum information. DNA's ability to store and transmit genetic information provides a natural analogy for how quantum information could be stored and processed. Understanding the quantum aspects of DNA's structure and behavior could pave the way for developing more effective and efficient DNA-based quantum computers. In essence, DNA-quantum computing is a specific application of the broader field of quantum biology, focusing on leveraging the unique properties of DNA for quantum information processing. It represents a promising area for developing novel quantum technologies and understanding the potential role of quantum mechanics in biological systems.

5.20   Difference Between Quantum-DNA Computing and DNA-Quantum Computing

It can be said that both quantum-DNA and DNA-quantum computing are quantum biology. But the quantum-DNA computing and DNA-quantum computing is not the same. Let's discuss some differences between these two new technologies.

From the design principle and the figures of a block diagram, it can easily observe that the construction procedure is almost opposite of one another. In quantum-DNA computing, the first part is the quantum part and the last part is the DNA part. So the inputs are qubits and the outputs are DNA sequences.

Another big difference is in the heat transfer circuit. In quantum-DNA computing, the heat emitted from the quantum part is transferred to the DNA circuit because the DNA circuit needs heat to perform perfectly. On the other hand, In DNA-quantum computing, the heat of the quantum circuit is managed by diminishing heat or by transferring the heat to some cold storage container.

In the quantum-DNA circuit, quantum cache memory is used to store qubits. But in the DNA-quantum circuit, DNA cache memory is used to store DNA sequences. In other words, "Quantum-DNA computing" and "DNA-quantum computing" are often used interchangeably, but they represent distinct approaches to computation. DNA-quantum computing focuses on leveraging DNA's structure and properties for quantum computing applications, potentially as a physical medium for qubits. Quantum-DNA computing, on the other hand, uses quantum computing principles to solve problems related to DNA, such as sequence analysis or drug discovery. Using DNA as a platform for quantum computation. This involves exploring ways to encode and manipulate quantum information using DNA molecules. DNA-quantum computing could offer unique advantages like molecular-scale qubits, self-assembly, and potential for biocompatible quantum sensors. This computing requires overcoming challenges in creating stable and controllable DNA-based qubits, as well as ensuring their scalability and coherence, such as using DNA origami to create

Table 5.1 Differences between DNA-quantum computing and quantum-DNA computing

Feature	DNA-Quantum Computing	Quantum-DNA Computing
Focus	DNA as a quantum platform	Quantum computers for DNA problems
Input	DNA molecules, DNA sequences	Problems related to DNA, quantum as a DNA platform, qubits
Output	Qubits	DNA sequences
Goal	Build a quantum computer using DNA	Use quantum computers to study DNA
Technology	DNA-based qubits, DNA origami, Quantum biotechnology	Quantum algorithms, quantum simulators
Applications	Quantum sensors, quantum biosensing, financial modeling, cybersecurity	DNA sequence analysis, drug discovery, materials science, optimization tasks, DNA-based quantum information processing

precise nanoscale structures for quantum components. Quantum-DNA computing uses quantum computers to solve problems in DNA science. This includes tasks like analyzing DNA sequences, predicting protein folding, and designing new drugs. For example, using quantum algorithms to find the smallest number of DNA strands needed for a particular task and using quantum simulations to model DNA replication or protein interactions. Table 5.1 presents the key differences between these two computing systems.

From the above discussion, it can be said that the quantum-DNA circuit and the DNA-quantum circuit are not the same. The design procedures, working principles, and the requirements all are different.

5.21 Summary

This chapter shows how to establish a cross-platform between quantum computing and DNA computing. At first, the challenges to develop quantum-DNA and DNA-quantum computing systems are shown and then solved them accordingly. To store quantum and DNA information, the quantum and the DNA cache memories can be utilized. To work in a cross-platform environment, data conversion is a must. NMR process can be used to convert DNA information into equivalent quantum bits (qubits). On the other side, the NMR relaxation is used to convert qubits to DNA sequences. This chapter has presented the basic organization of quantum-DNA computing and DNA-quantum computing or quantum biocomputing and Bio-quantum computing. The merits and demerits of quantum-DNA computing and DNA-quantum computing are also discussed, where advantages of quantum computing and DNA computing are combined. These two computing forms are combinedly called quantum biology. The computer that is attempted to build by using computations in quantum biology in this book is actually the cutting-edge quantum biocomputer which is the part of the nanotechnology. Nanotechnology refers to the branch of science and engineering devoted to designing, producing, and using structures, devices, and systems by manipulating atoms and molecules at nanoscale, i.e. having one or more dimensions of the order of 100 nanometres (100 millionth of a millimetre) or less. In the natural world, there are many examples of structures with one or more nanometre dimensions, and many technologies have incidentally involved such nanostructures for many years, but only recently has it been possible to do it intentionally. Many of the applications of nanotechnology involve new materials that have very different properties and new effects compared to the same materials made at larger sizes. This is due to the very high surface to volume ratio of nanoparticles compared to larger particles, and to effects that appear at that small scale but are not observed at larger scales. On the other hand, quantum biocomputing emerging field explores the possibility of using biological systems, like DNA or proteins, to perform computations based on quantum principles. It leverages the unique properties of biological molecules, such as their ability to store and process information, and the potential for quantum entanglement and superposition at the molecular level. In essence, nanotechnology provides the tools and techniques to build and manipulate the physical systems used in quantum computing and potentially biological computing, while quantum mechanics provides the underlying principles that govern the behavior of these systems at the nanoscale.

Bibliography

1. J. Anders, D.K.L. Oi, E. Kashefi, D.E. Browne, E. Andersson, Ancilla-driven universal quantum computation. Phys. Rev. A **82**(2), 020301 (2010)
2. D. Auguin, V. Catherinot, T.E. Malliavin, J.L. Pons, M.A. Delsuc, Superposition of chemical shifts in NMR spectra can be overcome to determine automatically the structure of a protein. Spectroscopy **17**(2–3), 559–568 (2003)
3. F. Hobo, M. Takahashi, H. Maeda, S 33 NMR cryogenic probe for taurine detection. Rev. Sci. Instruments **80**(3), 036106 (2009)
4. V.D. Kodibagkar, M.S. Conradi, Remote tuning of NMR probe circuits. J. Magn. Resonance **144**(1), 53–57 (2000)
5. L. Slocombe, J.S. Al-Khalili, M. Sacchi, Quantum and classical effects in DNA point mutations: Watson–Crick tautomerism in at and GC base pairs. Phys. Chem. Chem. Phys. **23**(7), 4141–4150 (2021)
6. N. Lambert, Y.-N. Chen, Y.-C. Cheng, C.-M. Li, G.-Y. Chen, F. Nori, Quantum biology. Nat. Phys. **9**(1), 10–18 (2013)
7. J. McFadden, J. Al-Khalili, Proc. R. Soc. A **474**(2220), 20180674 (2018)

Part II
Arithmetic Circuits in Quantum Biocomputing

Overview

Every second, millions of calculations take place in all living systems in human life. All the variants of a conventional computer are commonly part of these calculations. Arithmetic operations such as addition, subtraction, multiplication, etc. play a major role in computer performance and efficiency. Arithmetic operations are the basic of all mathematical structures. It is the root of statistics and the solution to many real-life problems in every aspect of life. Creating new computing platforms supported by traditional binary arithmetic and silicon-based technologies are not enough to fulfill the needs. As a result, a considerable amount of research effort has been devoted to the study of unconventional computer systems in order to explore more efficient arithmetic circuits as well as to improve computer technologies to facilitate progress. Quantum computing is a study in which calculations are a million times faster than conventional computers. IBM has developed quantum computers to solve complex problems that are unsolvable and unsolvable in most modern conventional computers. DNA calculation is a type of calculation in which many arithmetic operations can be performed at high speed and in a parallel way. DNA computers perform calculations using biological molecules instead of traditional silicon chips. So, with the help of quantum-DNA and DNA-quantum counting in arithmetic operations, problems with computational speed and a large amount of data manipulation can be easily solved. Time is the most valuable thing on earth. Time can be saved by the advancement of the combined technology of quantum computing and DNA computing for arithmetic operations which will be the best achievement of all times. So, the role of quantum biocomputing in arithmetic operation is appreciable and the development of both computing can give us more advantages in the future. It can be called computing with arithmetic operations in quantum biology too. This part will describe the arithmetic operations in quantum computing and DNA computing first. After that, this chapter will describe all the same arithmetic operations in quantum biocomputing which means in quantum-DNA computing and in DNA-quantum computing.

Chapter 6
Quantum Arithmetic Circuits

6.1 Introduction

In recent years, quantum computation has received a lot of attention because of its potential application to challenging problems in classical calculation. Despite the difficulties of deploying quantum devices in the lab, theoretical physicists have attempted to devise various quantum algorithm implementations. Following their path, some of the most prominent arithmetic operations in quantum computing have developed here.

The arithmetic operations which will be discussed in this chapter are given below:

1. Quantum Half-Adder
2. Quantum Full-Adder
3. Quantum Qubit Adder
4. Quantum BCD Adder
5. Quantum Carry-Skip Adder
6. Quantum Carry Look ahead Adder
7. Quantum Half-Subtractor
8. Quantum Full-Subtractor
9. Quantum Multiplier
10. Quantum Divider
11. Quantum Comparator

The general organizations and block diagram of arithmetic operations in quantum computing are presented in this chapter with the circuit architecture and working procedure.

© The Author(s), under exclusive license to Springer Nature Singapore Pte Ltd. 2025   99
H. M. Hasan Babu, *Quantum Biocomputing in Quantum Biology Volume I*,
https://doi.org/10.1007/978-981-97-7154-7_6

6.2   Quantum Half Adder

A half adder is a circuit that adds two-qubit digits to produce a sum and a carry. This adder can only add two qubits at a time. In other words, a quantum half adder is a quantum circuit that performs the addition of two single-bit quantum values (qubits), producing a sum and carry. In quantum computing, the half adder is implemented using quantum gates and qubits, such as Toffoli and CNOT gates, which are the quantum equivalents of classical logic gates. Quantum circuits can be built using these gates to perform the addition operation, similar to how classical half adders are built using XOR and AND gates. Quantum half adders are fundamental building blocks for more complex quantum arithmetic circuits, such as quantum full adders and multi-bit quantum adders. These adders are also used in quantum computing algorithms that require arithmetic operations on quantum data. The truth table for quantum half adder is given in Table 6.1.

From the above truth table, the equation of Sum (S) and Carry (C_{out}) are as follows:

$$S = A' B + A B'$$
$$= A \text{ XOR } B$$
$$C_{out} = A.B$$

Table 6.1   The truth table for quantum half adder

Inputs		Outputs	
$|A>$	$|B>$	$|S>$	$|Cout>$
$|0>$	$|0>$	$|0>$	$|0>$
$|0>$	$|1>$	$|1>$	$|0>$
$|1>$	$|0>$	$|1>$	$|0>$
$|1>$	$|1>$	$|0>$	$|1>$

Fig. 6.1   Block diagram of quantum half adder

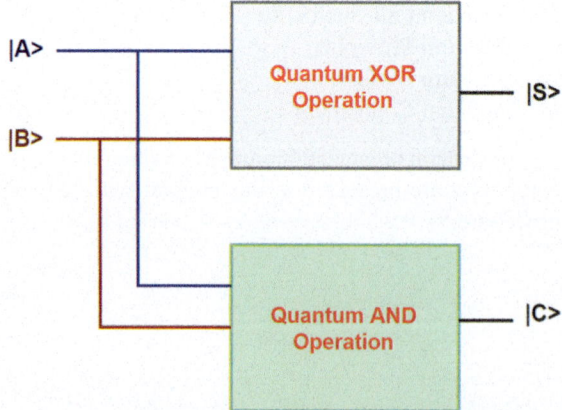

6.2.1 Block Diagram

The quantum circuit for a half adder is created by several quantum gates. To design a half adder, one XOR gate and one AND gate are required. In this quantum circuit, one XOR operation is needed to produce the SUM of the half adder. Additionally, to generate the carry qubit, one quantum AND operation is used. The block diagram of the quantum half adder is illustrated in Fig. 6.1.

6.2.2 Circuit Architecture

Algorithm 6.2.1 represents the overall procedures of the quantum-based half-adder operation. If the total number of inputs qubits is n then O (n) is the run time complexity of this algorithm. To perform one AND operation, and one XOR operation, in this work one quantum XOR and one quantum one operation are used. Quantum XOR operation is denoted by DO_Quant_XOR $(|A>, |B>)$, quantum AND operation are denoted by DO_Quant_AND $(|A>, |B>)$. "HALF_SUM" is used to represent the Half_Adder sum's output which is obtained after doing the quantum XOR operation. "HALF_Carry" is the carry of the half adder that is obtained after doing quantum AND operation.

Algorithm 6.2.1 Quantum Half Adder Algorithm

---
1. **Begin**
2. **while** i equals to 1 to n **do**
3. HALF_SUM<-Do_Quant_XOR(|Ai>, |Bi>)
4. HALF_Carry<-Do_Quant_AND(|Ai>, |Bi>)
5. **end while**

Figure 6.2 depicts the quantum circuit of the half-adder. In this half adder circuit, the XOR operation for $|A>$ and $|B>$ inputs is done by a quantum operation and creates an output of $|S>$. As for the output of $|C>$, both inputs $|A>$ and $|B>$ pass through a quantum AND operation.

6.2.3 Working Principle

The quantum half-adder needs two inputs, A and B. For various input qubits A and B, consider the following cases:

(i) When both A and B are "true" $|1>$, the output qubit of S is "false" $|0>$ and C is "true" $|1>$.

Fig. 6.2 Quantum
half-adder circuit

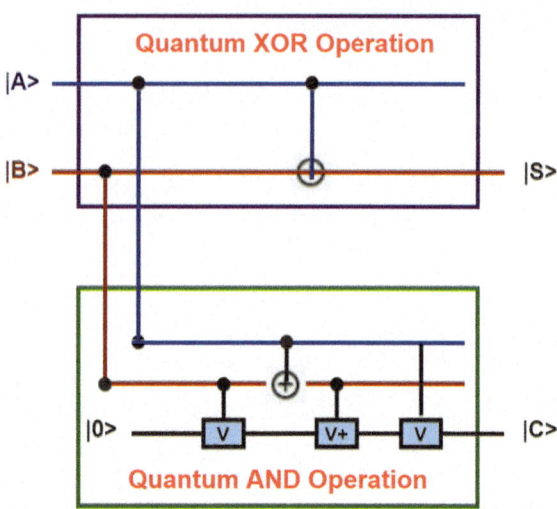

(ii) When both A and B are "false" $|0>$, the output qubit of both S and C are "false" $|0>$.

(iii) When A is "false" $|0>$, and B is "true" $|1>$, the output qubit of S is "true" $|1>$ and C is "false" $|0>$,

(iv) When A is "true" $|1>$ and B is "false" $|0>$, the output qubit of S is "false" $|0>$, and C is "true" $|1>$.

6.3   Quantum Full Adder

A full adder, a fundamental building block in classical computing, can be implemented in quantum computing using qubits and quantum gates. These adders can be implemented using various quantum gates, such as CNOT gates, Toffoli gates, and other quantum operations. A full adder is a circuit designed to overcome the disadvantages of the half adder circuit. It accepts three inputs and produces two outputs after adding them. The truth table (Table 6.2) is given below:

From the above truth table,

S = A′ B′ Cin + A′ B Cin′ + A B′ Cin′ + A B Cin
= Cin (A′ B′ + A B) + Cin′ (A′ B + A B′)
= Cin XOR (A XOR B)
Cout = A B + A Cin + B Cin (A + A′)
= A B Cin + A B + A Cin + A′ B Cin
= A B (1 +Cin) + A Cin + A′ B Cin
= A B + A Cin + A′ B Cin
= A B + A Cin (B + B′) + A′ B Cin

Table 6.2 The truth table for quantum full adder

Inputs			Outputs	
|A >	**|B >**	**|Cin >**	**|S >**	**|Cout >**
|0 >	|0 >	|0 >	|0 >	|0 >
|0 >	|0 >	|1 >	|1 >	|0 >
|0 >	|1 >	|0 >	|1 >	|0 >
|0 >	|1 >	|1 >	|0 >	|1 >
|1 >	|0 >	|0 >	|1 >	|0 >
|1 >	|0 >	|1 >	|0 >	|1 >
|1 >	|1 >	|0 >	|0 >	|1 >
|1 >	|1 >	|1 >	|1 >	|1 >

$$= A \, B \, Cin + A \, B + A \, B' \, Cin + A' \, B \, Cin$$
$$= A \, B \, (Cin + 1) + A \, B' \, Cin + A' \, B \, Cin$$
$$= A \, B + A \, B' \, Cin + A' \, B \, Cin$$
$$= AB + Cin \, (A' \, B + A \, B')$$

6.3.1   Block Diagram

The quantum circuit for a full adder is created by several quantum gates. To create a full adder, two XORs, two ANDs, and an OR quantum gates are required. In this quantum circuit, two quantum XOR operations are required to produce the SUM of the full adder. To produce the carry qubit, two quantum AND operations and a quantum OR operation are needed along with the quantum XOR operation. The block diagram of quantum full adder is illustrated in Fig. 6.3.

6.3.2   Circuit Architecture

Algorithm 6.3.2 represents the overall procedures of the quantum-based full-adder operation. If the total number of inputs bits is n then O (n) is the run time complexity of this algorithm. In this work, two quantum XORs, two quantum ANDs, and one quantum OR operations are used. Quantum XOR operation is denoted by DO_Quant_XOR (|A >, |B >), quantum AND operation are denoted by DO_Quant_AND (|A >, |B >), and quantum OR operation is denoted by DO_Quant_OR (|A >, |B >).

"FA_SUM" is used to represent the Full_Adder sum's output which is obtained after doing the second quantum XOR operation. "FA_Carry" is the final carry of the full adder is obtained after doing quantum OR operation.

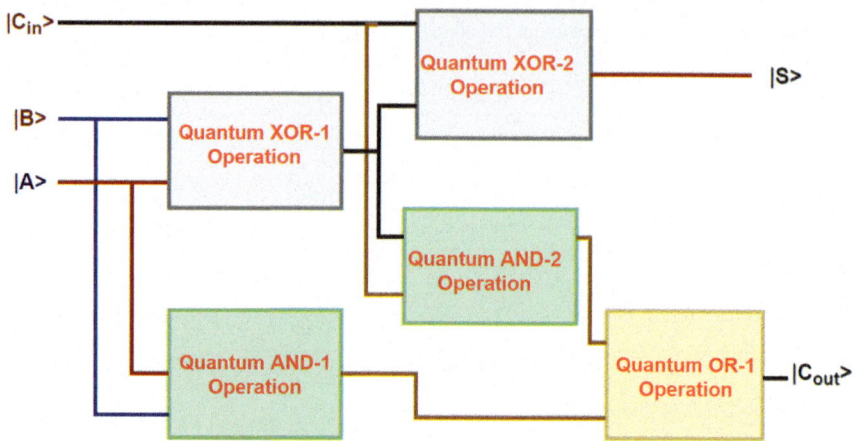

Fig. 6.3 Block diagram of quantum full adder

Algorithm 11.2: Quantum Full Adder Algorithm

1. Begin
2. **while** i equals to 1 to n **do**
3. |T1 > <-DO_Quant_XOR(|Ai >, |Bi >)
4. |T2 > <-Do_Quant_AND(|T1 >, |Ci >)
5. |T3 > <-Do_Quant_AND(|Ai >, |Bi >)
6. FA_SUM <-Do_Quant_XOR(|T1 >, |Ci >)
7. FA_Carry <-Do_Quant_OR(|T2 >, |T3 >)
8. **end while**

Figure 6.4 shows the quantum circuit of the full adder. In this full adder, the first XOR operation for |A > and |B > inputs is done by a quantum XOR operation and creates output |A > $\oplus$ |B >. This output, |A > $\oplus$ |B >, is used as one of the inputs of the second XOR operation and another input qubit comes from |Cin >. Finally, the output |S > qubit of the full adder is generated through this quantum XOR operation which represents the sum of the full adder.

In addition, two quantum AND operations are needed to get another output qubit of |C$_{out}$ >. First quantum AND operation occurs between input |A > and |B > and produces |A >.|B >. As for the second AND operation, one input comes from |A > $\oplus$ |B > and another input is directly provided from |Cin >. These two input qubits then generate (|A > $\oplus$ |B >).|Cin >. Finally, the outputs of both quantum AND operations pass through a quantum OR operation to produce the value of |Cout > qubit.

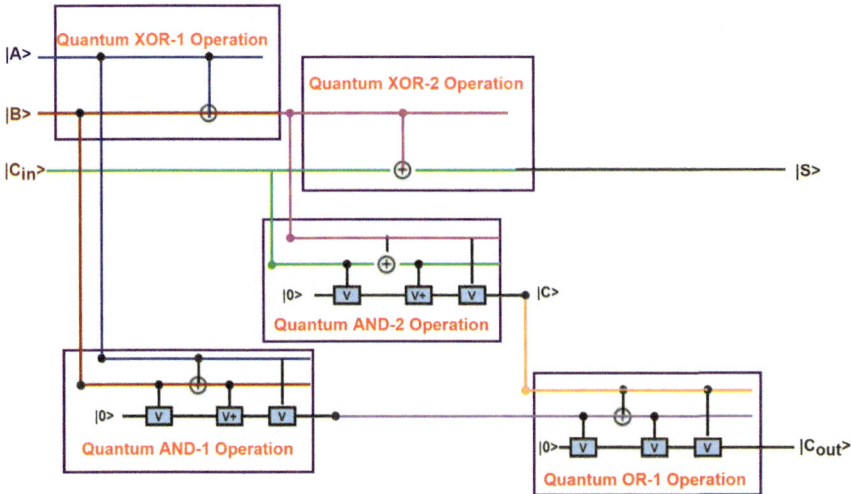

Fig. 6.4   Quantum full adder circuit

6.3.3   *Working Principle*

The quantum full adder needs three inputs, A, B, and Cin, and provides two outputs S and Cout. To understand the working principle of quantum full adder, consider the values of inputs A, B, and Cin are $|1>$, $|0>$, and $|1>$. Then

(i) The output in quantum XOR-1 is "true" $|1>$, in quantum AND-1 is "false" $|0>$, and in quantum AND-2 is "false" $|0>$.

(ii) Finally, the value of output S is "true" $|1>$, and Cout is "false" $|0>$.

Rest of the input combinations will work as the same way by following the truth table shown in Table 6.2.

6.4   Quantum Qubit Adder

A quantum adder is a circuit in quantum computing that performs addition operations using qubits. Quantum qubit adder needs quantum XOR operation, quantum AND operation, and quantum OR operation. In this section, quantum 2-qubit adder and quantum 4-qubit adder will be explained with the figures.

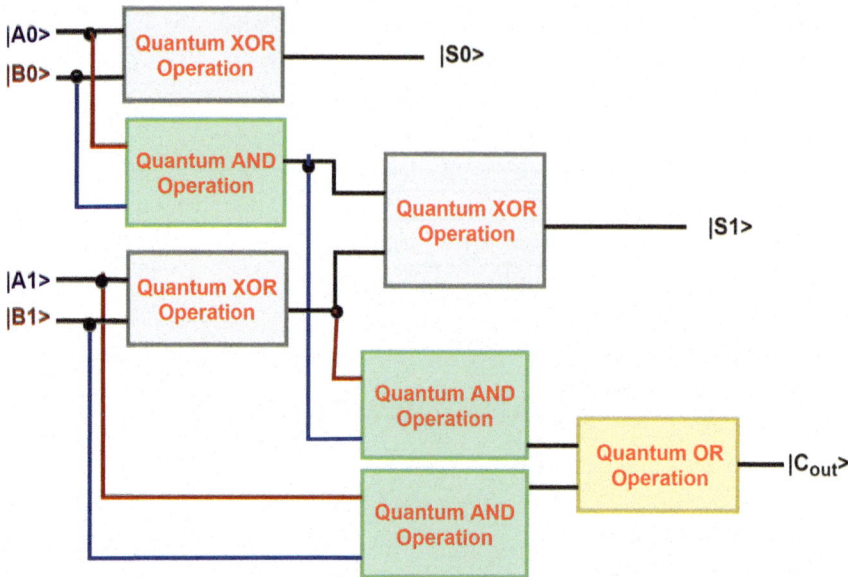

Fig. 6.5 Block diagram of 2-qubit quantum adder

6.4.1   Block Diagram

Figure 6.5 illustrates the block diagram of the 2-qubit quantum adder. Quantum 2-qubit adder circuit requires three quantum XOR operations, three quantum AND operations, and one quantum OR operation to add the inputs of 2-qubit |A > and |B >. All of these operations are executed using quantum gates.

And Fig. 6.6 demonstrates an overall block diagram of the 4-qubit quantum adder. Quantum 4-qubit adder circuit needs eight quantum XOR operations, eight quantum AND operations, and four quantum OR operations to complete the circuit.

It is observed from the above figure that the quantum 4-qubit adder needs 4 quantum full adder. So, the simplified block diagram can be drawn like the Fig. 6.7.

Figure 6.8 shows the general organizations of a quantum n-qubit adder.

So, n numbers of quantum full-adders are connected to each other to perform the addition process of two values of n-qubits.

6.4.2   Circuit Architecture

In the quantum 2-qubit adder circuit, |A0 > and |B0 > inputs perform a quantum XOR operation to generate the output of |S0 >. These inputs also execute another quantum AND operation whose output acts as an input for both quantum XOR-2

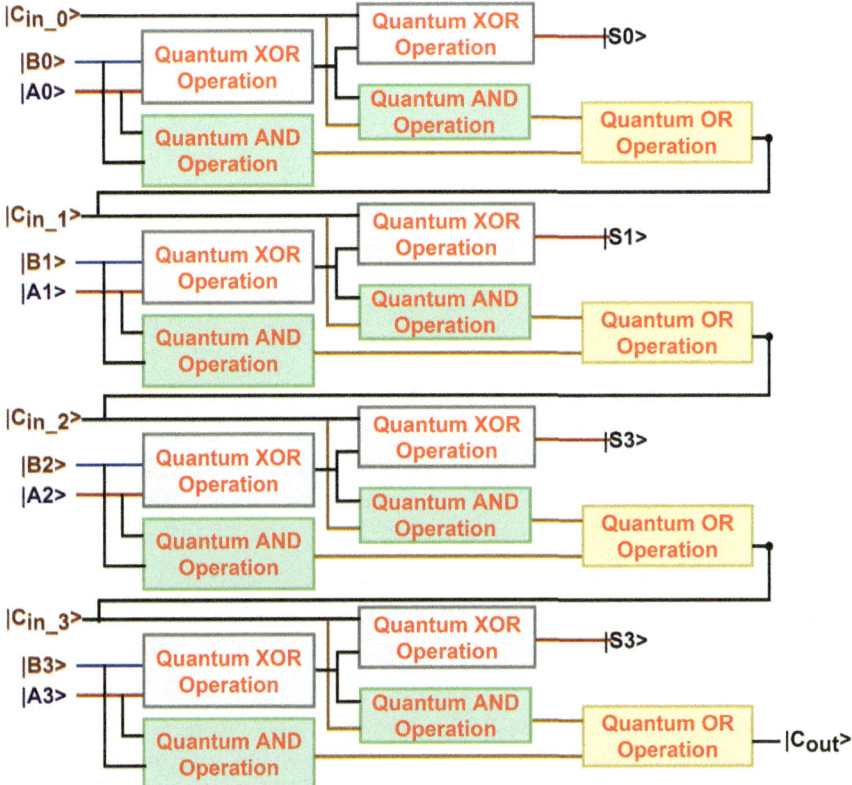

Fig. 6.6 Block diagram of 4-qubit quantum adder

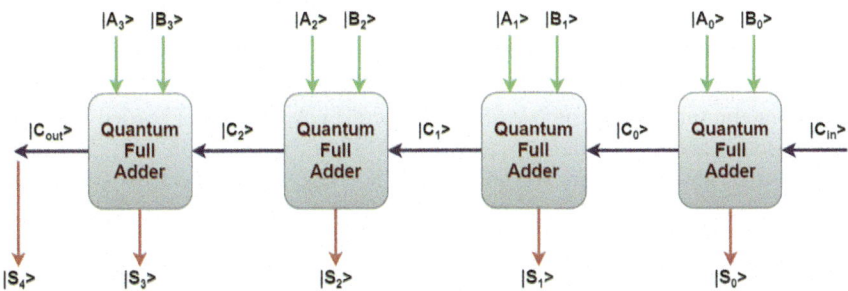

Fig. 6.7 The simplified block diagram of quantum 4-qubit adder

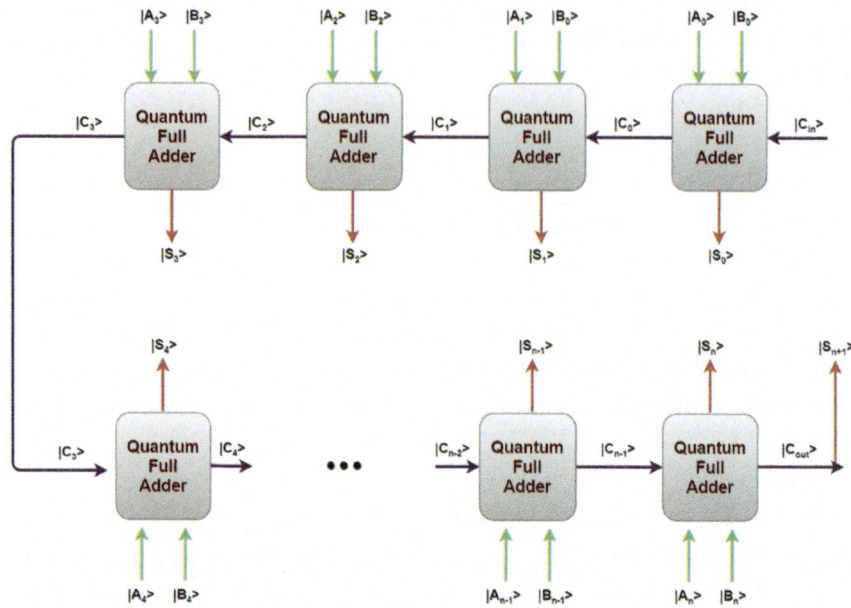

Fig. 6.8 The block diagram of quantum N-qubit adder

operation and quantum AND-2 operation, where the input qubits $|A1>$ and $|B1>$ pass through the quantum XOR-3 operation to produce $|A1> \oplus |B1>$. This $|A1> \oplus |B1>$ then go to the XOR-2 operation and AND-2 operation as the second input. Second XOR then generates the output of $|S1>$, and the output that quantum AND-2 operation produces, it goes to the quantum OR process whose other input comes from the quantum AND-3 operation of $|A1>$ and $|B1>$. Finally, the quantum OR operation gives the output of $|Cout>$. The whole architecture of the circuit is shown in Fig. 6.9.

As for the quantum 4-qubit adder, the overall circuit can be divided into four parts as each qubit of input $|A>$ performs a full adder operation with each qubit of $|B>$ that has the same position. Each qubits pair of $|A>$ and $|B>$ performs two sequential quantum XOR operations to give the result of $|S_0>$, $|S_1>$, $|S_2>$, and $|S_3>$, respectively. On the other side, the carry-out qubit of $|A0>$ and $|B0>$, which is the output of quantum OR-1 operation, is transferred as an input to the $|C_{in\_1}>$, an input of the second part. Similarly, the result of quantum OR-2 is provided to $|C_{in\_2}>$. After continuing the trend, finally, the quantum OR-4 operation generates the result of $|C_{out}>$. At the beginning, qubit $|0>$ is inserted in the $|C_{in\_0}>$ input line. Figure 6.10 shows the 4-qubit quantum adder circuit.

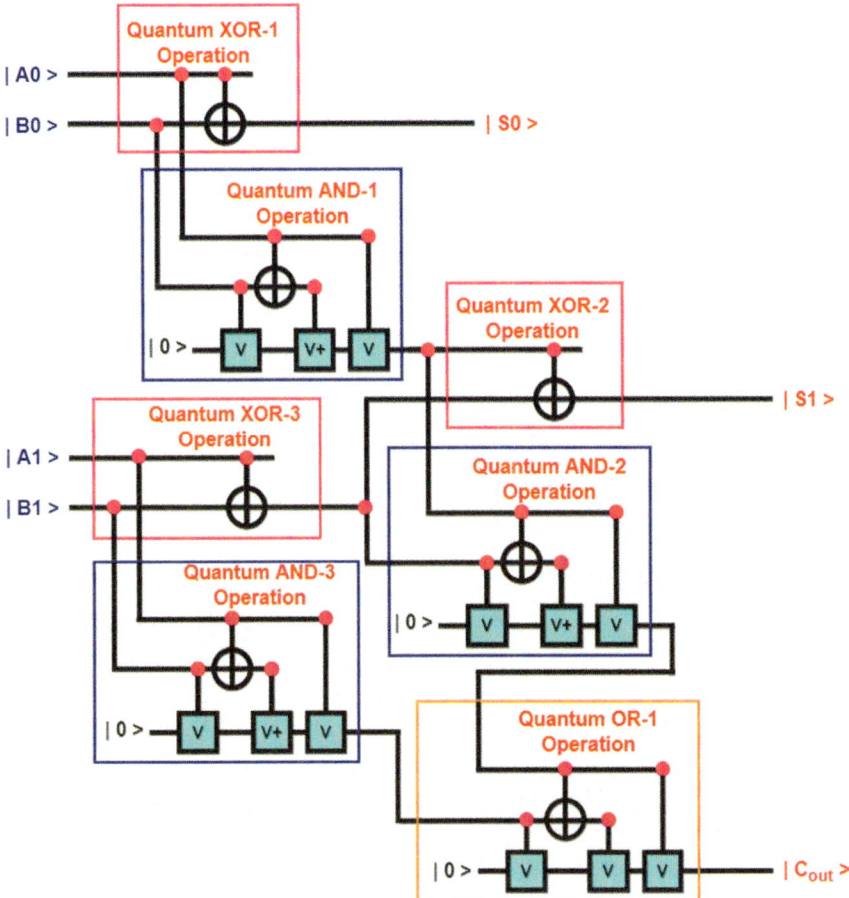

Fig. 6.9 2-Qubit quantum adder circuit

6.4.3 Working Principle

Quantum 2-Qubit Adder

The quantum 2-qubit adder needs two 2-qubit inputs, A and B and Cin, and provides three outputs S0, S1, and Cout. To understand the working principle of quantum 2-qubit adder, consider the values of input qubits A0, A1, and B0 and B1 are $|1>, |0>, |1>,$ and $|1>$. Then

(i) The output (S0) of quantum XOR-1 is "false" $|0>$
(ii) The output of S1 is "false" $|0>$ as quantum AND-1 is "true" $|1>$, and quantum XOR-2 is "true" $|1>$. Thus, the quantum XOR-3 will be "false" $|0>$.

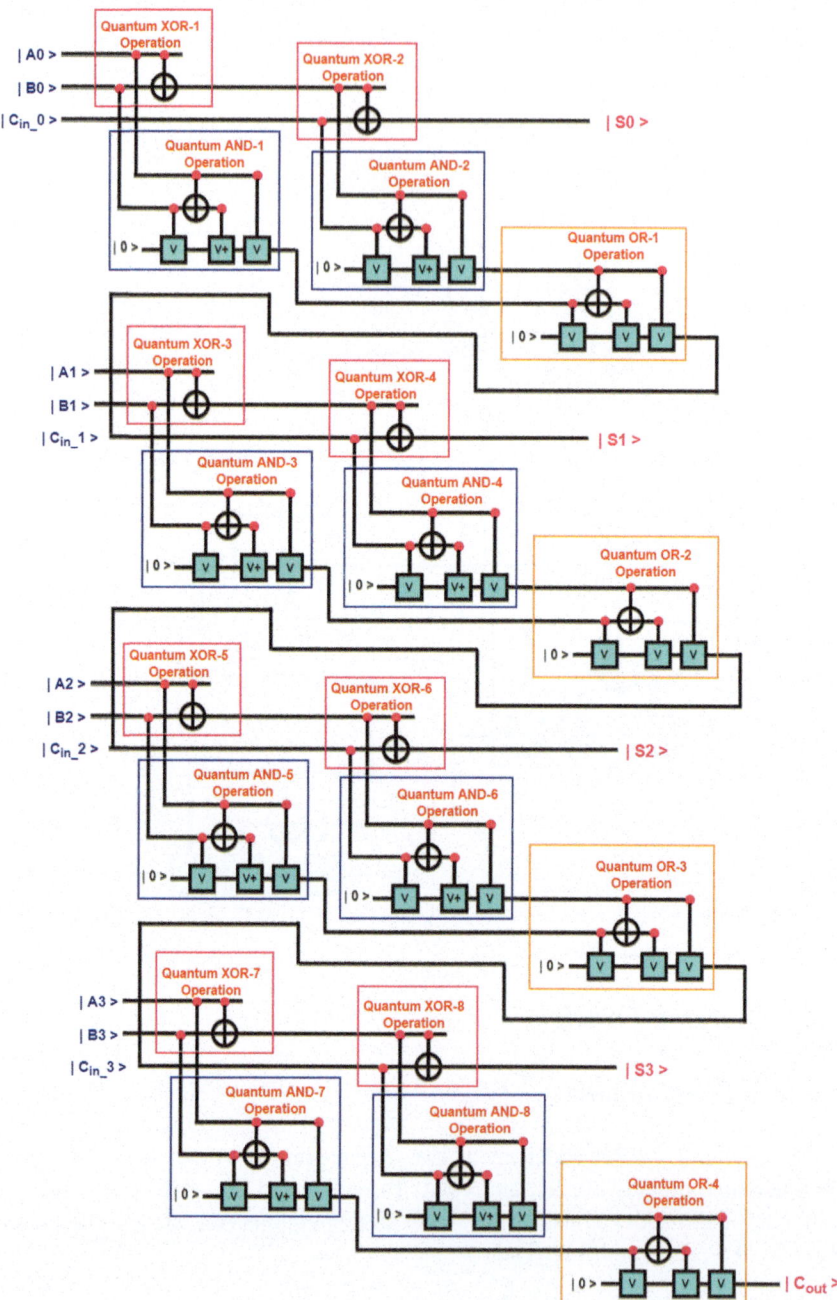

Fig. 6.10 4-Qubit quantum adder circuit

(iii) The output of Cout is "true" $|1>$ as quantum AND-3 is "true" $|1>$, and quantum AND-2 is "false" $|0>$. Thus, the quantum OR-1 will be "true" $|1>$.

Since the truth table and the working principle of quantum XOR, quantum AND, and quantum OR operations (from the previous chapter) are known to us, it is easy to obtain the rest of the output combinations for the different input combinations by following the circuit:

Quantum 4-Qubit Adder

Suppose the values of $|A3>$, $|A2>$, $|A1>$, and $|A0>$ are $|1>$, $|1>$, $|0>$, and $|1>$. On the other side, the values of $|B3>$, $|B2>$, $|B1>$, and $|B0>$ are $|1>$, $|1>$, $|1>$, and $|1>$. Then, the outputs of $|S3>$, $|S2>$, $|S1>$, and $|S0>$ are $|1>$, $|1>$, $|0>$, and $|0>$, respectively. Additionally, the output of $|Cout>$ is $|1>$.

Since the truth table and working principle of quantum XOR, quantum AND, and quantum OR operations (from the previous chapter) are known, it is easy to obtain the rest of the output combinations for the different input combinations by following the circuit.

6.5 Quantum BCD Adder

A Binary Coded Decimal (BCD) adder in quantum computing involves designing a circuit that performs addition of two BCD numbers, resulting in a BCD sum and a carry output. This is crucial for performing arithmetic operations on decimal values within the context of quantum computing. BCD adders are implemented using quantum gates like Toffoli gates (C3 gates) and CNOT gates, which are fundamental building blocks in quantum circuits. The BCD adder is utilized in PC frameworks and adding machines that perform mathematical tasks in the decimal number framework straightforwardly. The binary-coded form of decimal numbers is accepted by the BCD adder. A threshold of nine inputs and five outputs are required for the Decimal-Adder.

6.5.1 Block Diagram

For the construction of the BCD adder, two 4-qubit quantum adders are required, along with two quantum AND operations and two quantum OR operations. The block diagram of the BCD adder is given in Fig. 6.11.

6.5.2 Circuit Architecture

Figure 6.12 shows the quantum circuit of the BCD adder. Here, the four outputs of the first 4-qubit quantum adder directly transfer to the next 4-qubit quantum adder, as the inputs of $|A>$. However, in between, two quantum AND operations and two

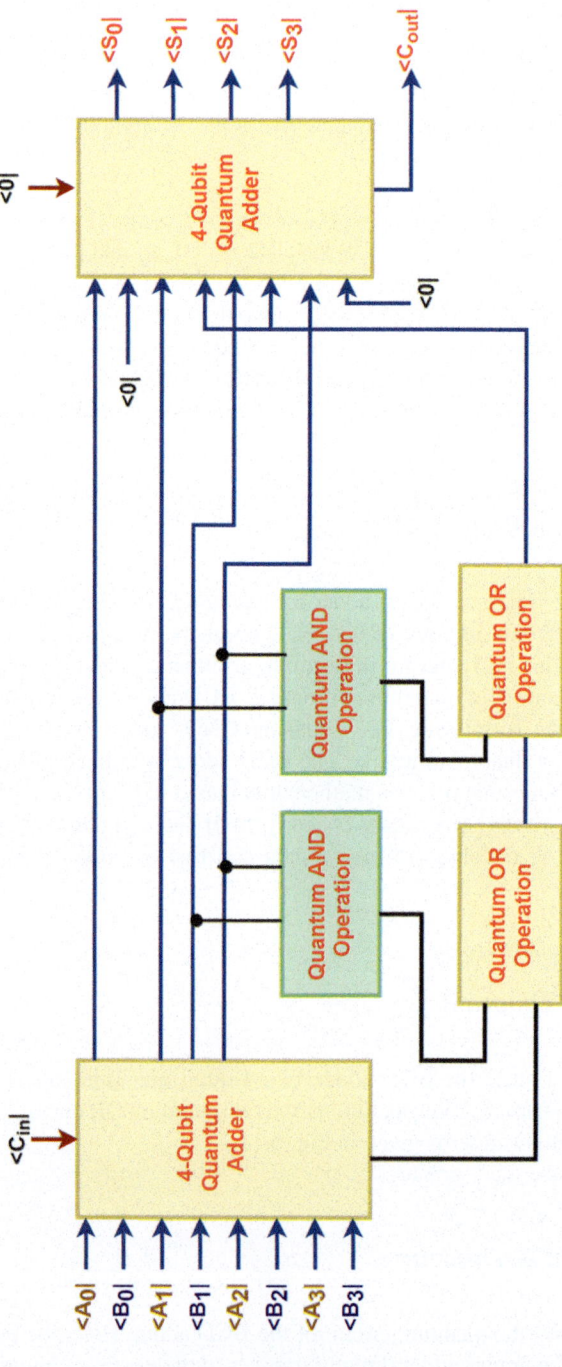

Fig. 6.11 Block diagram of BCD adder

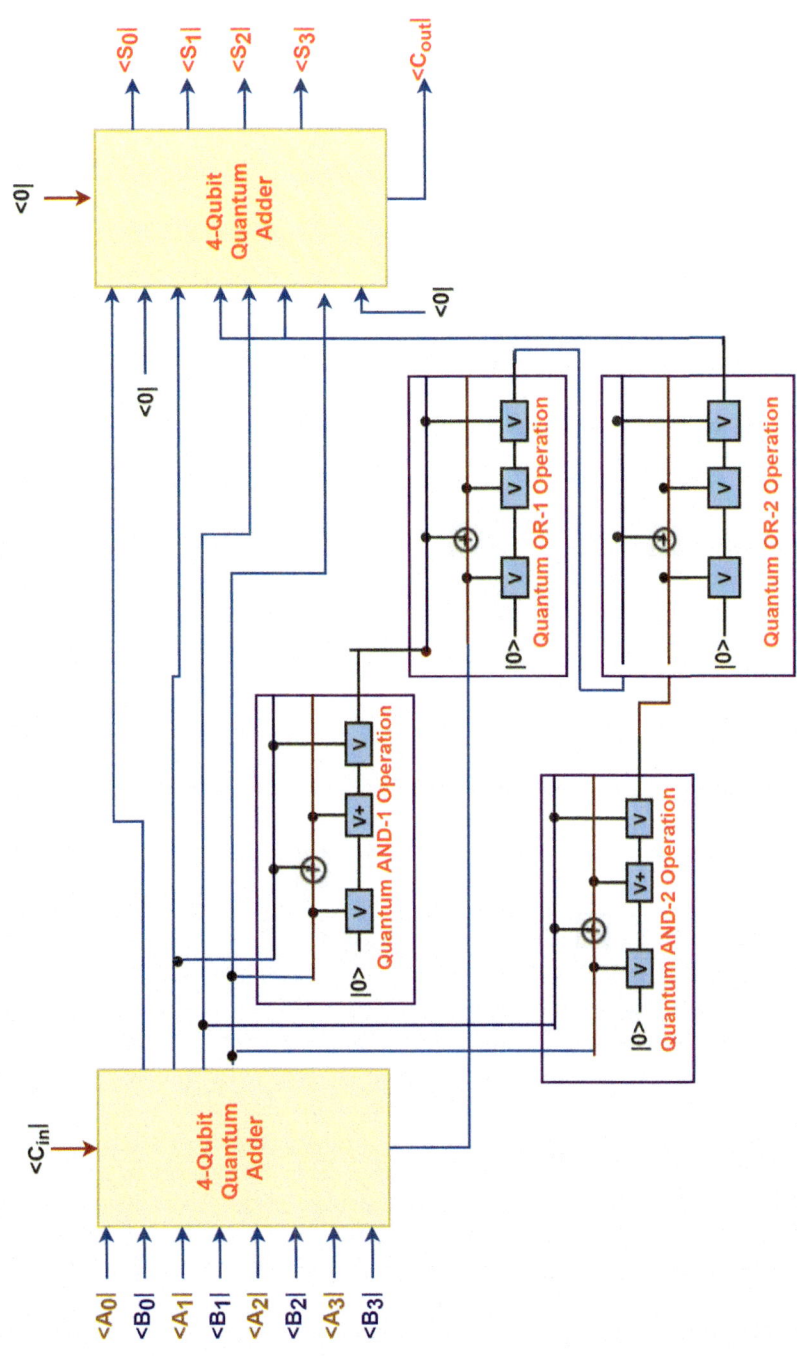

Fig. 6.12 Quantum BCD adder circuit

quantum OR operations are conducted using them. The values of $|S_1>$ and $|S_3>$ perform quantum AND-1 operation, whereas quantum AND-2 operation is operated on $|S_2>$ and $|S_3>$. In the next step, both the outputs of quantum AND operation, along with the $|C_{out}>$ of the first 4-qubit quantum adder, sequentially execute two quantum OR operations. Then, the result of the quantum OR-2 operation acts as the inputs for the $|B>$ of the last 4-qubit quantum adder.

6.5.3   *Working Principle*

The quantum BCD adder needs two 4-qubit adders where there are two 4-qubit inputs, A and B and a carry input, Cin, and in order to understand the working principle of quantum BCD adder, consider the values of input qubits $|A_0>$, $|A_1>$, $|A_2>$, and $|A_3>$ are $|0>$, $|1>$, $|1>$ and $|0>$, equal to decimal value '6' and $|B_0>$, $|B_1>$, $|B_2>$, and $|B_3>$ are $|1>$, $|0>$, $|1>$ and $|0>$, equal to decimal value '5'. Then

(i)   The output of the first 4-Qubit adder will be $|S_0> = |1>$, $|S_1> = |1>$, $|S_2> = |0>$, $|S_3> = |1>$ and $|C_{out}> = |0>$.

(ii)   Then, the results of two Quantum AND operations are $|S_2>$. $|S_3> = |0>$ and $|S_1>$. $|S_3> = |1>$.

(iii)   By continuing the above result, the final quantum OR operation generates "True" $|1>$.

(iv)   Finally, the outputs of the second 4-qubit adder are $|S_0> = |1>$, $|S_1> = |0>$, $|S_2> = |0>$, $|S_3> = |0>$ and $|C_{out}> = |1>$ which is the BCD number of 11.

In this way, it is easy to obtain different outputs for different input qubits combinations.

6.6   Quantum Carry-Lookahead Adder

The addition of two numbers is started in parallel adders when all the qubits of the augend and the addend must be available at the same time to conduct the computation. Because the carry output of each full adder stage is connected to the carry input of the next higher-order stage in a parallel adder circuit, it is also known as a ripple-carry type adder. In other words, a Quantum Carry-Lookahead Adder (QCLA) is a quantum circuit designed to perform addition with a lower depth than a quantum ripple-carry adder. It achieves this by using techniques to calculate carries across multiple qubits simultaneously. QCLAs are important because they can be used in various quantum algorithms, including those related to encryption and physics. A quantum carry-lookahead adder (QCLA) offers a way to perform addition with a critical path depth of $O(\log n)$, which is faster than a traditional ripple-carry adder that has a depth of $O(n)$. This faster addition can be beneficial in algorithms like Shor's algorithm, which uses modular multiplication that relies on addition. QCLAs are valuable for building quantum circuits for various algorithms, such as those used in quantum cryptography (e.g., Shor's algorithm). QCLAs offer faster addition times, which can be crucial for the performance of quantum algorithms. They can also reduce the overall depth of a quantum circuit, making it more efficient and potentially easier to implement on near-term quantum computers.

The quantum carry-lookahead adder simply finds all the carry outputs at a time before rather than one after another, like the quantum N-qubit adder. This method employs logic operations to examine the lower-order qubits of the augend and addend to determine whether or not a higher-order carry should be created.

For simplicity, all variables will be used as classical notation here. Two variables are defined as carry generate $\mathbf{G}_i$ and carry propagate $\mathbf{P}_i$ then,

$$P_i = A_i \oplus B_i \tag{6.1}$$

$$G_i = A_i B_i \tag{6.2}$$

The sum output and carry output can be expressed as

$$S_i = P_i \oplus C_i \tag{6.3}$$

$$C_{i+1} = P_i C_i + G_i \tag{6.4}$$

where $\mathbf{G}_i$ is a *carry generate* that provides the carry regardless of the input carry when both A_i and B_i are "$|1>$". $\mathbf{P}_i$ is a carry propagate that is linked to the carry propagation from C_i to C_{i+1}.

Then,

$$\begin{aligned} C_1 &= P_0 C_{in} + G_0 \\ C_2 &= P_1 C_1 + G_1 \\ &= (A_1 \oplus B_1) C_1 + A_1 B_1 \end{aligned} \tag{6.5}$$

$$\begin{aligned} C_3 &= P_2 C_2 + G_2 \\ &= (A_2 \oplus B_2) C_2 + A_2 B_2 \\ &= P_2 (P_1 C_1 + G_1) + G_2 \\ &= P_2 P_1 C_1 + P_2 G_1 + G_2 \end{aligned} \tag{6.6}$$

$$\begin{aligned} C_4 &= P_3 C_3 + G_3 \\ &= (A_3 \oplus B_3) C_2 + A_3 B_3 \\ &= P_3 (P_2 P_1 C_1 + P_2 G_1 + G_2) + G_3 \\ &= P_3 P_2 P_1 C_1 + P_3 P_2 G_1 + P_3 G_2 + G_3 \end{aligned} \tag{6.7}$$

Therefore, it is evident that the carry output of every quantum full-adder operation will depend only on the value of C_1. Following equation helps to calculate C_n.

$$C_n = P_{n-1}P_{n-2}\ldots P_2P_1C_1 + P_{n-1}P_{n-2}\ldots P_3P_2G_1$$
$$+ P_{n-1}P_{n-2}\ldots P_4P_3G_2 \qquad\qquad (6.8)$$
$$+ \cdots + P_{n-1}P_{n-2}G_{n-3} + P_{n-1}G_{n-2} + G_{n-1}$$

So, at first, it is required to calculate the value of C_1, then it will be able to execute all the next operations concurrently, which will definitely give a huge jump to reduce the computational execution time for larger qubits.

6.6.1 Block Diagram

To construct the quantum carry-lookahead adder, it is required to design extra circuitry with the quantum parallel adder, which will determine all the carry outputs without the value of the carry inputs generated from the previous operations. The value of $|C_i>$ can be calculated by a combination of quantum AND and quantum OR operations. The combinational circuit is designed to determine the value of $|C_i>$ and after that, the carry values will be used to determine the sum outputs.

The general architecture of the quantum 4-qubit carry-lookahead adder is shown in Fig. 6.13.

Figure 6.14 shows the block diagram for the quantum N-qubit carry-lookahead adder, where the addition can be performed of n-qubit inputs, and again they all will be executed at the same time after determining the value of $|C_1>$ which will reduce the total computational time.

Figure 6.15 shows the block diagram to estimate the value and $|C_2>$ which is estimated by Eq. (6.5).

Figure 6.16 shows the block diagram to estimate the value and $|C_3>$ which is estimated by Eq. (6.6).

It is easy to construct the block diagram for the value of other $|C_i>$, in the same manner.

6.6.2 Circuit Architecture

In this section, the construction of a circuit diagram for the 3-qubit quantum carry-lookahead adder will be presented. To construct a 3-qubit quantum carry-lookahead adder, it is needed to determine the value of $|C_2>$ and the value of $|C_3>$. The circuit diagram to determine the value of $|C_2>$ is shown in Fig. 6.17.

Figure 6.18 shows the operational circuit diagram of the quantum 3-qubit carry-lookahead adder, which is constructed using several numbers of quantum XOR, quantum AND, and quantum OR operations to perform the addition operation of two input values of 3-qubit.

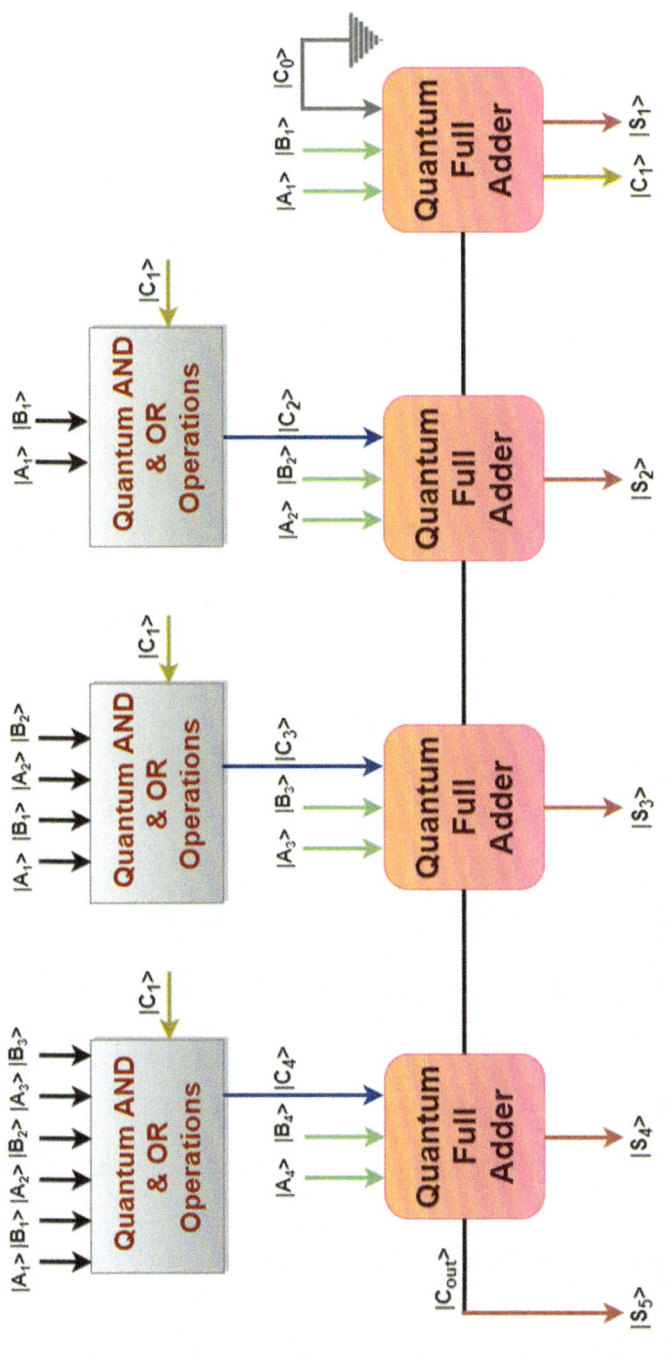

Fig. 6.13 The general architecture of quantum 4-qubit carry-lookahead adder

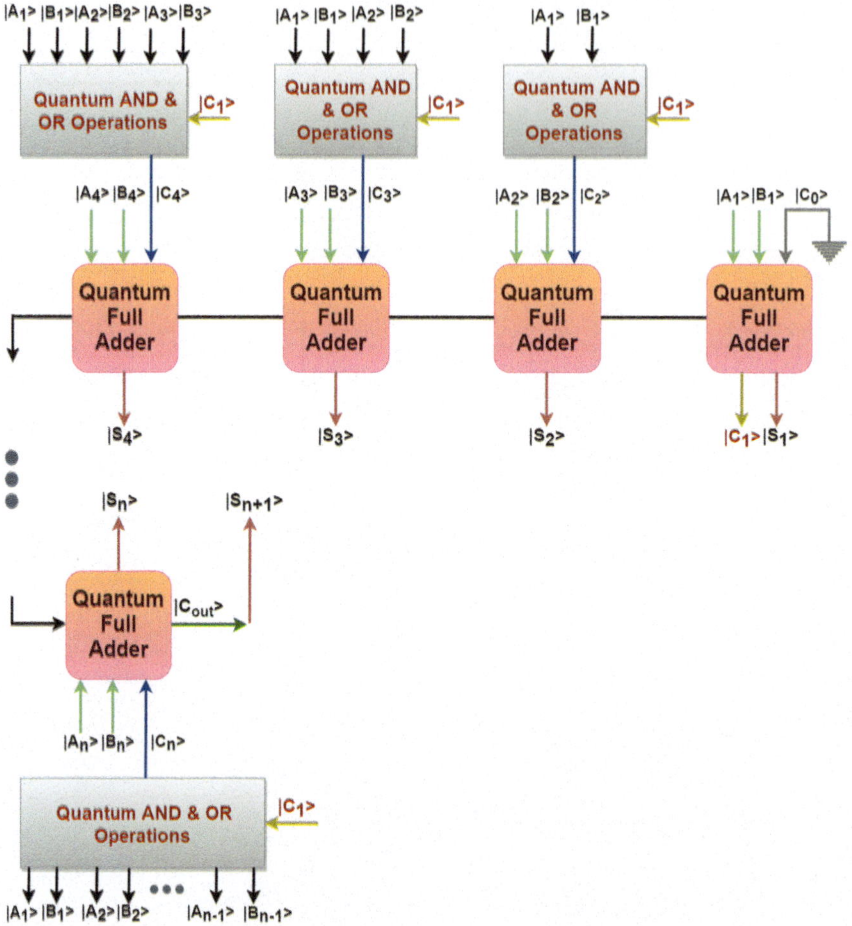

|A1> |B1>|A2>|B2> |A3>|B3>
Quantum AND & OR Operations
|C1>
|A1> |B1> |A2> |B2>
Quantum AND & OR Operations
|C1>
|A1> |B1>
Quantum AND & OR Operations
|C1>
|A4> |B4> |C4>
|A3> |B3> |C3>
|A2> |B2> |C2>
|A1>|B1> |C0>
Quantum Full Adder
Quantum Full Adder
Quantum Full Adder
Quantum Full Adder
|S4>
|S3>
|S2>
|C1>|S1>
|Sn>
|Sn+1>
Quantum Full Adder
|Cout>
|An>|Bn> |Cn>
Quantum AND & OR Operations
|C1>
|A1>|B1> |A2>|B2> |An-1> |Bn-1>

Fig. 6.14 The general architecture of quantum N-qubit carry-lookahead adder

|P1>
|C1>
|G1>
QAND Operation
QOR Operation
|C2>

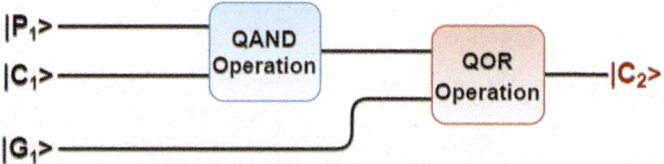

Fig. 6.15 The block diagram for deriving the value of $|C_2>$

It is necessary to check carefully to understand and determine the values of $|C_1>$, $|C_2>$, and $|C_3>$.

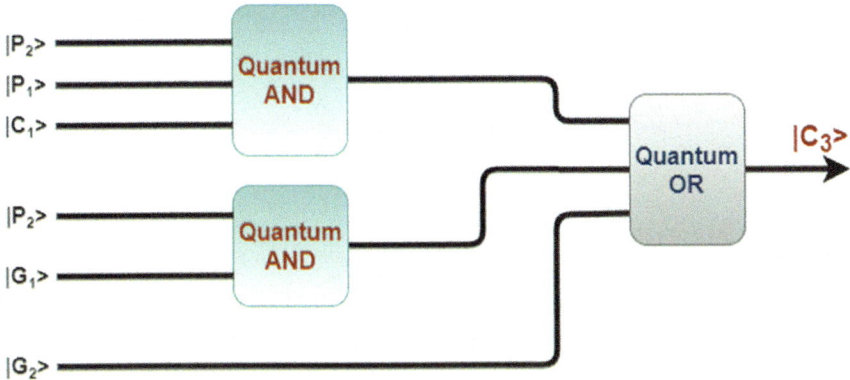

Fig. 6.16 The block diagram for deriving the value of $|C_3>$

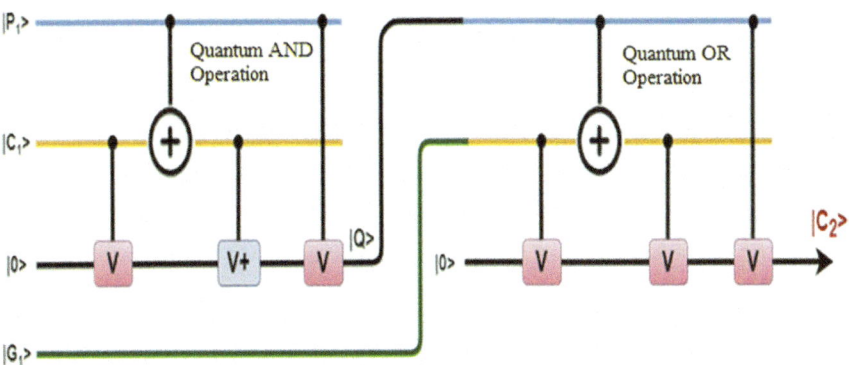

Fig. 6.17 The circuit diagram to derive the value of $|C_2>$

6.6.3 Working Principle

The quantum carry-lookahead adder's working principles are very simple. Suppose, $|A> = |110>$ and $|B> = |101>$.

1. The 1st addition operation will be performed between $|A_1> = |0>$, and $|B1> = |1>$. And for the 1st operation the value is $|C_0> =0>$. The 1st quantum XOR will produce $|1>$ and the 2nd quantum XOR will also produce $|1>$. Therefore the value of $|S_1>$ will be $|1>$ and the carry $|C_1>$ will be generated using the quantum AND and quantum OR operations, which will be $|0>$. Thus the $|C_1> = |0>$ will be used to determine all the values of other $|C_i>$.
2. The value of $|P_1>$ will be $|1>$, and the value of $|G_1>$ will be $|0>$. Therefore, the value of $|C_2> = |0> (P_1C_1 + G_1)$.
3. Now the 2nd addition operation will be performed between $|A_2> = |1>$, and $|B_2> = |0>$. And $|C_2> = |0>$. Therefore, the value is $|S_2> = |1>$. Hence, the value of $|C_3>$ can be determined.

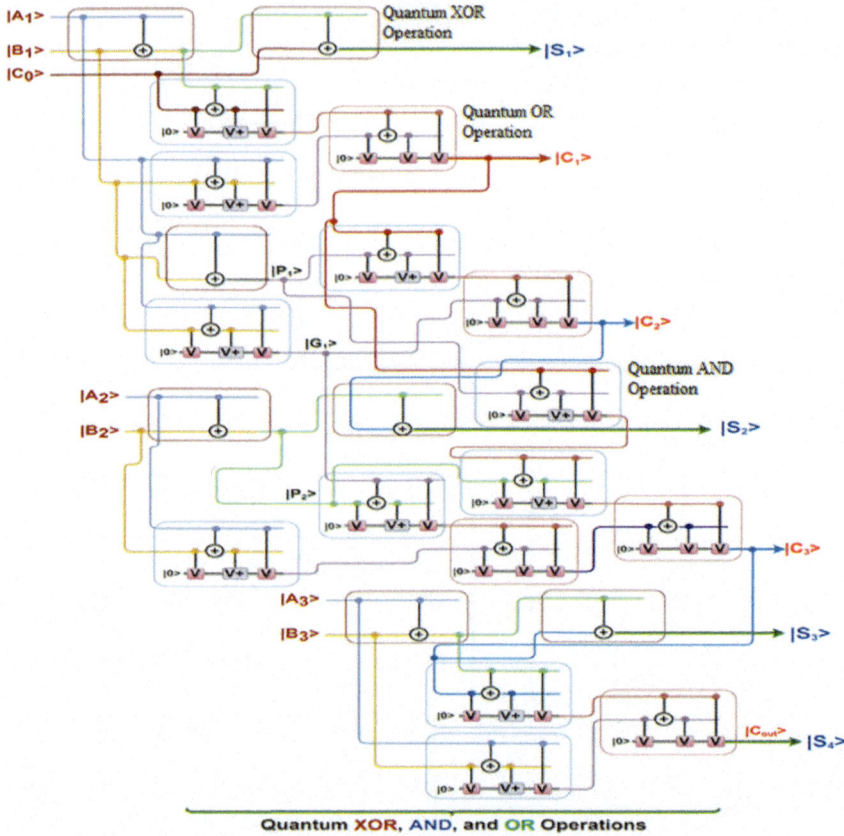

Fig. 6.18 The circuit diagram of 3-qubit quantum carry-lookahead adder

4. Here, the value of $|P_2>$ will be $|1>$, and the value of $|G_2>$ will be $|0>$ again. Therefore, the value of $|C_3>$ will be $|0>$ (from Eq. 6.6).
5. Now the 3rd addition operation will be performed between $|A_3> = |1>$, and $|B_3> = |1>$. And $|C_3> = |0>$. Therefore, the value is $|S_3> = |0>$. And the value of $|C_4>$ will be the most significant qubit of the result.
6. Therefore, the value of $|S_4>$ will be $|1>$.

It can be tested for different 3-qubit inputs. Identify what will be the output at each operation in the circuit of quantum 3-qubit carry-lookahead adder.

6.7 Quantum Carry-Skip Adder

A quantum carry-skip adder is a quantum adder implementation. When compared to other adders, it improves the latency of a quantum n-qubit adder (ripple-carry adder) with minimal effort. By combining many carry-skip adders to construct a quantum block-carry-skip adder, the worst-case delay can be reduced.

The quantum carry-skip adder skips a number of qubits (often called block) if the addition process of qubit does not generate any carry output, which is equivalent to $|1>$. It bypasses the 1st carry qubit of the block when it is found that there exists no operation which generates carry output $|1>$. That is why it is often called the quantum bypass adder. In other words, a Quantum Carry-Skip Adder (QCSA) is a type of adder circuit that operates on quantum bits (qubits) and uses the principles of a carry-skip adder to perform addition. It aims to reduce the time and resources needed for quantum arithmetic operations by allowing carry bits to bypass certain adder stages under specific conditions. A QCSA translates the principles of the classical carry-skip adder into the realm of quantum computing, leveraging the unique properties of qubits to perform arithmetic operations. QCSAs are designed to minimize the number of qubits and quantum gates required for addition, which is crucial for practical quantum algorithms due to the limited resources available on current quantum computers. In essence, a QCSA is a quantum circuit that emulates the carry-skip adder's ability to bypass certain adder stages based on the inputs, enabling faster and more resource-efficient quantum arithmetic operations.

Figure 6.19 shows how a carry is propagated through the operations. Here, BP stands for block propagate, which determines whether a carry will generate as $|1>$ or not. If there will be no carry output as $|1>$, a multiplexer will bypass the 1st carry input $|C_0>$ as the input for the next quantum full-adder operation. If there will be generated a carry as $|1>$ within these four quantum full-adder operations, this carry-skip adder will work just like the quantum parallel adder.

The block propagate BP can be determined as follows:

$BP_i = |P_0>. |P_1>. |P_2> ... |P_i>$

where $|P_i> = (A_i$ Quantum XOR $B_i)$

6.7.1 Block Diagram

In this section, a 4-qubit quantum carry-skip adder will be designed. For 4-qubit quantum carry-skip adder, four quantum full adders are needed. The output of these quantum full adders will go through a quantum AND operation, and finally a quantum multiplexer will be used to generate the final output. Figure 6.20 shows the block diagram of a 4-qubit quantum carry-skip adder.

From the Fig. 6.20, the block propagate will be generated first. If $BP = |1>$ then the circuit will bypass the first carry qubit to the next quantum full-adder using a quantum multiplexer.

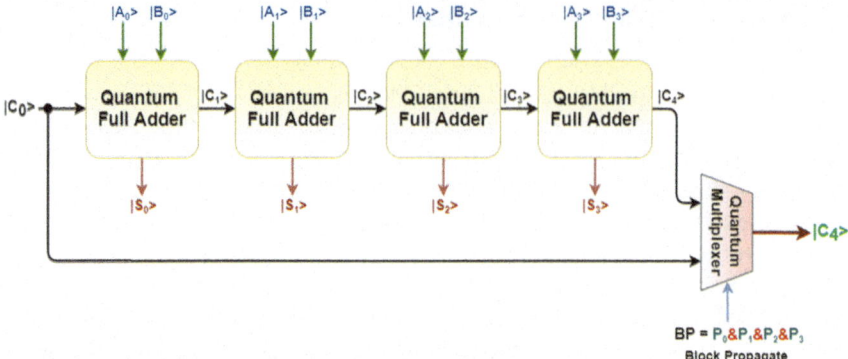

Fig. 6.19 Basic structure of a 4-qubit quantum carry-skip adder

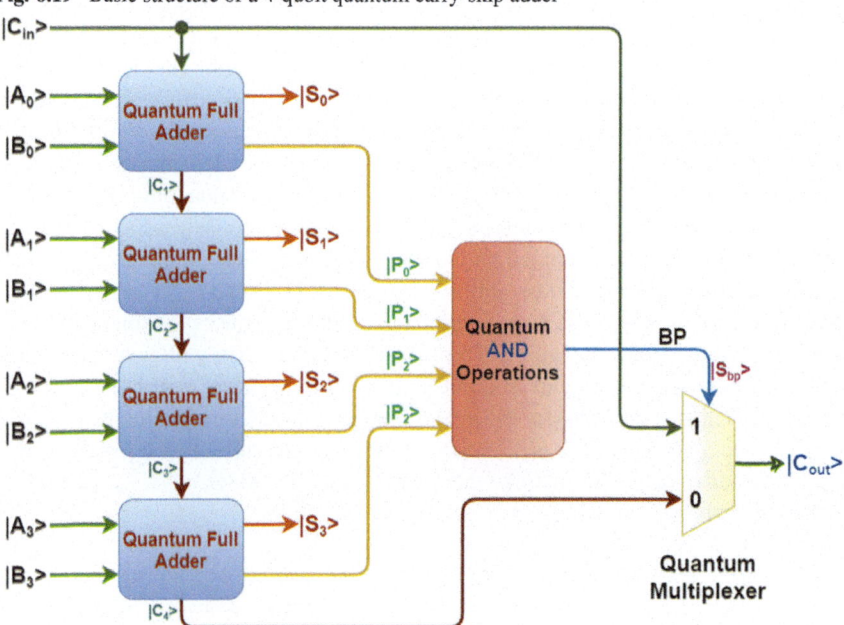

Fig. 6.20 The block diagram of a 4-qubit quantum carry-skip adder

6.7.2   *Circuit Architecture*

It is easy to construct the circuit of a 4-qubit carry-skip adder from the block diagram. Figure 6.21 shows the circuit diagram of the quantum 4-qubit carry-skip adder.

To construct the 4-qubit quantum carry-skip adder, four quantum full-adders, four additional quantum XOR and quantum AND operations are required to determine the block propagate, and a 2-to-1 quantum multiplexer is needed which will bypass the block propagate.

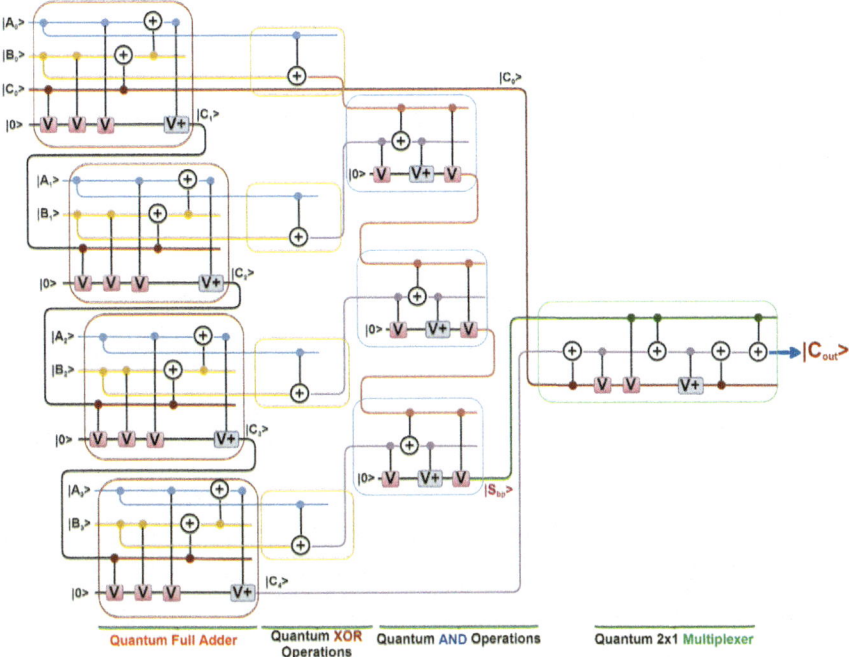

Fig. 6.21 The circuit diagram of a 4-qubit quantum carry-skip adder

6.7.3 Working Principles

To make this idea easier as always, the working procedure of the quantum carry-skip adder will be explained through example.

Suppose, perform the addition of two input qubits $|A>$ and $|B>$, where $|A> = |01101011>$ and $|B> = |10010100>$.

$$|A>:|01101011>$$
$$|B>:|10010100>$$

$$|S>:|11111111>$$

In such cases, the carry does not generate. Therefore the determining carry for such cases will increase the execution speed.

For the above example, the block propagation (BP) can be calculated as follows:

$BP = \prod_{i=0}^{n-1} P_i$

And, $|P_i> = |A_i>$ Quantum XOR $|B_i>$

If BP = $|1>$, then the carry of the last adder operation will be the carry output of the 1st adder operation. This means no carry is generated as $|1>$. Let's calculate BP for the above example.

BP = $|1>.|1>.|1>.|1>.|1>.|1>.|1>.|1> = |1>$

Therefore, no carry will be generated as $|1>$. Thus, the last carry output will be $|0>$ which was the 1st carry input of this operation.

For the 4-qubit circuit, suppose $|A> = |1011>$, and $|B> = |0100>$. So, $|A_0> = |1>$, $|A_1> = |1>$, $|A_2> = |0>$, and $|A_3> = |1>$. And $|B_0> = |0>$, $|B_1> = |0>$, $|B_2> = |1>$, and $|B_3> = |0>$. The operations can be described as follows:

1. The 1st quantum full-adder will receive inputs $|A_0> = |1>$, $|B_0> = |0>$, and $|C_0> = |0>$. Remember that the carry will only be determined if the value of BP is not $|1>$. Therefore the sum will be $|1>$, and the value of $|P_0> = |1>$.
2. Then, for the 2nd quantum full-adder operation, input qubits are $|A_1> = |1>$ and $|B_1> = |0>$. Therefore, $|P_1> = |1>$.
3. Now, for the 3rd quantum full-adder operation, input qubits are $|A_2> = |0>$ and $|B_2> = |1>$. Therefore, $|P_2> = |1>$.
4. Finally, for the 4th quantum full-adder operation, input qubits are $|A_3> = |0>$ and $IB_3> = |1>$. Therefore, $|P_3> = |1>$.

Now, all the values of P_i are calculated. So the block propagates will be BP = $|1>$.

Therefore, no carry was generated as $|1>$. And that is why the sum of the operations will be the values of all $|P_i>$'s for each A_i and B_i.

The value of BP will activate the quantum multiplexer. Consequently, the carry output for $|C_4>$ will be the value of $|C_0>$ for this example.

This circuit can be tested for input qubits that generates carry qubit as $|1>$ in any of the quantum full adder operations. For example, $|A> = |1011>$, and $|B> = |1001>$, and see how the circuit gives the result.

For this example, BP will be $|0>$, and that is why the circuit will work exactly like the quantum parallel adder.

6.8 Quantum Half Subtractor

The quantum half subtractor can also subtract two qubit numbers. It is equipped with two inputs and two outputs. This circuit is used to subtract two qubit numbers with single qubits, $|A>$ and $|B>$. The quantum half subtractor has two output states: 'D' and 'Bout.' In other words, a quantum half subtractor is a quantum circuit designed to perform the subtraction of two single qubits. A quantum half subtractor is constructed using quantum gates, such as CNOT and Toffoli gates, which manipulate qubits to perform the subtraction operation. The quantum half subtractor can be implemented by applying a Toffoli gate followed by a CNOT gate. These gates

Table 6.3 Quantum half subtractor truth table	Inputs		Outputs	
	|A >	**|B >**	**|D >**	**|Bout >**
	|0 >	|0 >	|0 >	|0 >
	|0 >	|1 >	|1 >	|1 >
	|1 >	|0 >	|1 >	|0 >
	|1 >	|1 >	|0 >	|0 >

are used to manipulate the qubits representing the inputs and produce the outputs. Quantum subtractors are essential components in building quantum computers and other computational units. They are used in quantum arithmetic operations and can be cascaded to perform subtraction on larger numbers. Truth table for the quantum half subtractor is given in Table 6.3.

From the above truth table, we get

$$D = A'B + A B'$$
$$= A \text{ XOR } B$$
$$B_{out} = A'B$$

6.8.1 Block Diagram

The quantum circuit for a half subtractor is created by several quantum gates. To design a half subtractor, one XOR gate, one NOT gate, and one AND gate are required. In this quantum circuit, one quantum XOR operation is needed to produce the difference of the half adder. Additionally, to generate the borrow qubit, one quantum NOT and quantum AND operations are used. The block diagram of the quantum half subtractor is illustrated in Fig. 6.22.

6.8.2 Circuit Architecture

Figure 6.23 depicts the quantum circuit of the half subtractor. In this half subtractor circuit, the quantum XOR operation for \|A > and \|B > inputs is done by a quantum operation and creates an output of \|D >. As for the output of \|Bout >, first \|A > is performed a quantum NOT operation, then both input qubits \|A'> and \|B > pass through a quantum AND operation.

6.8.3   *Working Principle*

The quantum half-adder needs two input qubits, A and B. For various input qubit values of A and B, consider the following cases:

 (i) When both A and B are "true" |1 >, the output sequences of both D and Bout are "false" |0 >.
 (ii) When both A and B are "false" |0 >, the output qubits of both D and Bout are "false" |0 >.
(iii) When A is "false" |0 > and B is "true" |1 >, the output qubit of D is "true" |1 > and Bout is "true" |1 >.
(iv) When A is "true" |1 > and B is "false" |0 >, the output qubit of D is "true" |1 > and Bout is "false" |0 >.

6.9   Quantum Full Subtractor

The quantum full subtractor is a circuit that performs subtraction on three input qubits: minuend, subtrahend, and borrow in. The difference and borrow output qubits are generated by the full subtractor. In other words, a quantum full subtractor is a quantum implementation of the classical full subtractor, utilizing quantum gates and principles to perform binary subtraction in the quantum realm. A quantum circuit designed to achieve the same subtraction operation but using quantum logic gates. It's often used in quantum algorithms that require subtraction, such as Shor's algorithm. Quantum full subtractors are built using quantum logic gates such as Toffoli gates, Feynman gates, and NOT gates, which are the fundamental building blocks of quantum circuits. Table 6.4 shows the truth table of the quantum full subtractor.

Fig. 6.22 Block diagram of quantum half subtractor

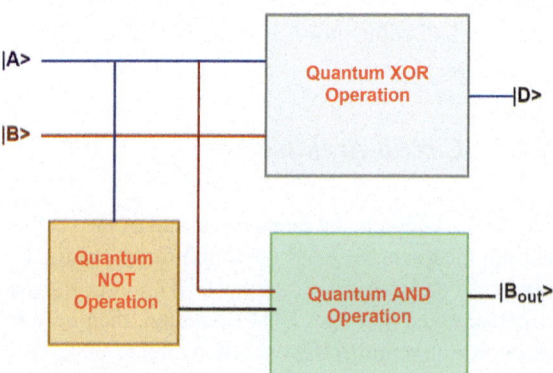

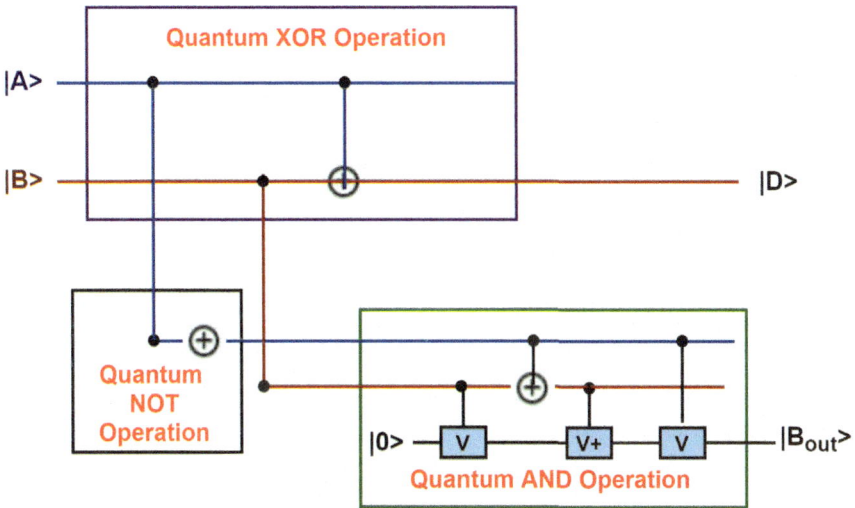

Fig. 6.23 Quantum half subtractor circuit

Table 6.4 A truth table of a quantum full subtractor

Inputs			Outputs	
|A >	**|B >**	**|Bin >**	**|D >**	**|Bout >**
|0 >	|0 >	|0 >	|0 >	|0 >
|0 >	|0 >	|1 >	|1 >	|1 >
|0 >	|1 >	|0 >	|1 >	|1 >
|0 >	|1 >	|1 >	|0 >	|1 >
|1 >	|0 >	|0 >	|1 >	|0 >
|1 >	|0 >	|1 >	|0 >	|0 >
|1 >	|1 >	|0 >	|0 >	|0 >
|1 >	|1 >	|1 >	|1 >	|1 >

From the above truth table, we get

$$
\begin{aligned}
\mathbf{D} &= A'B'Bin + A'BBin' + AB'Bin' + ABBin \\
&= Bin\,(A'B' + AB) + Bin'(AB' + A'B) \\
&= Bin\,(A\ XNOR\ B) + Bin'(A\ XOR\ B) \\
&= Bin\,(A\ XOR\ B)' + Bin'(A\ XOR\ B) \\
&= Bin\ XOR\ (A\ XOR\ B) \\
&= (A\ XOR\ B)\ XOR\ Bin \\
Bout &= A'B'Bin + A'BBin' + A'BBin + ABBin \\
&= Bin\,(AB + A'B') + A'B\,(Bin + Bin') \\
&= Bin\,(A\ XNOR\ B) + A'B \\
&= Bin\,(A\ XOR\ B)' + A'B
\end{aligned}
$$

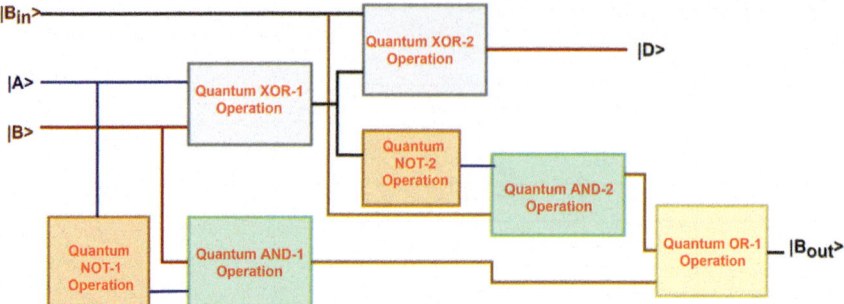

Fig. 6.24 Block diagram of quantum full subtractor

6.9.1   Block Diagram

The quantum circuit for a full subtractor is created by several quantum gates. To design a full subtractor, two XOR, two AND, two NOT, and an OR quantum gates are required. In this quantum circuit, two quantum XOR operations are required to produce the difference ($|D>$) of the full subtractor, and as for $|Bout>$, two quantum AND operations, two quantum NOT, and one quantum OR operation are needed along with the quantum XOR operation. The block diagram of the quantum full subtractor is illustrated in Fig. 6.24.

6.9.2   Circuit Architecture

Figure 6.25 shows the quantum circuit of the full subtractor. In this full subtractor, the first XOR operation for $|A>$ and $|B>$ inputs is done by a quantum XOR operation and creates output $|A> \oplus |B>$. This output, $|A> \oplus |B>$, is then used as one of the inputs of the second XOR operation, where another input qubit comes from $|Bin>$. Finally, the output $|D>$ qubit of the full subtractor is generated through this quantum XOR operation which presents the difference of the full subtractor.

In addition, two quantum AND operations are needed to get another output qubits of $|Bout>$. The first AND operation occurs between the negation of $|A>$ input qubit and $|B>$ input qubit and produces $|A'>.|B>$. As for the second AND operation, one input comes from the NOT operation of $|A> \oplus |B>$ and another input is directly provided from $|Bin>$. These two input qubits then generate $(|A> \oplus |B>').|Bin>$. Finally, the outputs of both quantum AND operations pass through a quantum OR operation to produce the value of $|Bout>$ qubit.

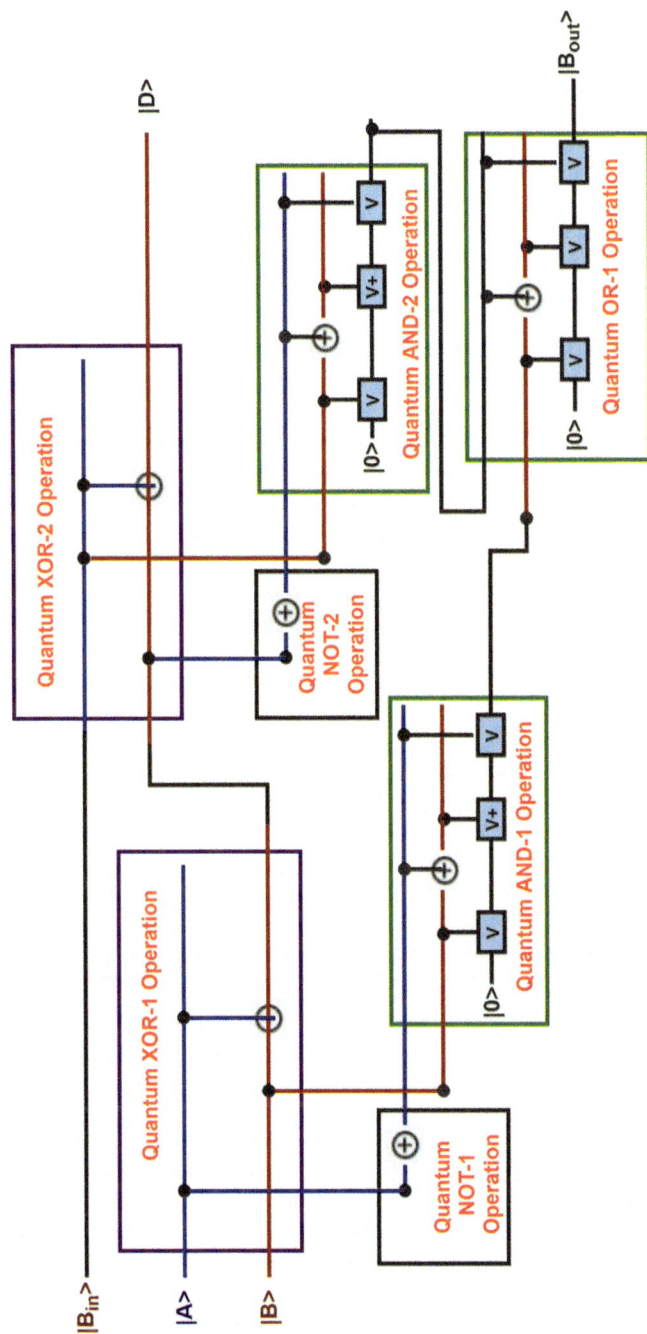

Fig. 6.25 Quantum full subtractor circuit

6.9.3 Working Principle

The quantum full subtractor needs three inputs, A, B, and Bin, and provides two outputs D and Bout. To understand the working principle of quantum full subtractor, consider the qubit values of inputs A, B, and Bin which are $|1>$, $|0>$, and $|1>$, respectively. Then,

(i) The output in quantum XOR-1 is "true" $|1>$, in quantum AND-1 is "false" $|0>$ and in quantum AND-2 is "false" $|0>$.
(ii) Finally, the value of output D is "false" $|0>$ after performing XOR-2 operation, where and Bout is "false" $|0>$.

According to the truth table as shown in Table 6.4, different input combinations will produce different output values. The working principles of quantum XOR operation, quantum NOT operation, quantum AND operation, and quantum OR operation are described in Chap. 2 and it is easy to perceive the working principles for the rest of the input and output combinations of the truth table by using the circuit architecture as shown in Fig. 6.25.

6.10 Quantum Multiplier

A quantum multiplier is a quantum circuit that multiplies two input values. The two numbers are called multiplicand and multiplier respectively, and the result is called a product.

Both the multiplicand and the multiplier might be of different qubit sizes. The qubit size of the product is determined by the qubit size of the multiplicand and multiplier. The product's qubit size is equal to the sum of the multiplier's and multiplicand's qubit sizes. In other words, a quantum multiplier is a circuit or algorithm used in quantum computing to perform multiplication operations. It's a fundamental building block for quantum algorithms that require arithmetic operations. Quantum multipliers can be designed using techniques like the quantum Fourier transform (QFT) or the exponent adder method, offering different trade-offs in terms of complexity and efficiency. Quantum multipliers are essential for performing arithmetic operations, which are necessary for many quantum algorithms. Quantum multipliers require qubits and quantum gates for their implementation. The number of qubits and gates depends on the specific design and the size of the numbers being multiplied. Quantum multipliers are used in various quantum algorithms, including those related to cryptography, optimization, and scientific simulations.

There are two processes involved in the quantum qubits multiplication of numbers with more than one qubit. The first stage is to do a single qubit-wise quantum multiplication, which is known as partial product, and the second step is to aggregate all partial products into a single product.

Using the quantum AND operations, it is possible to get partial products or single-qubit products. However, the quantum full-adders and the quantum half-adders are required to add these partial products.

6.10.1 Block Diagram

A 2×2 quantum multiplier will take two 2-qubit input values and will produce the product value of them.

Figure 6.26 shows the block diagram of the quantum 2×2 multiplier.

Suppose multiplication will be performed within two 2-qubit inputs $|A >$ and $|B >$.

Here, $|A > = |A_1 A_0>$, and $|B > = |B_1 B_0>$.

The quantum multiplication operation between them can be performed as follows: (To keep it simple here, qubit notations are not used.)

$|A >$: A_1 A_0 (multiplicand)
$|B >$: B_1 B_0 (multiplier)
Quantum AND $A_1.B_0$ $A_0.B_0$
    $A_1.B_1$ $A_0.A_1$ **X**
Quantum OR $|C_2>$ $A_1.B_1$ $A_1.B_0 A_0.B_0$
    $+ |C_1> + A_0.B_1$
$|P >$: $|P_3> |P_2> |P_1 > |P_0>$

The above calculation shows how a 2×2-qubit quantum multiplication is performed, where two 2-qubit input $|A > = |A_1 A_0>$, and $|B > = |B_1 B_0>$ using 2×2 quantum multiplier produces a product of $|P > = |P_3 P_2 P_1 P_0>$. At first, the quantum AND operation is performed as shown in the calculation. Then, the quantum AND results are added in the way the above calculation showed to get the expected result.

6.10.2 Circuit Architecture

At first, the quantum AND operation circuits are needed to perform quantum AND operations among them (as the calculation showed). Afterwards, it is needed to add

Fig. 6.26 The general working block of a quantum 2×2 multiplier

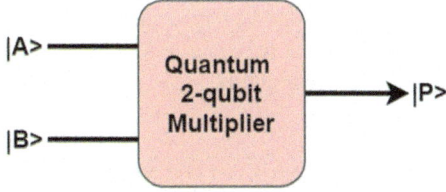

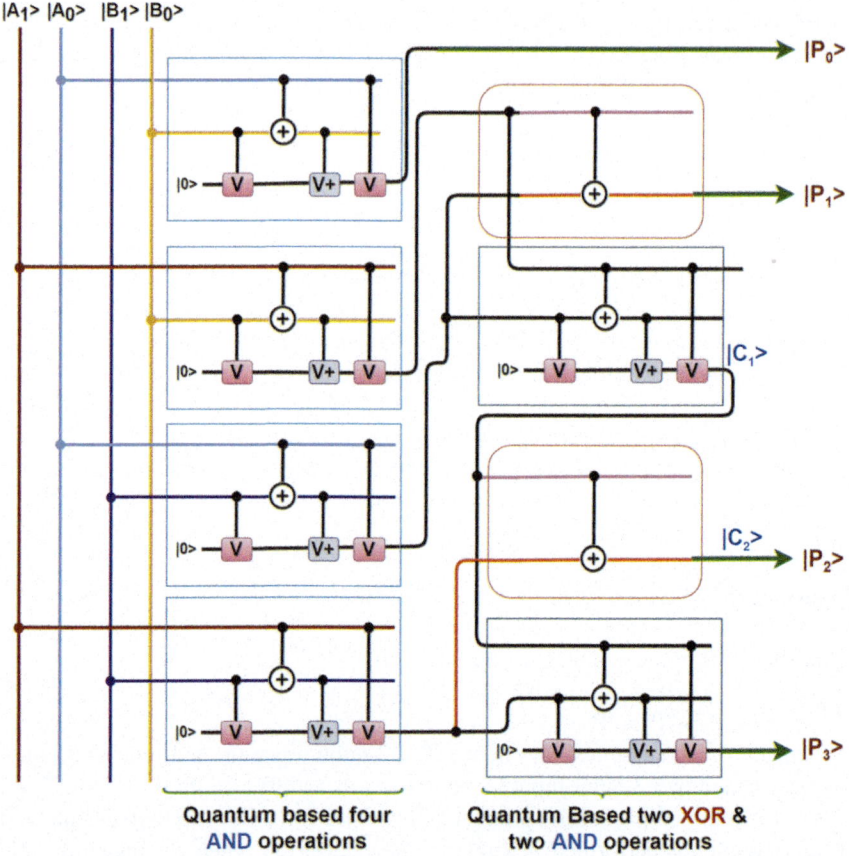

Fig. 6.27 The circuit diagram of a quantum 2 × 2 multiplier

two input values. Therefore, the quantum half-adders are needed to perform the addition operations.

And while performing the addition process, a carry qubit will be generated, and propagated the carry to the following add operations. However, the final carry qubit will be the most significant qubit of the output value.

The following figure shows the circuit diagram of the quantum 2 × 2 multiplier.

The circuit in Fig. 6.27 consists of four quantum AND operational circuits, two quantum OR, and two quantum XOR operational circuits. And they are combined in a way so that they can produce the expected output qubits, which is the product of the two input qubits.

6.10.3 Working Principle

Suppose, two input values, where $|A> = |11>$, and $|B> = |10>$. So, the values of $|A_0> = |1>$, and $|A_1> = |1>$. And the values of $|B_0> = |0>$, and $|B_1> = |1>$.

1. The 1st quantum AND operation will receive inputs $|A_0> = |1>$, and $|B_0> = |0>$. This will produce the output $|0>$, which is the value of $|P_0>$.
2. The 2nd quantum AND operation will receive inputs $|A_1> = |1>$, and $|B_0> = |0>$. This will produce the output $|0>$ and this qubit will work as an input qubit of the 1st quantum XOR operation and the 5th quantum AND operation.
3. The 3rd quantum AND operation will receive input qubits $|A_0> = |1>$, and $|B_1> = |1>$. This will produce the output $|1>$ and this qubit will be worked as an input qubit of the 1st quantum XOR operation and the 5th quantum AND operation as shown in the circuit.
4. The 4th quantum AND operation will receive input qubits $|A_1> = |1>$, and $|B_1> = |1>$. This will produce the output $|1>$ and this qubit will work as an input qubit of the 2nd quantum XOR operation and the 6th quantum AND operation as shown in the circuit.
5. Now, the 1st quantum XOR operation will receive input qubits $|0>$ (from Step 2) and $|1>$ (from Step 3). Thus, it will produce $|1>$ as output. Therefore, the value of $|P_1>$ is $|1>$ which will be finally obtained.
6. The 5th quantum AND operation will receive input qubits $|0>$ (from Step 2) and $|1>$ (from Step 3). Thus, it will produce $|0>$ as output, which is the carry qubit and it will work as an input for the next quantum XOR and quantum AND operations.
7. Now, the 2nd quantum XOR operation will receive input qubits $|1>$ (from Step 4) and $|0>$ (from Step 6). Thus, it will produce $|1>$ as output. Therefore, the value of $|P_2>$ is $|1>$ which is finally obtained.
8. One more operation is performed here now. The 6th quantum AND operation will receive input qubits $|1>$ (from Step 4) and $|0>$ (from Step 6). Thus, it will produce $|0>$ as output. And this output value will be the value of $|P_3>$. Therefore, the most significant qubit of the product result is $|0>$.

Consequently, the product of $|A> = |11>$, and $|B> = |10>$ is $|P> = |P_3 P_2 P_1 P_0> = |0110>$.

By taking some other 2-qubit values for $|A>$ and $|B>$, it is easy to describe the circuit operational behavior for the corresponding input.

6.11 Quantum Divider

In the context of quantum computing, a divider refers to a quantum circuit or algorithm that performs integer division. Division is a fundamental arithmetic operation, and its quantum implementation is crucial for various algorithms, especially those involving numerical computations. A quantum divider is a quantum circuit designed to perform division. It can be implemented using various quantum gates and algorithms, such as the long division algorithm, or using quantum Fourier transforms. Efficient quantum division is essential for various quantum algorithms, including those used in factoring numbers, simulating quantum systems, and solving other complex problems. Researchers are actively working on optimizing quantum dividers to reduce the number of qubits and gates required for the operation. This is crucial for enabling the implementation of complex algorithms on noisy intermediate-scale quantum devices, which have limited qubit resources. Quantum division is an important mathematical calculation but often overlooked in quantum arithmetic operations. Although the quantum mathematical division is not very difficult, but it can be challenging to grasp the other quantum arithmetic operations. This is due to the fact that all other quantum arithmetic operations are comparable, yet quantum division is a bit of an oddity. Besides, the circuit structure is so much complex than the others. And there are no specific rules that must be followed when performing quantum mathematical division. Furthermore, the circuit structure is significantly more complicated than the others.

6.11.1 Definition

A quantum divider will divide an n-qubit input value (called the dividend) by an n-qubit input value (called the divisor). After performing the operation the quantum divider will produce an n-qubit quotient and an n-qubit remainder. The following calculation will make it easy to understand:

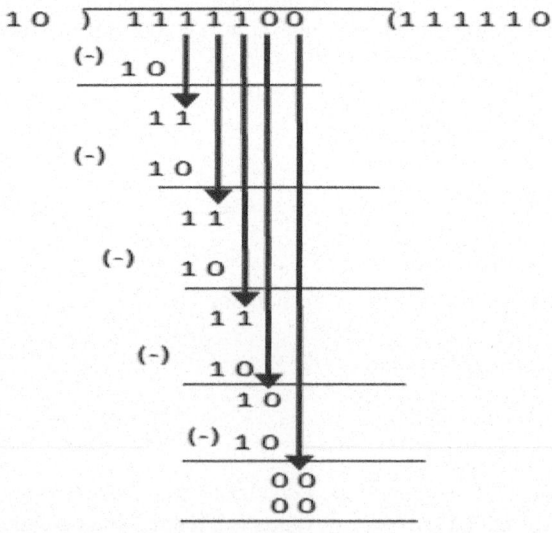

Fig. 6.28 The input-output mapping of the quantum divider

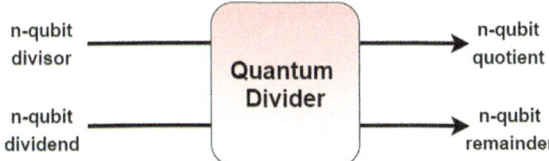

Here, the dividend is a 7-qubit value - |1111100 >. But why the other values are not 7-qubit? Well, they are! But it is not necessary to show them in that manner.

Here, the divisor is |0000010 >. As the result of the division operation, the quotient is |0111110 > and the remainder value is |0000000 > which are shown in the calculation. Therefore, it can be said that the qubit size of the divisor, quotient, and remainder will depend on the qubit size of the dividend. Figure 6.28 shows the input-output mapping of the quantum divider.

6.11.2   Block Diagram

This section will describe how to perform the operation of a quantum divisor. Firstly, it is needed to make the divisor value equivalent to the dividend value by adding the corresponding number of |0 > to the most significant qubit in divisor input. Then, it is easy to design the quantum divider for the 2-qubit dividend, the 4-qubit dividend, and the n-qubit dividend.

Figure 6.29 shows the general architecture of a 2-qubit quantum divider (here 2-qubit means the dividend value of the operation is in size of two qubits). Figure 6.30 shows the general architecture of a 4-qubit quantum divider (here, 4-qubit means the dividend value of the operation is in size of four qubits). And, Fig. 6.31 shows the general architecture of an n-qubit quantum divider (here n-qubit means the dividend value of the operation is in size of n-qubit).

From Fig. 6.29, it is clear that to construct a 2-qubit quantum divider, four quantum subtractors, three quantum 2-to-1 multiplexers, and two quantum NOT operations are required which will produce a 2-qubit quotient value and a 2-qubit remainder.

It can easily be mentioned that the construction of the quantum divider for more qubit value inputs is very difficult as because of the input qubit value in size, the circuit becomes more complicated. The number of resources increases at a large scale. Therefore, the design also becomes much more complicated.

Look how the architecture changes with just a slight change in input qubit (2-qubit to 4-qubit). Now, sixteen quantum full-subtractor, thirteen quantum 2-to-1 multiplexers, and four quantum NOT operations are needed to generate the 4-qubit quotient and the 4-qubit remainder. N numbers of quantum NOT operations, n^2 numbers of quantum full subtractors, and $[n(n-1)+1]$ numbers of quantum 2-to-1 multiplexers are needed to perform its activity. Figure 6.31 shows the architecture of the quantum n-qubit divider.

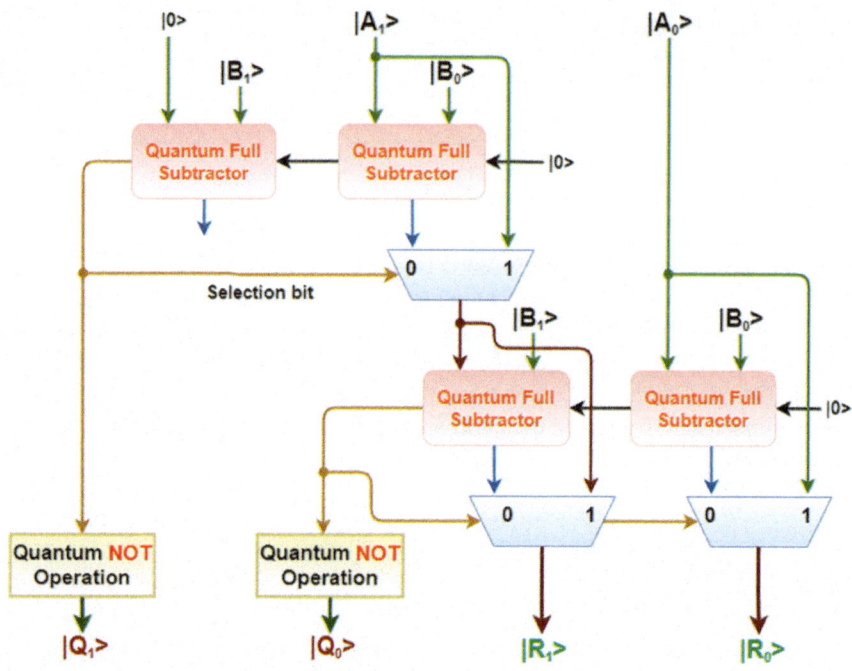

Fig. 6.29 The general architecture of the quantum 2-qubit divider

Note that, in all the general architectures, the divisor is the values of $|B>$, the dividend is the values of $|A>$ and the output values of $|Q>$ refer to the quotient values, and the output of $|R>$ refers to the values of the remainder.

6.11.3 Circuit Architecture

Well, the general architecture of the quantum divider for 2-qubit, 4-qubit, and n-qubit are designed. Now, it is shown how to construct the quantum operational circuit diagram? Figure 6.32 shows the operational circuit diagram of the quantum 2-qubit divider.

6.11.4 Working Principle

Suppose the 2-qubit dividend is $|A>$ and the 2-qubit divisor is $|B>$, where $|A> = |11>$, and $|B> = |10>$. Therefore, $|A_0> = |1>$, and $|A_1> = |1>$. And

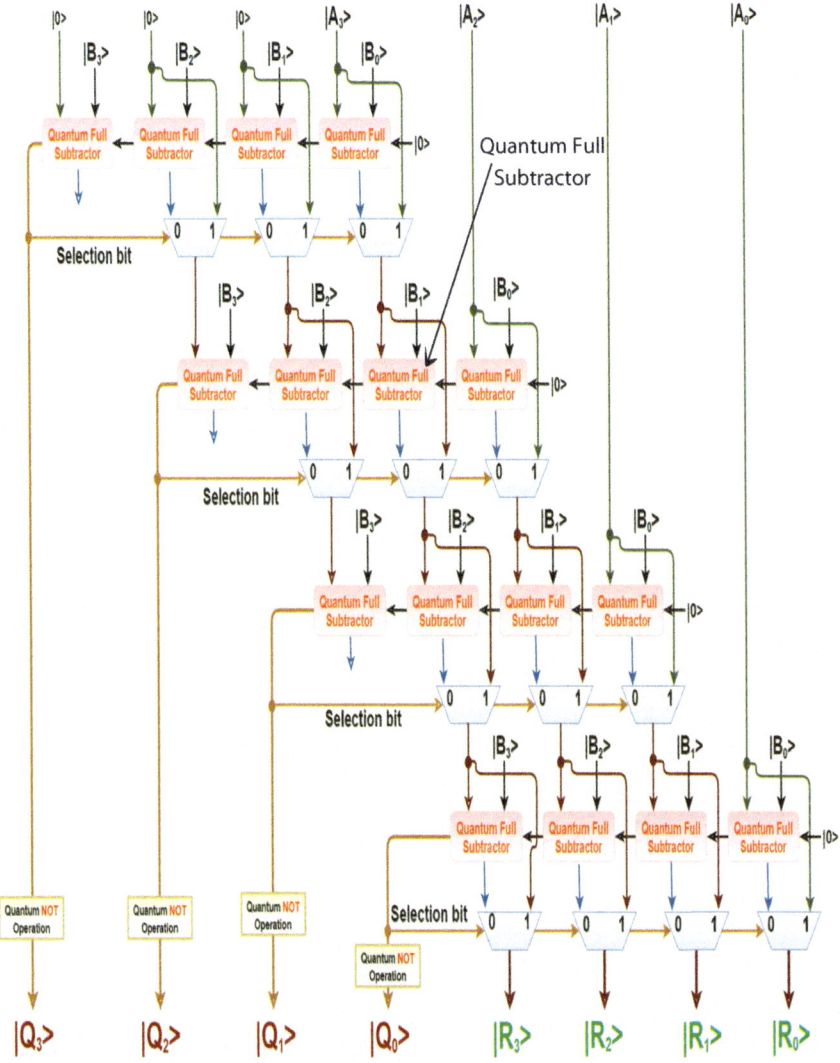

Fig. 6.30 The general architecture of the quantum 4-qubit divider

$|B_0> = |0 >$, and $|B_1> = |1 >$. These inputs will generate the quotient $|Q > = |Q_1 Q_0>$, and the remainder as $|R > = |R_1 R_0>$.

The behavior of the quantum 2-qubit divider circuit for these qubits is given below.

1. The operation of the quantum full-subtractor will start from the right side, and the operation of the quantum multiplexer will start from the left side of the figure as shown in Fig. 6.32. The 1st quantum full-subtractor will get inputs $|B_0> = |0 >$, and $|A_1> = |1 >$ and the other two inputs of the full-subtractor are two ancilla

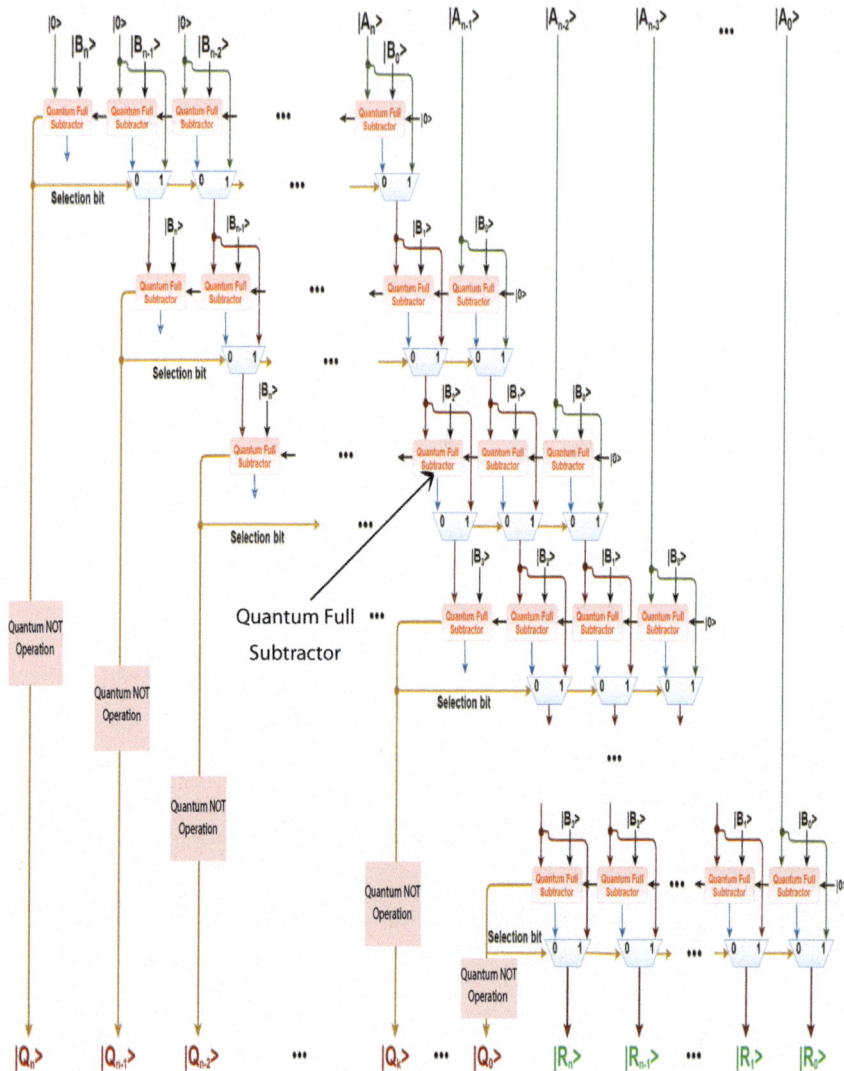

Fig. 6.31 The general architecture of the quantum N-qubit divider

qubits as $|0>$. Therefore, the difference output will be $|1>$, which will work as the input of the 1st quantum multiplexer. And the borrow output will be $|0>$, which will work as an input of the 2nd quantum full-subtractor.

1. The 2nd quantum full-subtractor will get inputs $|B_1> = |1>$, $|0>$ from Step 1, and the other two inputs of the full-subtractor are two ancilla qubits as $|0>$. Therefore the borrow output will be $|1>$, which will work as the selection input

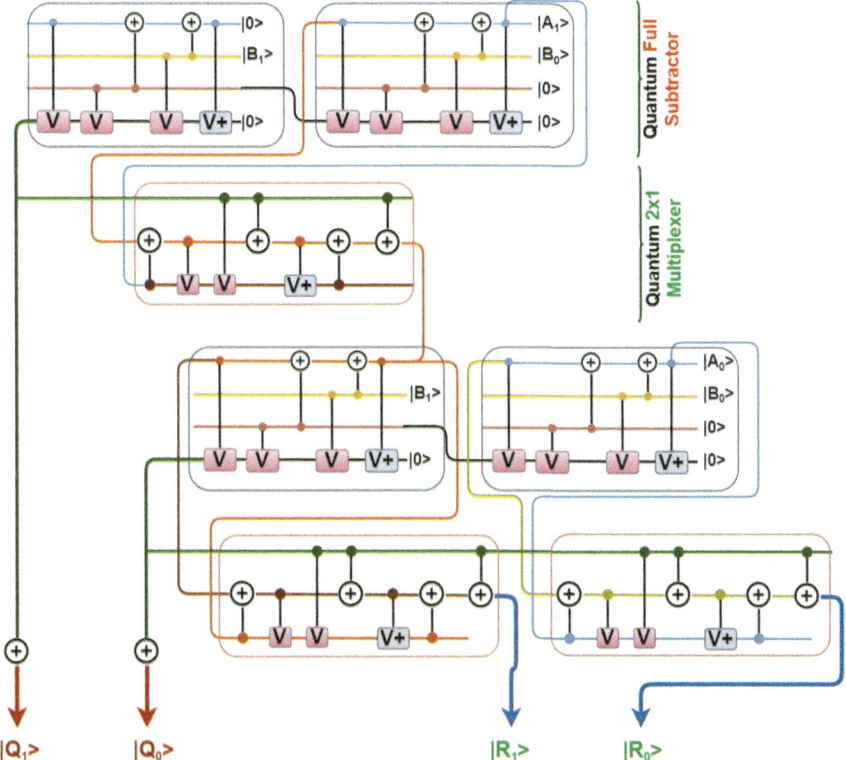

Fig. 6.32 The circuit diagram of the quantum 2-qubit divider

of the 1st quantum multiplexer and the quantum NOT of this borrow qubit $|1>$ will be the value of $|Q_1>$. So, $|Q_1> = |0>$.

2. Now, consider the 1st quantum multiplexer operation. It gets the selection inputs $|1>$ (from Step 2), and also $|1>$ (from Step 1), and the value of $|A1> = |1>$. Therefore, the output will be $|1>$. This output will work as an input of the 4th quantum full-subtractor and the 2nd quantum multiplexer.

3. The 3rd quantum full-subtractor will get input $|B_0> = |0>$, and $|A_0> = |1>$. The other two inputs of the full-subtractor are two ancilla qubits as $|0>$. Therefore, the difference output will be $|1>$, which will work as the input of the 3rd quantum multiplexer. The borrow output will be $|0>$ which will work as an input of the 4th quantum full-subtractor.

4. The 4th quantum full-subtractor will get inputs $|B_1> = |1>$, $|0>$ from Step 4, and $|1>$ from Step 3. Therefore, the borrow output will be $|0>$, which will work as the selection inputs of the 2nd and 3rd quantum multiplexers, and the quantum NOT of this borrow qubit $|0>$ will be the value of $|Q_0>$. So, $|Q_0> = |1>$. And the difference output will also be $|0>$, which will work as an input of the 2nd quantum multiplexer.

5. The quotient part is completed. The 2nd quantum multiplexer will get the selection input $|0>$ (from Step 5), another input $|0>$ from Step 5, and $|1>$ from Step 3. Thus, the output will be $|0>$. So, the value is $|R_1> = |0>$.
6. Finally, the 3rd quantum multiplexer will get the selection input $|0>$ (from Step 5), another input $|1>$ from Step 3, and $|1>$ from the value of $|A_0>$. Thus, the output will be $|1>$. Therefore, the value is $|R_0> = |1>$.

So, it is clear that, after performing the division operation between the dividend $|A> = |11>$ and the divisor $|B> = |10>$, we get the quotient $|Q> = |01>$, and the remainder $|R> = |01>$.

6.12 Quantum Comparator

A comparator that compares two input signals and gives output as which input is larger or smaller or equal. In other words, a quantum comparator is a quantum circuit designed to compare two quantum states, typically represented by qubits or bit strings. It determines if the two states are equal or, if not, which is larger. These comparators are crucial in quantum algorithms, allowing for conditional logic and are used in areas like quantum machine learning and image processing. Quantum comparators are fundamental building blocks for quantum algorithms that require conditional statements or comparisons between quantum states. They are used in various quantum algorithms, including: i) Quantum machine learning; ii) Quantum image processing; iii) Quantum search algorithms; iv) Probabilistic comparisons, etc. Quantum 1-qubit comparator takes two single-qubit inputs, $|A>$ and $|B>$. The values of the inputs can be $|0>$ and $|1>$. This circuit gives three outputs $|X>$, $|Y>$, and $|Z>$, where $|X>$ is true when $|A>$ is smaller than $|B>$, $|Y>$ is true, when $|A>$ is equal to $|B>$ and $|Z>$ is true, when $|A>$ is greater than $|B>$. Here, Table 6.5 shows the truth table of the 1-qubit comparator.

Table 6.5 Truth table of a 1-qubit comparator

Inputs		Outputs		
$|A>$	$|B>$	$|X>$	$|Y>$	$|X>$
		A < B	A= B	A > B
$|0>$	$|0>$	$|0>$	$|2>$	$|0>$
$|0>$	$|1>$	$|2>$	$|0>$	$|0>$
$|1>$	$|0>$	$|0>$	$|0>$	$|2>$
$|1>$	$|1>$	$|0>$	$|2>$	$|0>$

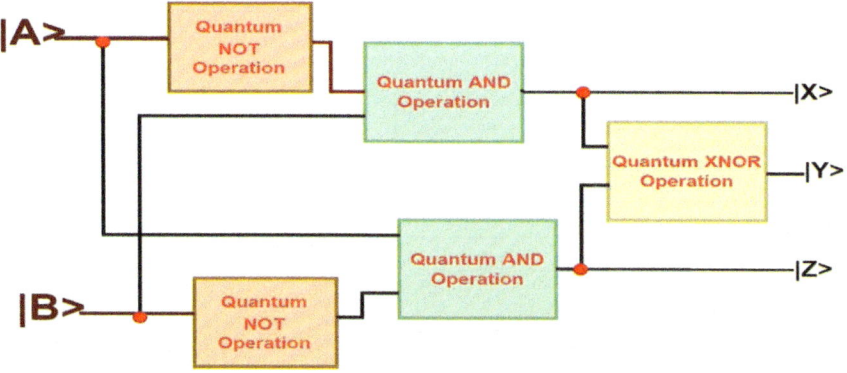

Fig. 6.33 Block diagram of quantum 1-qubit comparator

From the truth table, the equations for $|X>$, $|Y>$, and $|Z>$ are as follows:

$|X> = A^0.B^1$
$|Y> = A^0.B^0 + A^1.B^1 = A\ XNOR\ B$
$|Z> = A^1.B^0$

6.12.1 Block Diagram

Now, using the equations, the quantum 1-qubit comparator is designed. Figure 6.33 displays the block diagram of the quantum 1-qubit comparator

6.12.2 Circuit Architecture

Quantum 1-qubit comparator (Fig. 6.34) needs quantum AND operations, two quantum NOT operations, and one quantum OR operation. Quantum AND-1 is connected to $|A'>$ and $|B>$ and it performs $|A'>. |B>$ which is the result of $|Z>$. In the same way, the quantum AND-2 provides the output of $|X>$. To get the value of $|Y>$, the both outputs of AND operation act as the input of the quantum OR operation

6.12.3 Working Principle

The quantum comparator needs two inputs, $|A>$ and $|B>$. For various values of inputs A and B, consider the following cases:

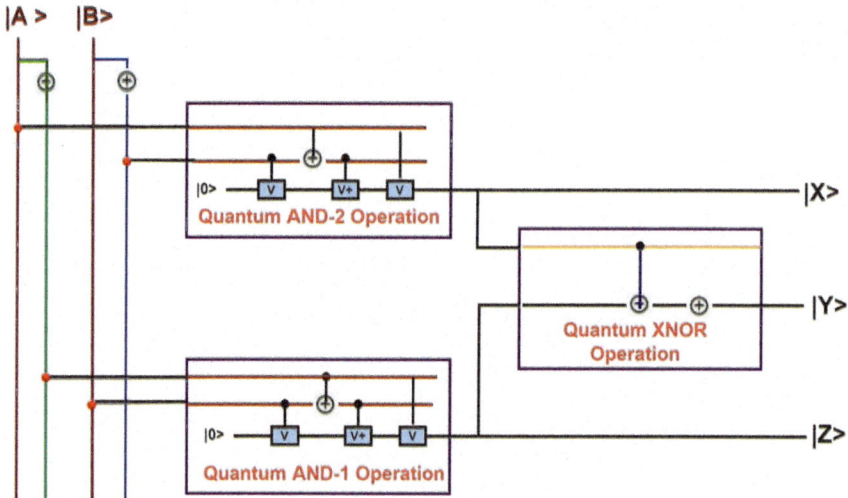

Fig. 6.34 Quantum 1-qubit comparator circuit

  (i) When both |A > and |B > are "false" |0 >, the outputs of both |Y > would be
      "true" |1 > and |X > and |Z > will be "false" |0 > as both AND operations will
      produce |0 >.

 (ii) When both |A > is "false" |0 > and |B > is "true" |1 >, the outputs of both |X
      > would be "true" |1 > and |Y > and |Z > will be "false" |0 >.

(iii) When |A > is "false" |0 > and |B > is "true" |1 >, the outputs of both |Z >
      would be "true" |1 > and |X > and |Y > will be "false" |0 >.

 (iv) When both |A > and |B > are "true" |1 >, the outputs of both |Y > would be
      "true" |1 > and |X > and |Z > will be "false" |0 > as both AND operations will
      produce |0 >.

6.13 Summary

Quantum carry ripple adder or quantum N-qubit adder is the chain of quantum
full-adders, where each quantum full adder's output carry qubit is connected to the
carry input of the next higher-order quantum full-adder in the chain. An n-qubit
adder requires n quantum full-adders to operate. To perform a 3-qubit addition, the
quantum carry-lookahead adder takes only 70 microseconds. And the quantum carry-
lookahead adder is the fastest adder. The quantum carry-skip adder skips the carry
qubit, when it is found that there is no carry generating, which is equivalent to
|1 >. It performs well in the best cases. Many information about quantum arithmetic
operations are discussed in this chapter. So, all important quantum logic operations

are discussed in detail, and their block diagrams with the circuit architectures are presented in the chapter.

Bibliography

1. G. Florio, D. Picca, *Quantum Implementation of Elementary Arithmetic Operations* (2004). arXiv preprint quant-ph/0403048
2. T. Ghosh, S. Sarkar, B.K. Behera, P.K. Panigrahi, *Masking of Quantum Information into Restricted Set of States* (2019). arXiv preprint arXiv:1910.00938
3. L.B. Levitin, T. Toffoli, Z. Walton, *Operation Time of Quantum Gates* (2002). arXiv preprint quant-ph/0210076
4. S.T. Marella, H.S.K. Parisa, in *Quantum Computing and Communications*. Introduction to Quantum Computing (IntechOpen, 2020)
5. Engin Şahin, Quantum arithmetic operations based on quantum Fourier transform on signed integers. Int. J. Quant. Inf. **18**(06), 2050035 (2020)
6. Hafiz Md. Hasan Babu, *Quantum Computing: A Pathway to Quantum Logic Design*, IOP (Institute of Physics) Publishing (2020), Bristol, UK
7. Hafiz Md. Hasan Babu, *Reversible and DNA Computing*, Wiley Publishers (2021), UK
8. Hafiz Md. Hasan Babu, *VLSI Circuits and Embedded Systems*, CRC Press (2022), USA
9. Md. Jahangir Alam, Guoqing Hu, Hafiz Md. Hasan Babu and Huazhong Xu, Control Engineering Theory and Applications, CRC Press (2022), USA
10. Hafiz Md. Hasan Babu, *Multiple-Valued Computing in Quantum Molecular Biology*, Volume I, CRC Press (2023), USA
11. Hafiz Md. Hasan Babu, *Multiple-Valued Computing in Quantum Molecular Biology*, Volume II, CRC Press (2023), USA
12. Hafiz Md. Hasan Babu, *DNA Logic Design: Computing with DNA*, World Scientific Publishing Company (2024), Singapore

Chapter 7
Arithmetic Operations in Quantum-DNA Computing

7.1 Introduction

This is the new technology to be introduced in this chapter which is called quantum biocomputing or quantum-DNA computing or DNA-quantum computing. The combination of quantum computing and DNA computing is called quantum-DNA computing or DNA-quantum computing systems. This cutting-edge technology based idea is introduced here for the first time in modern history. Arithmetic operations are the fundamental mathematical calculations used in everyday life and in more complex mathematical concepts. These operations include addition, subtraction, multiplication, and division, and are the foundation upon which many other mathematical concepts are built. Arithmetic operations form the basis of many branches of mathematics, such as algebra, calculus, and statistics. They play a similar role in the sciences, like physics and economics. Arithmetic is present in many aspects of daily life, for example, to calculate change while shopping or to manage personal finances. An arithmetic circuit in quantum-DNA computing will contain two parts; one is quantum part and another is DNA part. The inputs are qubit and the outputs are DNA molecular sequences. Some necessary issues like data conversion, data management, heat transfer, or management are also needed in these quantum-DNA circuits, which will be discussed in this chapter.

So the topics to be discussed are given below:

1. Quantum-DNA Half Adder
2. Quantum-DNA Full Adder
3. Quantum-DNA qubit Adder
4. Quantum-DNA BCD Adder
5. Quantum-DNA Carry look ahead Adder
6. Quantum-DNA Carry-Skip Adder
7. Quantum-DNA Half Subtractor
8. Quantum-DNA Full Subtractor
9. Quantum-DNA Multiplier

H. M. Hasan Babu, *Quantum Biocomputing in Quantum Biology Volume I*,
https://doi.org/10.1007/978-981-97-7154-7_7

Table 7.1 Truth table of a quantum-DNA half adder

Inputs		Outputs	
$|A\rangle$	$|B\rangle$	S	Cout
$|0\rangle$	$|0\rangle$	TGGATC	TGGATC
$|0\rangle$	$|1\rangle$	ACCTAG	TGGATC
$|1\rangle$	$|0\rangle$	ACCTAG	TGGATC
$|1\rangle$	$|1\rangle$	TGGATC	ACCTAG

10. Quantum-DNA Divider
11. Quantum-DNA Comparator

7.2 Quantum-DNA Half Adder

A quantum-DNA half adder is a circuit that adds two qubits to produce a sum and a carry-in DNA molecular sequences. Quantum biocomputing utilizes quantum information processing techniques to simulate and model biological systems. Half-adders can be employed in this context for tasks like simulating molecular interactions or analyzing complex biological data. In essence, the quantum half-adder is a quantum version of the classical half-adder, implemented using quantum gates and qubits to perform the basic addition operation in the quantum domain, which is relevant in quantum biocomputing for various applications. This type of adder can only add two qubits at a time. The truth table of the quantum-DNA half adder is given in Table 7.1

From the above truth table, the following equations of Sum (S) and Carry (Cout) are obtained:

$$S = A' B + A B'$$
$$= A \text{ XOR } B$$
$$Cout = A B$$

7.2.1 Block Diagram

The quantum-DNA circuit for a half adder is created by a number of quantum operations and DNA operations. To design a half adder, one XOR gate, and one AND gate are required. In this quantum-DNA circuit, one quantum XOR operation is needed to produce the SUM of the half adder. Additionally, to generate the carry qubit, one DNA AND operation is used. The block diagram of the quantum-DNA half-adder is illustrated in Fig. 7.1.

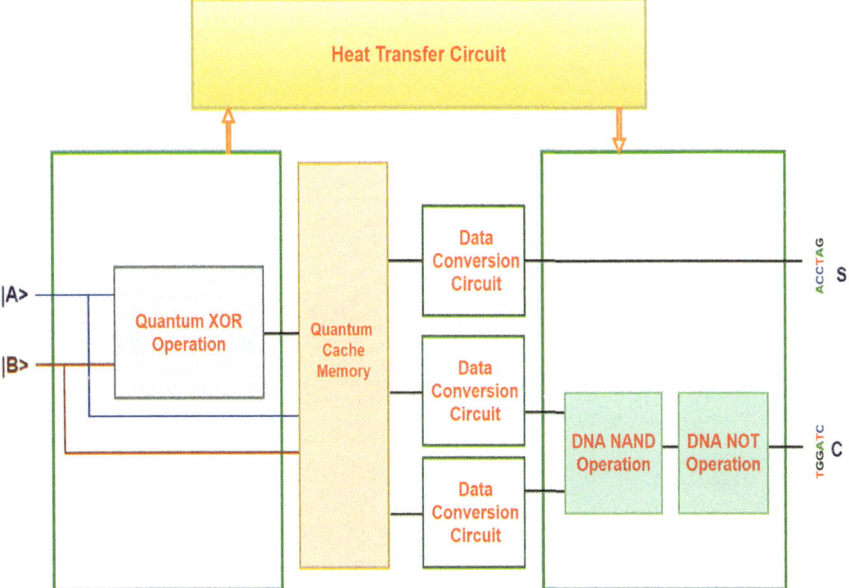

Fig. 7.1 Block diagram of quantum-DNA half adder

7.2.2 Circuit Architecture

Figure 7.2 depicts the DNA circuit of the half adder. In this half adder circuit, inputs are qubits |A> and |B>. The XOR operation for |A> and |B> inputs are done by a quantum operation and creates an output of S, which is a DNA sequences and for the output of C, both inputs A and B pass through a DNA AND operation. Between these quantum and DNA parts, there must be a quantum cache memory to store the qubits temporarily and an NMR relaxation (data conversion) to convert the qubits into DNA sequences, which are needed for quantum-DNA platform.

7.2.3 Working Principle

The quantum-DNA half adder needs two inputs, A and B. For various values of inputs A and B, consider the following cases:

(i) When both |A> and |B> are "true" |1>, the output qubit of S is "false" **TGGATC** and C is "true" **ACCTAG.**

(ii) When both |A> and |B> are "false" |0>, the output qubits of both S and C are "false" **TGGATC.**

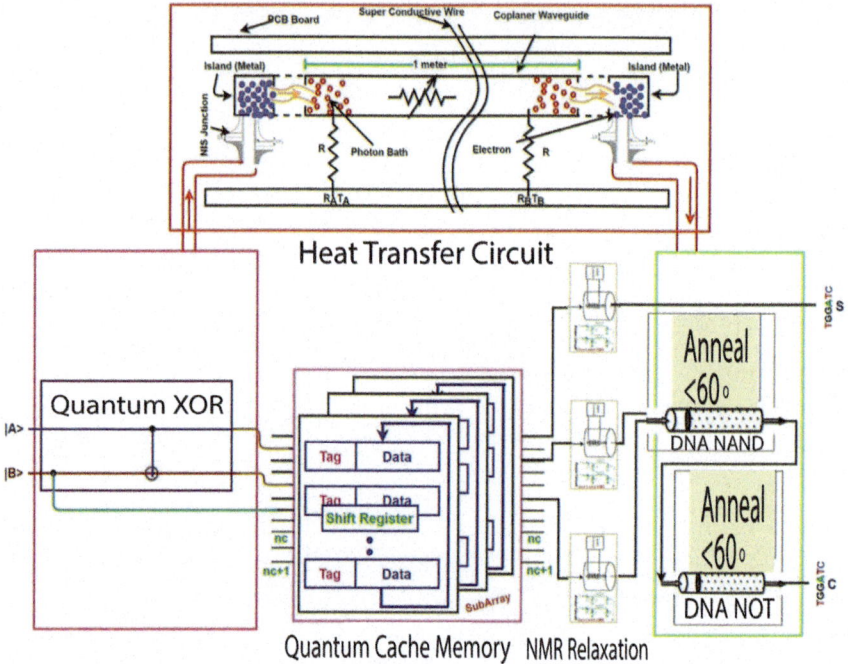

Fig. 7.2 Quantum-DNA half adder circuit

(iii) When |A> is "false" |0>, and |B> is "true" |1>, the output qubit of S is "true" **ACCTAG** and C is "false" **TGGATC.**

(iv) When |A> is "true" |1> and |B> is "false" |0>, the output qubit of S is "false" **TGGATC** and C is "true" **ACCTAG.**

7.3 Quantum-DNA Full Adder

A quantum-DNA full adder is a combinational circuit designed to overcome the drawback of the quantum-DNA half adder circuit. Quantum biocomputing is a field that explores using quantum computing principles to model and solve biological problems. Quantum full adders, like other quantum algorithms, can be used in quantum biocomputing for tasks like simulating biochemical reactions, analyzing genomic data, or designing new drugs. Here, this circuit accepts three inputs and produces two outputs as shown in the truth Table 7.2. In the truth table, it is seen that the inputs are qubits and the outputs are DNA sequences.

Table 7.2 Truth table for quantum-DNA full adder

Inputs			Outputs	
IA>	IB>	ICin>	S	Cout
I0>	I0>	I0>	TGGATC	TGGATC
I0>	I0>	I1>	ACCTAG	TGGATC
I0>	I1>	I0>	ACCTAG	TGGATC
I0>	I1>	I1>	TGGATC	ACCTAG
I1>	I0>	I0>	ACCTAG	TGGATC
I1>	I0>	I1>	TGGATC	ACCTAG
I1>	I1>	I0>	TGGATC	ACCTAG
I1>	I1>	I1>	ACCTAG	ACCTAG

From the above truth table, we get

$S = A' B' Cin + A' B Cin' + A B' Cin' + A B Cin$
$= Cin (A' B' + A B) + Cin' (A' B + A B')$
$= Cin \; XOR \; (A \; XOR \; B)$
$Cout = A B + A Cin + B Cin (A + A')$
$= A B Cin + A B + A Cin + A' B Cin$
$= A B (1 + Cin) + A Cin + A' B Cin$
$= A B + A Cin + A' B Cin$
$= A B + A Cin (B + B') + A' B Cin$
$= A B Cin + A B + A B' Cin + A' B Cin$
$= A B (Cin + 1) + A B' Cin + A' B Cin$
$= A B + A B' Cin + A' B Cin$
$= AB + Cin (A' B + A B')$

7.3.1   Block Diagram

The quantum-DNA circuit for a full adder is created by a number of quantum operations and DNA operations; and in between these two types of operations, a quantum cache memory and data conversion technique are used. To create a full adder, two XOR, two AND, and an OR operations are mainly required. In this quantum-DNA circuit, one quantum XOR operation, two quantum AND operations, and one DNA XOR and one DNA OR operations are required to produce SUM and the carry molecular sequences. The block diagram of the quantum-DNA full adder is illustrated in Fig. 7.3.

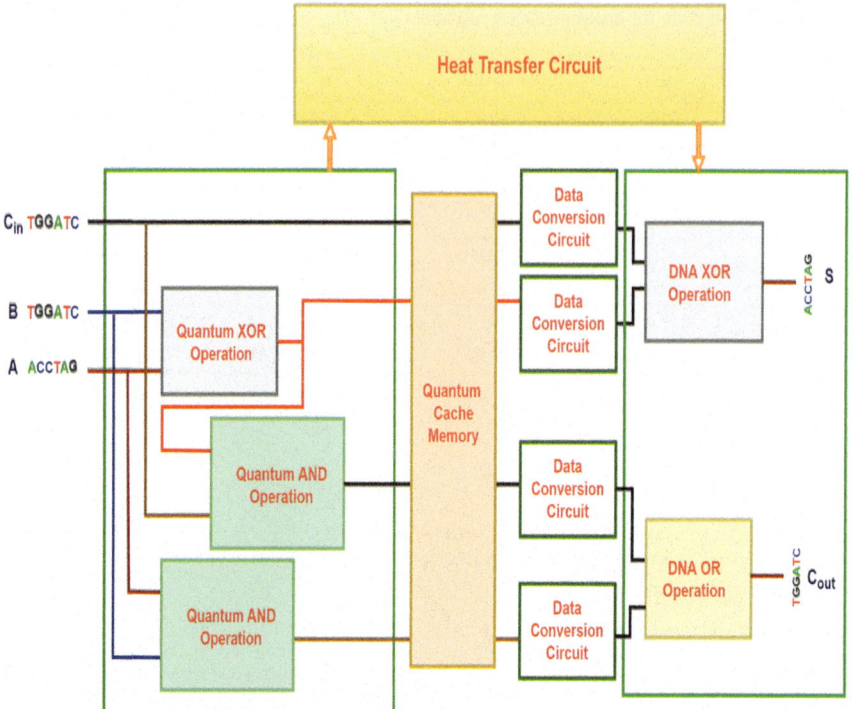

Fig. 7.3 Block diagram of quantum-DNA full adder

7.3.2 Circuit Architecture

Figure 7.4 shows the quantum-DNA circuit of the full adder. In this full adder, the first XOR operation for A and B input sequences is done by a DNA XOR operation and creates output A ⊕ B. This output, A ⊕ B, is used as one of the inputs of the second XOR operation and another input sequence comes from Cin. Finally, the output S of the full adder is generated through this DNA XOR operation, which represents the sum of the full adder.

In addition, two DNA NAND operations and two DNA NOT operations are needed to get another output sequence of Cout. At first NAND operation occurs between the inputs A and B, and then the output passes through a NOT operation to produce |A>.|B>. For the second NAND operation, one input comes from A ⊕ B and another input is directly come from Cin. The output of these two input sequences goes to a NOT operation to generate (|A> ⊕ |B>).|Cin>. Finally, the outputs of both DNA NOT operations pass through a DNA OR operation to produce the value of |Cout>.

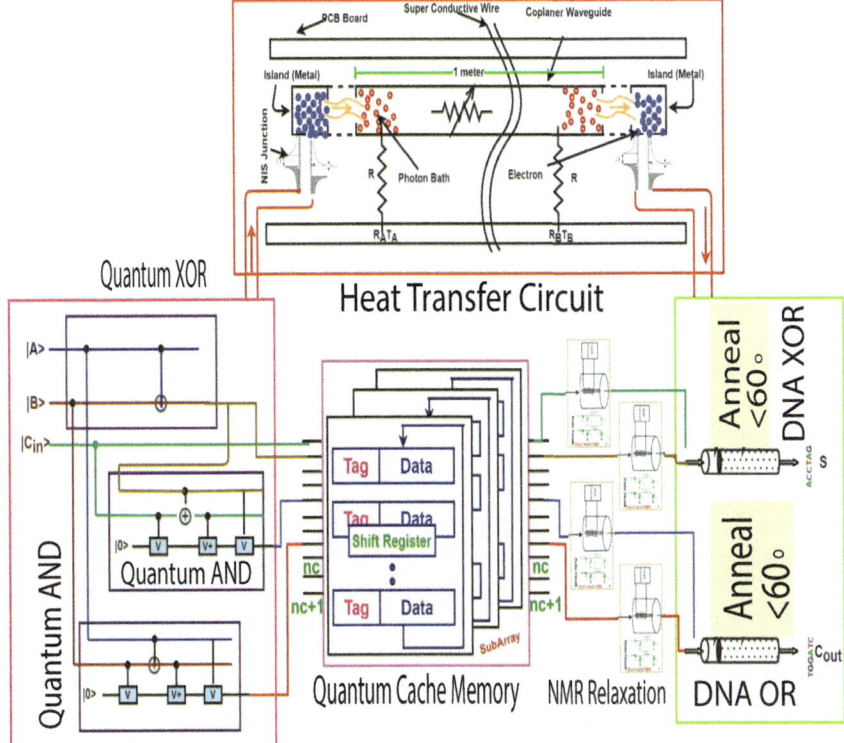

Fig. 7.4 Quantum-DNA full adder circuit

7.3.3 Working Principle

The quantum-DNA full adder needs three inputs, A, B, and Cin and provides two outputs S and Cout. In order to understand the working principle of quantum-DNA full adder, consider the values of input qubits $|A>$, $|B>$ and $|Cin>$ are $|1>$, $|0>$, and $|1>$, respectively. Then

(i) The output of quantum XOR-1 is "true" $|1>$, quantum AND-1 is "false" $|0>$, and quantum AND-2 is "false" $|0>$.
(ii) Finally, the value of output S is "true" **ACCTAG**, and Cout is "false" **TGGATC.**

7.4   Quantum-DNA Qubit Adder

The Quantum-DNA 2-qubit adder circuit requires three XOR operations, three AND operations, and one OR operation to add the inputs of 2-qubit $|A>$ and $|B>$. All of these operations are executed using quantum and DNA operations.

In the context of quantum biocomputing, a qubit adder refers to a quantum circuit designed to perform binary addition using qubits. These adders are essential for executing complex algorithms on quantum computers, including those used in biological applications. In essence, qubit adders are the building blocks for quantum arithmetic operations, enabling quantum computers to perform calculations relevant to various fields, including quantum biocomputing. In quantum biocomputing, qubit adders can be used in various applications, such as:

i) Drug Discovery: Quantum adders can be used in simulations of molecular interactions, which are essential for drug discovery.
ii) Protein Folding: Simulating protein folding, a complex biological process, also requires quantum arithmetic operations.
iii) Biological Algorithms: Many biological algorithms rely on mathematical operations, and quantum adders can enable efficient implementation of these algorithms.

7.4.1   Block Diagram

Here, one quantum XOR, three quantum AND, two DNA XOR, and one DNA OR are needed. Figure 7.5 illustrates the block diagram of the quantum-DNA 2-qubit adder.

7.4.2   Design and Working Procedures of 2-Qubit Quantum-DNA Qubit Adder

To construct the **Quantum-DNA 2-qubit adder circuit**, a total of five components are needed (i.e., quantum system, quantum cache memory, data conversion circuits, DNA system, and the heat transfer circuit). The quantum system includes four quantum operations (three quantum AND operations and one quantum XOR operation), and the DNA system includes three DNA operations such as two DNA XOR operations and one DNA OR operation. The output qubits from the quantum system are stored into the quantum cache memory temporarily. From the cache memory, the qubits are passing through the data conversion units (here NMR relaxation process) to convert the quantum qubits into the equivalent DNA information.

Then, the DNA information works as the input to the DNA system. The DNA operations produce the final output. The whole architecture is shown in Fig. 7.6.

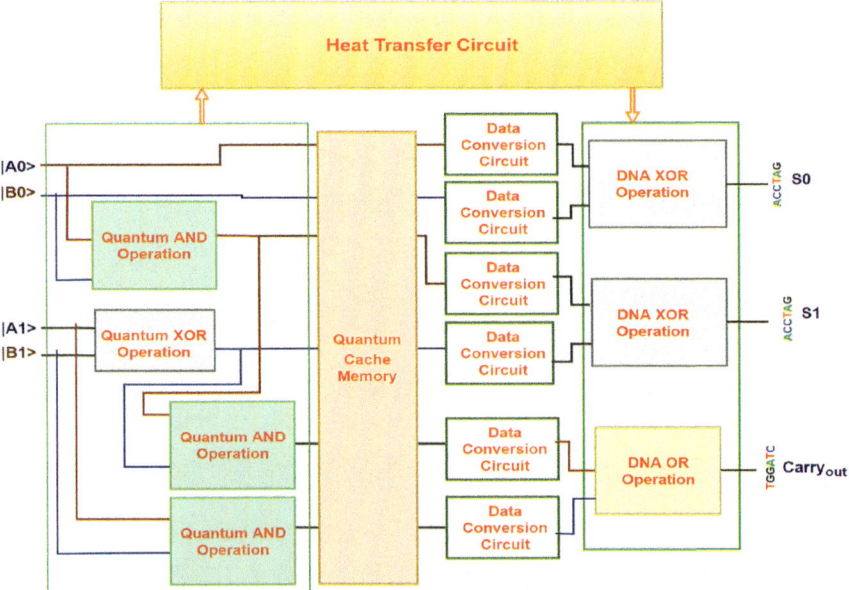

Fig. 7.5 Block diagram of 2-qubit quantum-DNA adder

The input qubits |A0> and |B0> are passed through the NMR relaxation and are converted into their equivalent DNA information. In the DNA system, those two input sequences produce the output **S0** by performing DNA XOR operation (DNA XOR-1 in the figure). Again |A0> and |B0> perform quantum AND-1 operation that produces the carry output of |A0> and |B0>. This carry output works as carry input to the next operation and go to the DNA system as well to perform the DNA XOR-2 operation that produces the output **S1** (the sum of |A1> and |B1>).

By performing quantum AND-2 and AND-3 operations, the outputs are stored in the quantum cache memory and eventually passed to the data conversion unit. They produce the final carry output by performing the DNA OR-1 operation of the addition operation, which is also called **S2.**

7.4.3 Design and Working Procedures of 4-Qubit Quantum-DNA Qubit Adder

Here, the quantum system includes sixteen quantum operations (eight quantum AND operations, four quantum OR operations, and four quantum XOR operations), and the DNA system includes four DNA XOR operations. The output qubits from the quantum system are stored into the quantum cache memory temporarily. From the cache memory, the qubits are passing through the data conversion units (here NMR

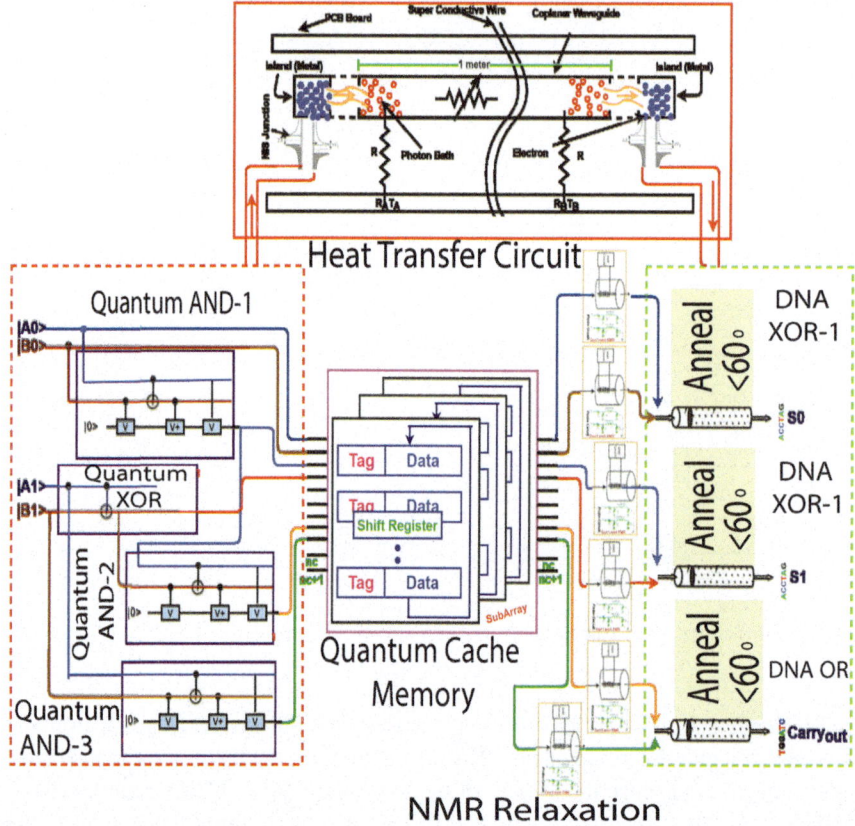

Fig. 7.6 2-Qubit quantum-DNA adder circuit

relaxation process) to convert the quantum bits into the equivalent DNA information. Besides, the produced massive heat from the quantum system is transferred through the heat conductance circuit. The required heat in the DNA system can also be taken from the quantum system. The circuit design architecture of the quantum-DNA 4-qubit adder is shown in Fig. 7.7.

The working procedure of the quantum-DNA 4-qubit adder is not very difficult to comprehend. By looking more attentively, it can be understood that the circuit design architecture of the quantum-DNA 4-qubit adder consists of four quantum full-adder operations, but the last XOR operation of each quantum full-adder is performed in the DNA system.

Therefore, any two 4-qubit numbers can be added through this quantum-DNA 4-qubit adder easily. The quantum system will perform all the operations of the 4-qubit parallel adder except for the last XOR operation of each full adder. The four DNA XOR operations will produce the four sum outputs **S0**, **S1**, **S2**, and **S3** accordingly.

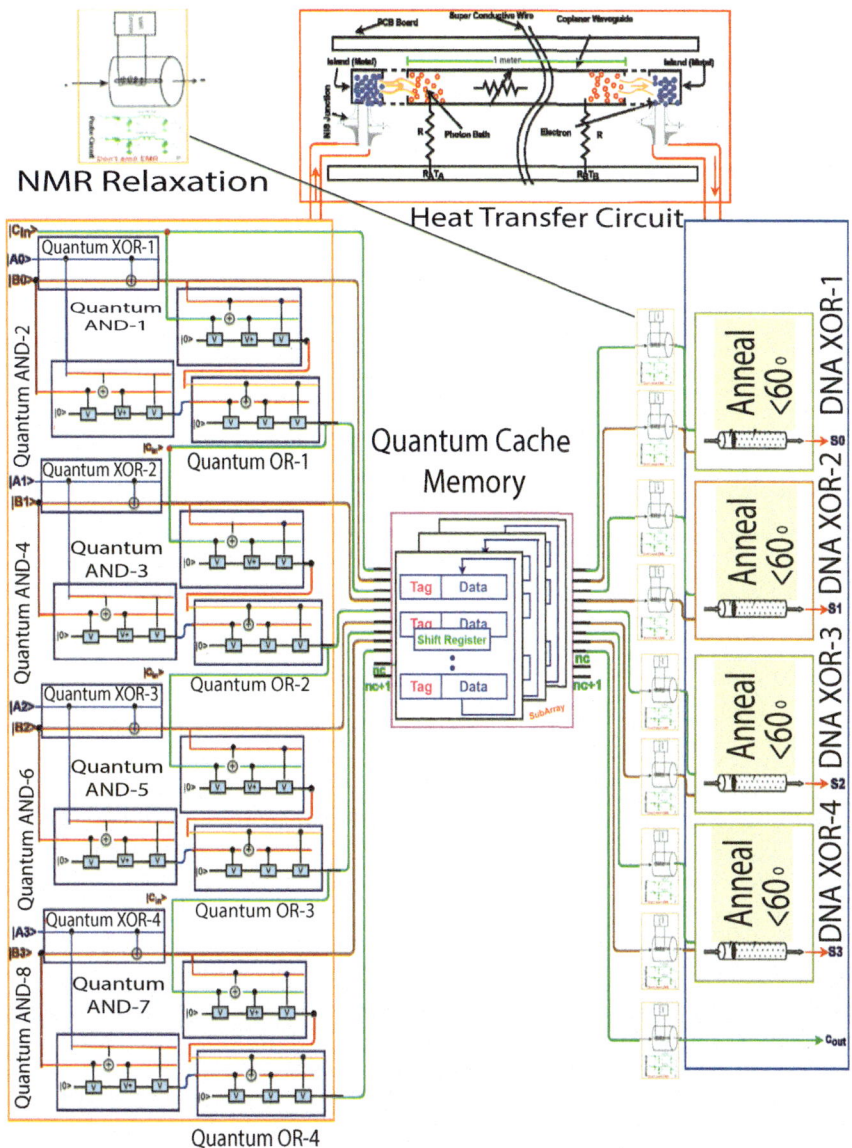

Fig. 7.7 4-Qubit quantum-DNA adder circuit

Point to be noted that, the most significant bit of the sum output of the addition operation will be achieved by the last quantum OR operation (quantum OR-4 in the figure), and the output of the quantum OR-4 will be converted into the equivalent

DNA sequence by the NMR relaxation process, and this DNA sequence will be the final carry output of the addition operation that can also be represented as **S4**.

7.5   Quantum-DNA BCD Adder

For the construction of the quantum-DNA BCD adder, one 4-qubit quantum adder and one 4-molecular DNA adder are required along with two quantum AND operations and two quantum OR operations. A Binary-Coded Decimal (BCD) adder in quantum biological computing refers to the use of BCD addition, a method for performing arithmetic operations on numbers represented in BCD format, within the context of quantum algorithms or simulations applied to biological systems. Quantum algorithms, potentially using BCD addition, could be developed to analyze and process large datasets in areas like genomics, proteomics, and drug discovery. In summary, BCD adders, while traditionally used in digital logic, have found a niche in the realm of quantum computing, particularly for applications in quantum simulation, algorithm design, and reversible circuit implementations in the context of biological systems.

7.5.1   Block Diagram

The block diagram of the quantum-DNA BCD adder is given in Fig. 7.8. Here all inputs are qubits and the outputs are DNA sequences. There is a heat transfer circuit to transfer accessive heat produced by the quantum circuit to the DNA circuit where the extra heat is required.

The general circuit of the DNA 4-molecular adder is shown in Fig. 7.9.

From the Fig. 7.9, to perform a 4-molecular DNA addition, four DNA full-adders are needed, where the carry output of each full-adder will work as the carry input to the very next DNA full-adder. And each full adder will generate the sum output of the addition of given inputs.

Figure 7.10 shows the DNA circuit of the full adder. In this full adder, the first XOR operation for A and B input sequences is done by a DNA XOR operation and generates the output A $\oplus$ B. This output, A $\oplus$ B, is used as one of the inputs of the second XOR operation and another input sequence comes from Cin. Finally, the output S of the full adder is generated through this DNA XOR operation, which represents the sum of the full adder.

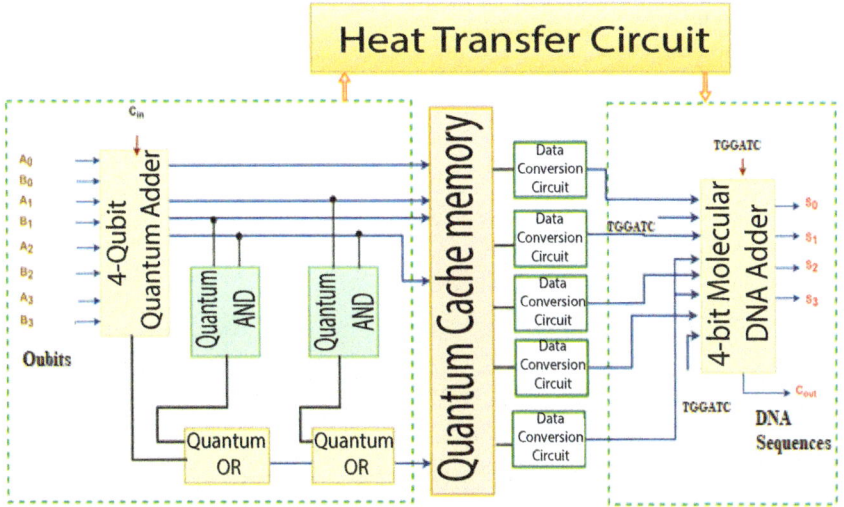

Fig. 7.8 Block diagram of a quantum-DNA BCD adder

7.5.2 Circuit Architecture

Figure 7.11 shows the quantum-DNA circuit of the BCD adder. Here, the four outputs of the first 4-qubit quantum adder directly transfer to the next 4-Molecular DNA adder, as the inputs of |A>. However, in between, two quantum AND operations and two quantum OR operations are performed using them. The values of |S1> and |S3> perform the quantum AND-1 operation, whereas quantum AND-2 operation is operated on |S2> and |S3>. In the next step, both outputs of the quantum AND operations along with the |Cout> of the first 4-qubit quantum adder, sequentially execute two quantum OR operations. Then, the result of the quantum OR-2 operation acts as the inputs for the |B> of the last 4-qubit quantum adder.

7.5.3 Working Principle

The quantum-DNA BCD adder needs two 4-qubit adders where there are two 4-qubits inputs, |A> and |B> and a carry input, |Cin> and in order to understand the working principle of quantum-DNA BCD adder, consider the values of input qubits |A0>, |A1>, |A2>, and |A3> corresponding to |0>, |1>, |1> and |0>, which are equal to decimal value '6' and |B0>, |B1>, |B2>, and |B3> corresponding to |1>, |0>, |1> and |0>, which are equal to decimal value '5'. Then,

(i) The output of the first 4-qubit adder will be |S0> = |1>, |S2> = |1>, |S2> = |0>, |S3> = |1> and |Cout> = |0>.

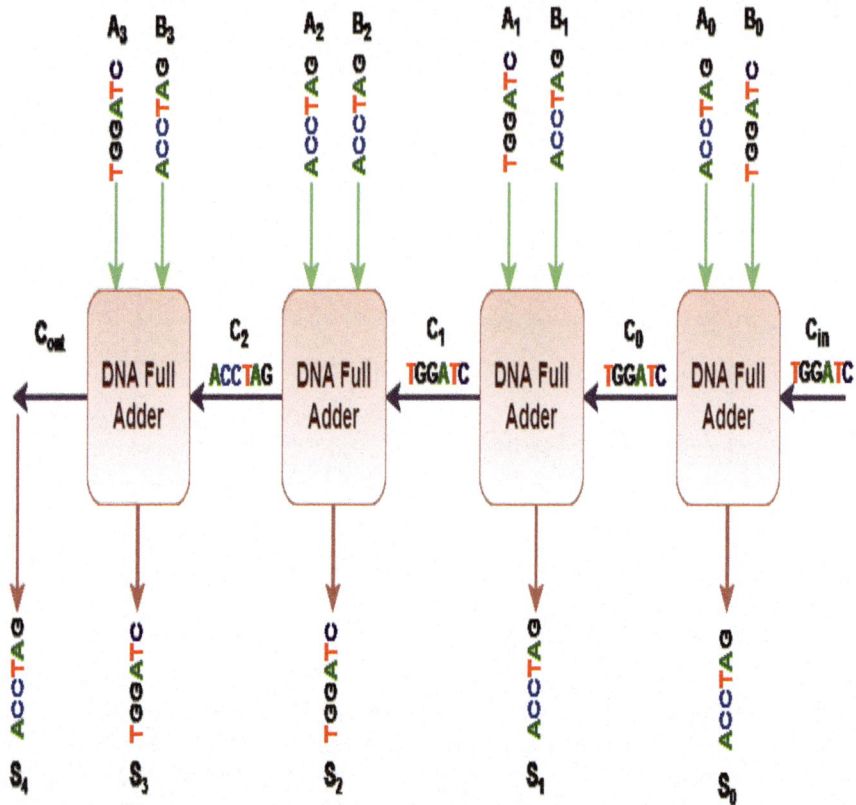

Fig. 7.9 4-Molecular DNA adder circuit

(ii) Then, the results of two quantum AND operations are $|S2>$. $|S3> = |0>$ and $|S1>$. $|S3> = |1>$.

(iii) By continuing the above result, the final quantum OR operation generates "True" $|1>$.

(iv) Finally, the outputs which are produced by the second 4-qubit adder are S0 = **ACCTAG,** S2 = **TGGATC,** S2 = **TGGATC,** S3 = **TGGATC,** and Cout = **ACCTAG**. This is the BCD number of 11.

7.6   Quantum-DNA Carry-Lookahead Adder

Remember the logical expressions to compute the carry by only using the carry output of the 1st addition operation. To determine those carry outputs which will be needed to design combinational circuits in both quantum and DNA systems. It can also be used in the quantum-DNA system. Figure 7.12 shows the general organizations

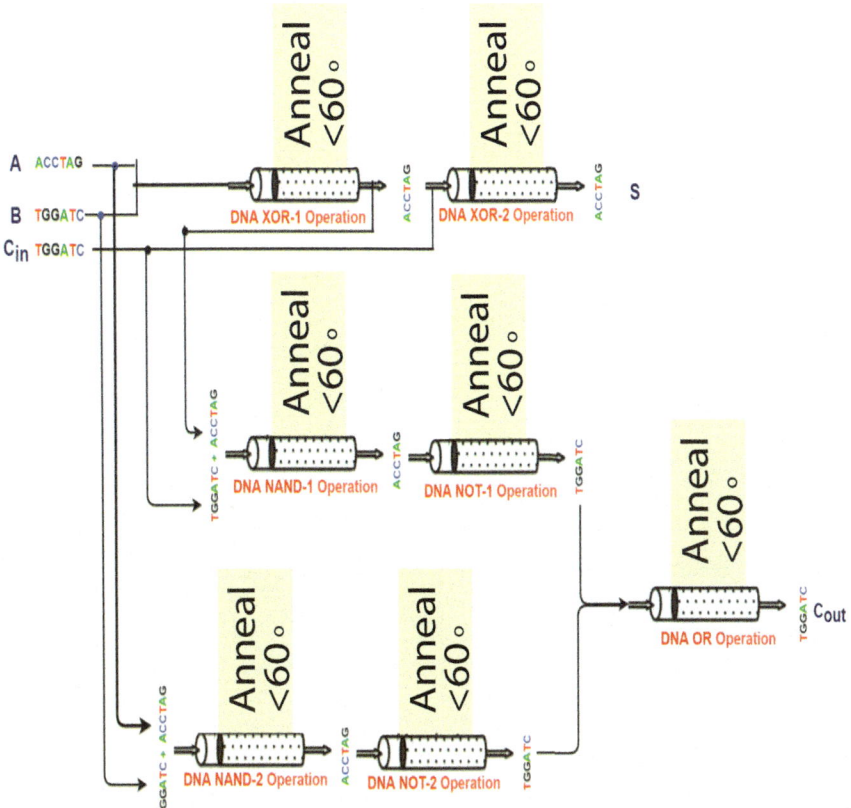

Fig. 7.10 DNA full adder circuit

of a 4-qubit quantum-DNA carry-lookahead adder with a basic idea, where each component is from the quantum-DNA system.

Figure 7.13a shows the design architecture of the combinational circuit which determines the value of C_2. Let's assume that there is a quantum cache memory where the qubits are stored as shown in Fig. 7.13b.

And Fig. 7.13b shows the general architecture of the combinational circuit which determines the value of C_3. In the same way, it can design the combinational circuit to determine the value of C_i using the logical expression.

Therefore, it is possible to construct the architecture of the n-qubit quantum-DNA carry-lookahead adder using the basic method.

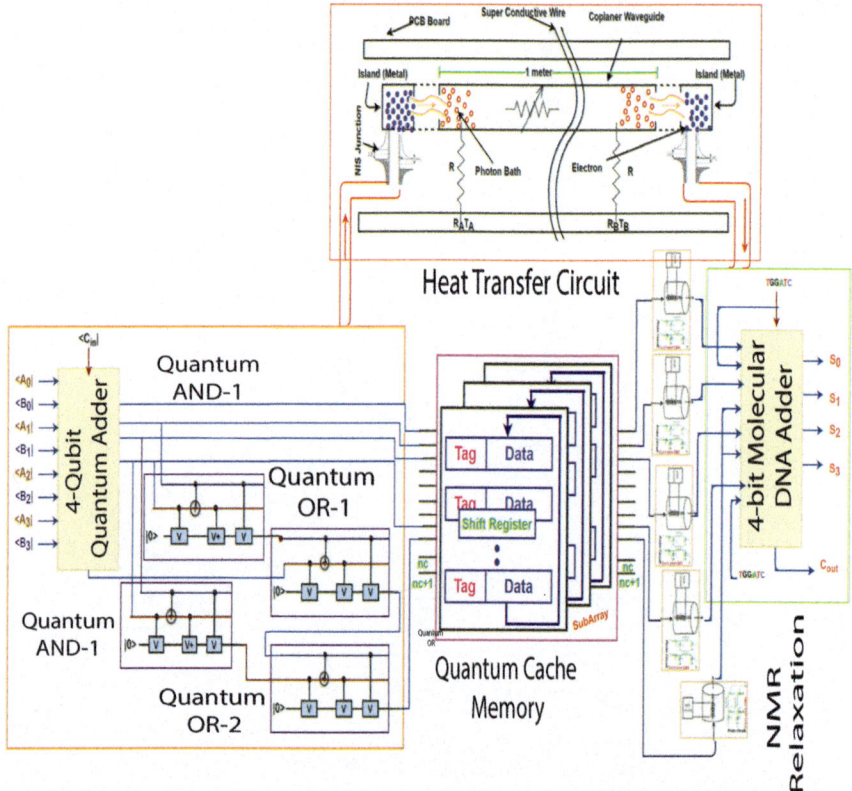

Fig. 7.11 Quantum-DNA BCD adder circuit

7.6.1 Block Diagram

Figure 7.14 shows the general organization of a 4-qubit quantum-DNA carry-lookahead adder. Here, among 4-qubit input addition operations, the three operations are performed in the quantum system, and the last addition operation is performed in the DNA system. The quantum cache memory stores the qubit values of the most significant qubits of the inputs and the carry value obtained by the combinational operations. The heat transfer circuit transfers the heat from the quantum system and the DNA system can get the necessary heat from the quantum system. Figure 7.15 shows the circuit diagram of a quantum-DNA 3-qubit carry-lookahead adder.

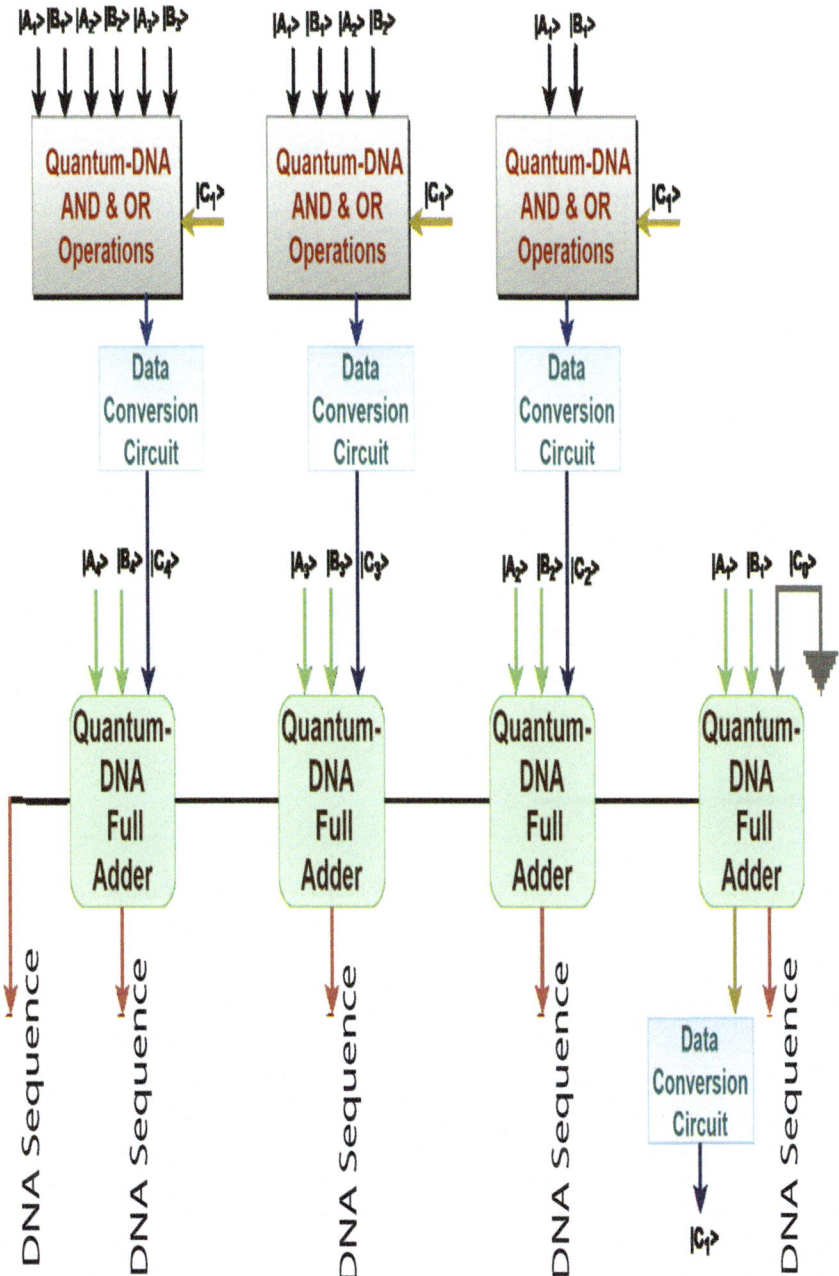

Fig. 7.12 The general organizations of a quantum-DNA 4-qubit carry-lookahead adder

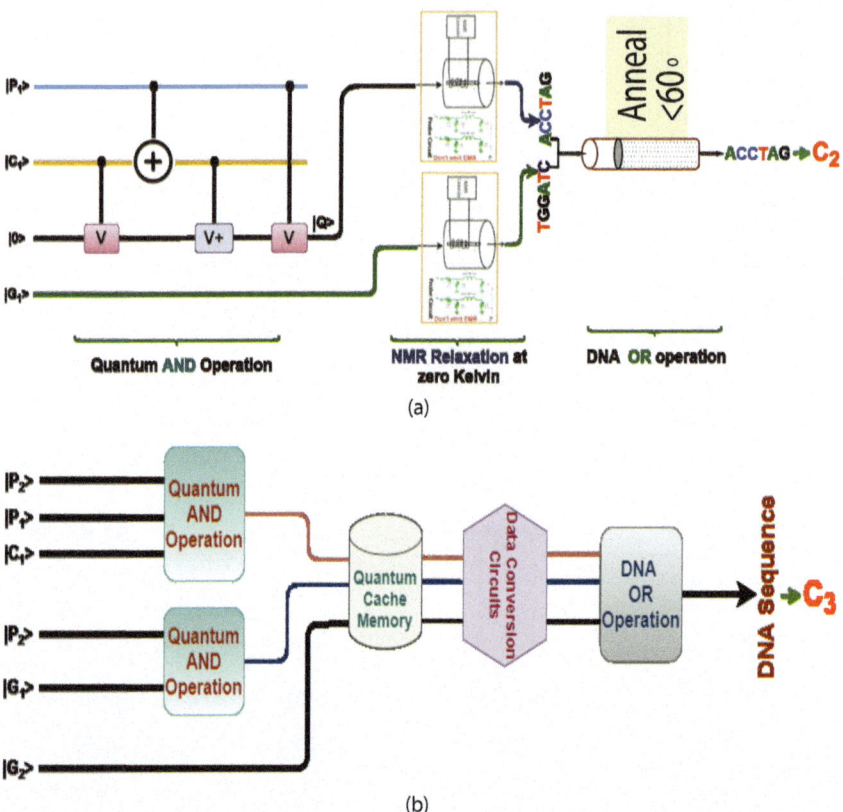

Fig. 7.13 The architecture of a quantum-DNA combinational circuit to determine the value of C_2 and C_3

7.6.2 Working Principles

The working procedures of the first part of the quantum-DNA system are the quantum system, which is the same as the working procedures of the quantum carry-lookahead adder. The circuit diagram shows that the 1st 2-qubit addition and the carry input qubit for the last addition operation are performed in the quantum system. Then, the carry qubit along with the other two input qubits are stored in the quantum cache memory. From the cache memory, they are passing through the trap ion to become the input of the DNA full-adder when the DNA system is ready. Then, the DNA system will perform the DNA full-adder operation and it will generate the final sum outputs.

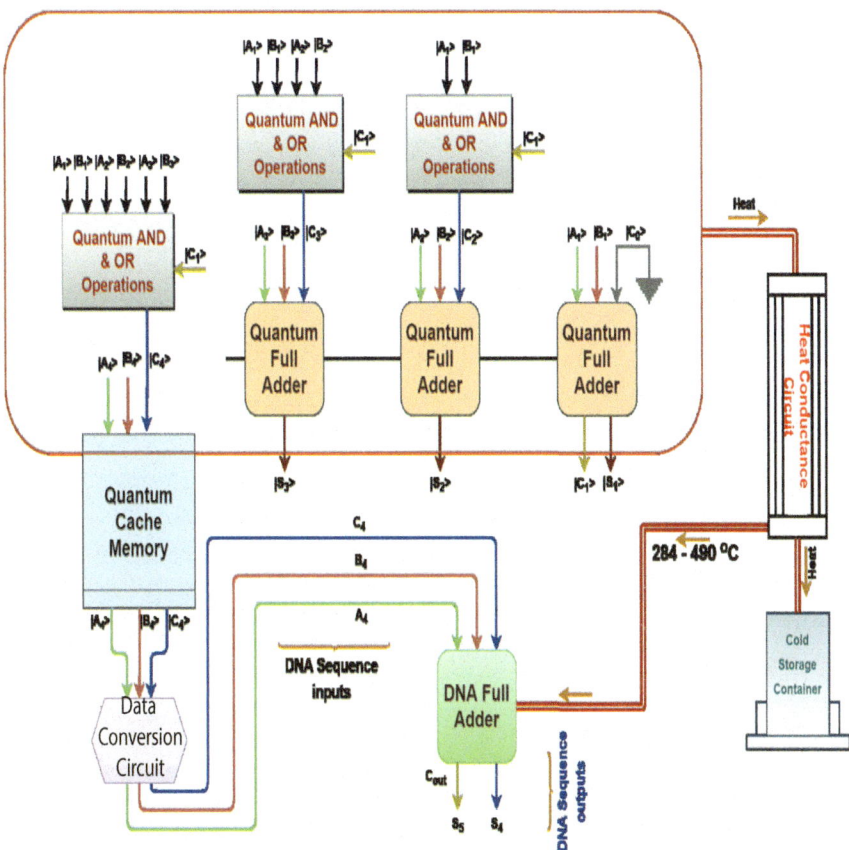

Fig. 7.14 The general organization of quantum-DNA 4-qubit carry-lookahead adder

7.7 Quantum-DNA Carry-Skip Adder

It is known how carry-skip adder works, how to design it in both the quantum system and the DNA system. A Carry-Skip adder, also known as a carry-bypass adder, is a method of implementing an adder that reduces the delay associated with ripple-carry adders. While it's a classical digital logic design, it can be adapted for use in quantum computing and even explored in the context of quantum biological computing. In summary, while the carry-skip adder is primarily a classical digital logic design, its principles can be extended to quantum computing, potentially leading to faster and more efficient quantum adders. The adaptation of these concepts could also have applications in the emerging field of quantum biological computing, particularly for tasks involving complex biological data analysis or simulations. Carry-skip adders or their quantum counterparts could potentially be used in quantum biological computing for tasks like:

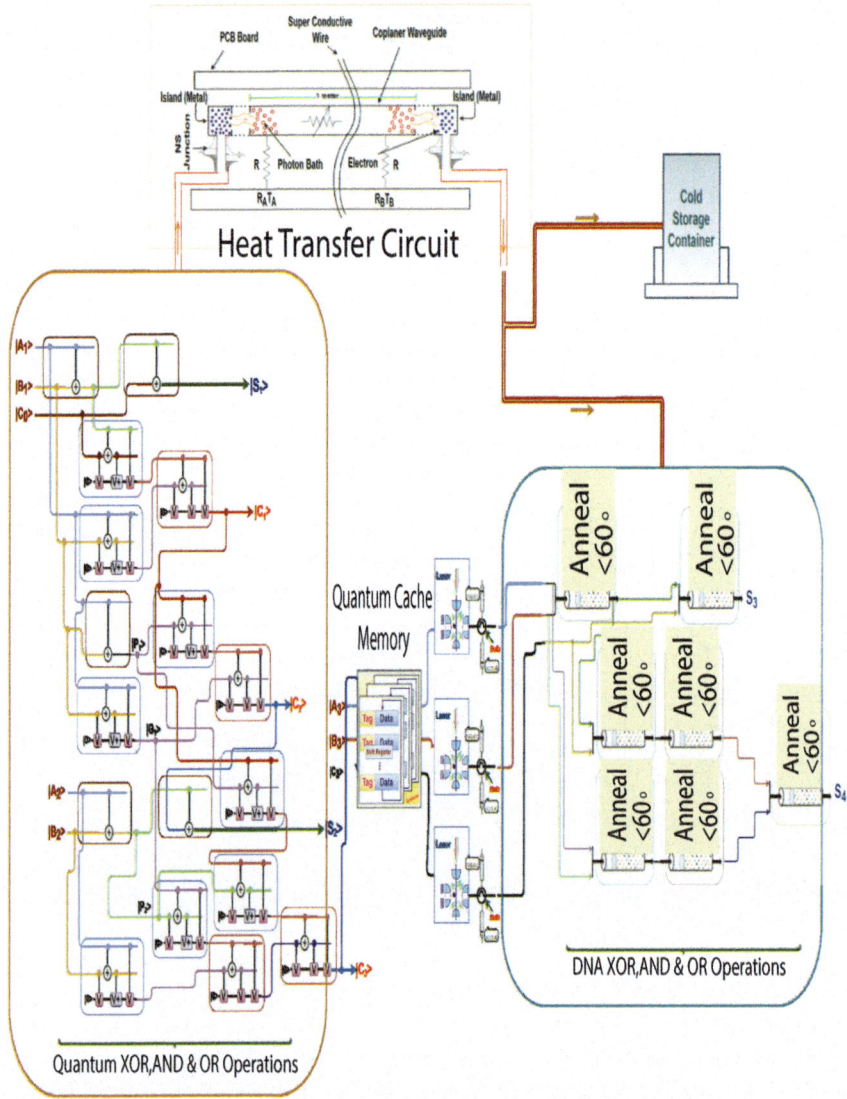

Fig. 7.15 The circuit architecture of quantum-DNA 3-qubit carry-lookahead adder

i) DNA Sequence Analysis: Analyzing the vast amount of data in DNA sequences, potentially requiring faster addition operations; and
ii) Protein Folding Simulation: Modeling the complex folding of proteins, which may involve numerous calculations.

In the carry-skip adder, a block of full-adders is created, and then determine the block propagate before operating the actual full-adder operations. If the block propagates

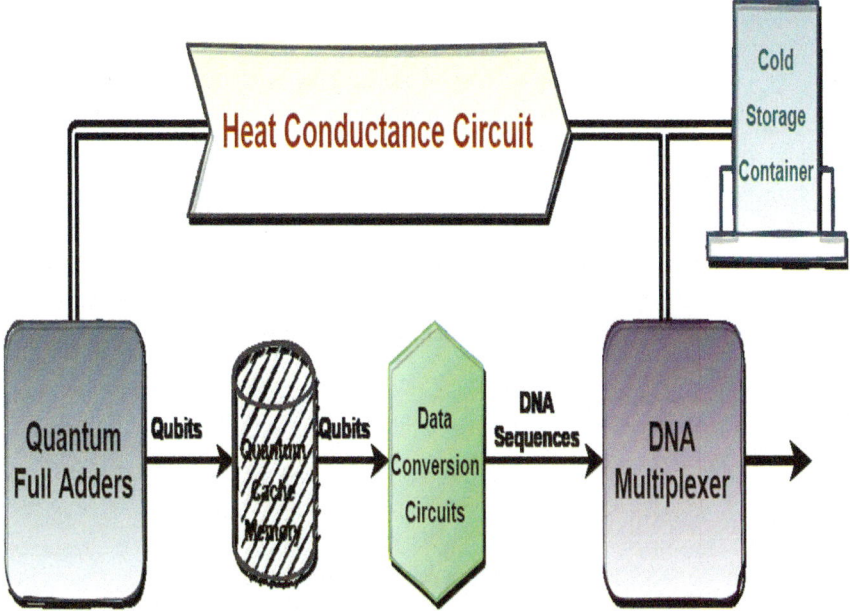

Fig. 7.16 The front-view of the quantum-DNA carry-skip adder

BP which is equal to $|1>$ (for quantum computing), then the output carry will be the same as the carry input given to the block.

Here, the block propagate is as follows:

$BP_i = |P_0> . |P_1> . |P_2> . .... . |P_i>;$

where $|P_i> = (A_i$ **Quantum XOR** $B_i)$

This is how to reduce the time delay to propagate the carry from one block to another. For more explanation, the carry-skip adder is exactly the same as the parallel adder when the BP is not equal to $|1>$. Therefore, it is required to design n numbers of full-adders of an n-qubit block. And if BP is equal to $|1>$ the value is bypassed to the carry input qubit of the block to the carry output using a 2-to-1 multiplexer.

7.7.1 Block Diagram

The general overview of the circuit diagram for the quantum-DNA carry-skip adder is depicted in Fig. 7.16.

From the above figure, it is observed that the quantum-DNA carry-skip adder is divided into two parts such as one part is in the quantum system which consists of connected parallel adders and the second part is in the DNA system, which consists of a 2-to-1 DNA multiplexer that will bypass the carry output of the block.

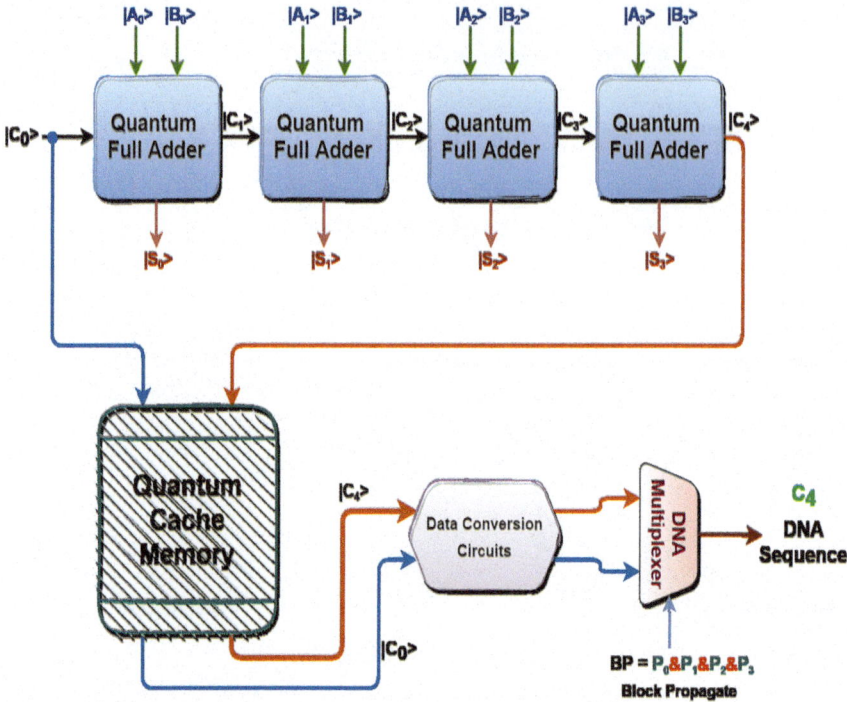

Fig. 7.17　The front-view of the quantum-DNA 4-qubit carry-skip adder

And a heat transfer circuit is connected to the quantum system to transfer the massive produced heat into the cooler. Moreover, the necessary heat can be transferred to the DNA system to operate it.

Figure 7.17 shows another front-view of a 4-qubit quantum-DNA carry-skip adder where the quantum system includes the four quantum full-adders. The first carry input to the block and carry output of the block are stored in a quantum cache memory so that they can be passed to the next operations.

The data conversion unit converts the data as DNA sequences and delivers the DNA sequences to the DNA system as the inputs, which is a DNA 2-to-1 multiplexer. And the DNA multiplexer does the rest of the operations.

In Fig. 7.18, an inside view of the quantum-DNA 4-qubit carry-skip adder is shown. The value of P_i is determined first, and the block propagation is determined by the quantum AND operations accordingly. The value of block propagation BP, and the two carry qubits C_0 and C_4 (Fig. 7.17 shows) will be stored in the quantum cache memory. And the data conversion unit will convert them accordingly. The converted values will be the input of the DNA multiplexer and will generate one DNA sequence as output. Note that the block propagate BP is working as the select input of the DNA multiplexer. If BP is ACCTAG (i.e., 1), then the multiplexer will bypass the value of C_0 (1st carry input to the block) as output. Otherwise, the value of C_4 goes through the DNA multiplexer.

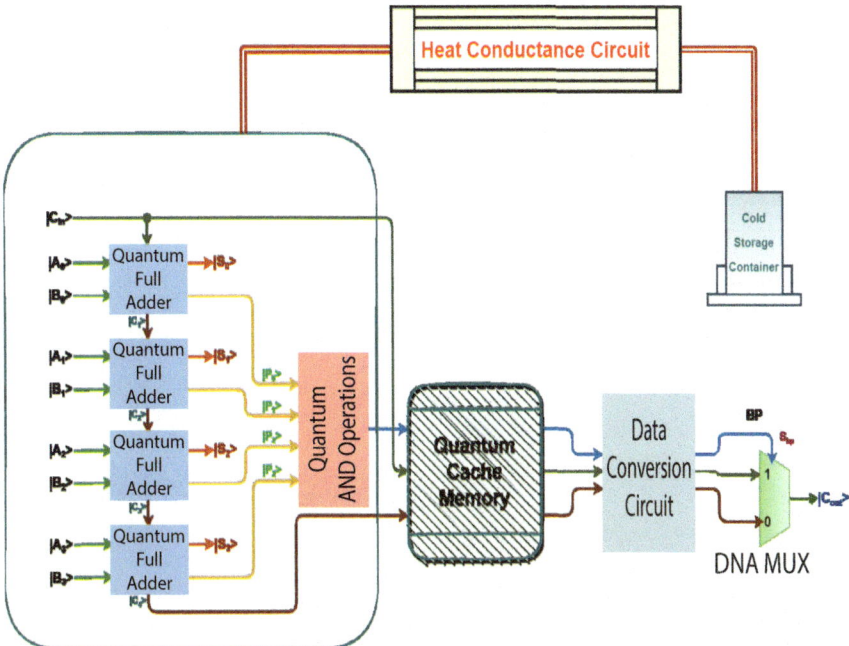

Fig. 7.18 The internal organizations of the quantum-DNA 4-qubit carry-skip adder

7.7.2 Circuit Architecture

Figure 7.19 shows the circuit diagram of the quantum-DNA 4-qubit carry-skip adder. The design procedure is the same as explained in the previous section. The quantum system has four quantum full-adders, the four quantum XOR operations to determine the value of P_i. It also has three quantum AND operations to get the output of BP. The three-qubit values (values of C0, C4, and BP) are stored in the quantum cache memory. Here again, trap ions are used to convert them into the corresponding DNA sequences. So, the stored qubits are passing through the trap ion circuits and become the equivalent DNA sequences. In the DNA system, the rest of the operations are performed according to DNA computing. DNA multiplexer passes the output according to the select input sequence, which is the value of BP. Thus, the bypassed output will be the carry input for the next cascaded block if exists.

7.7.3 Working Principles

The working procedures of the quantum-DNA carry-skip adder include four main units: Quantum system, data storing to cache memory, data conversion, and the DNA system. The operations in the quantum system were discussed in Chap. 6. And the

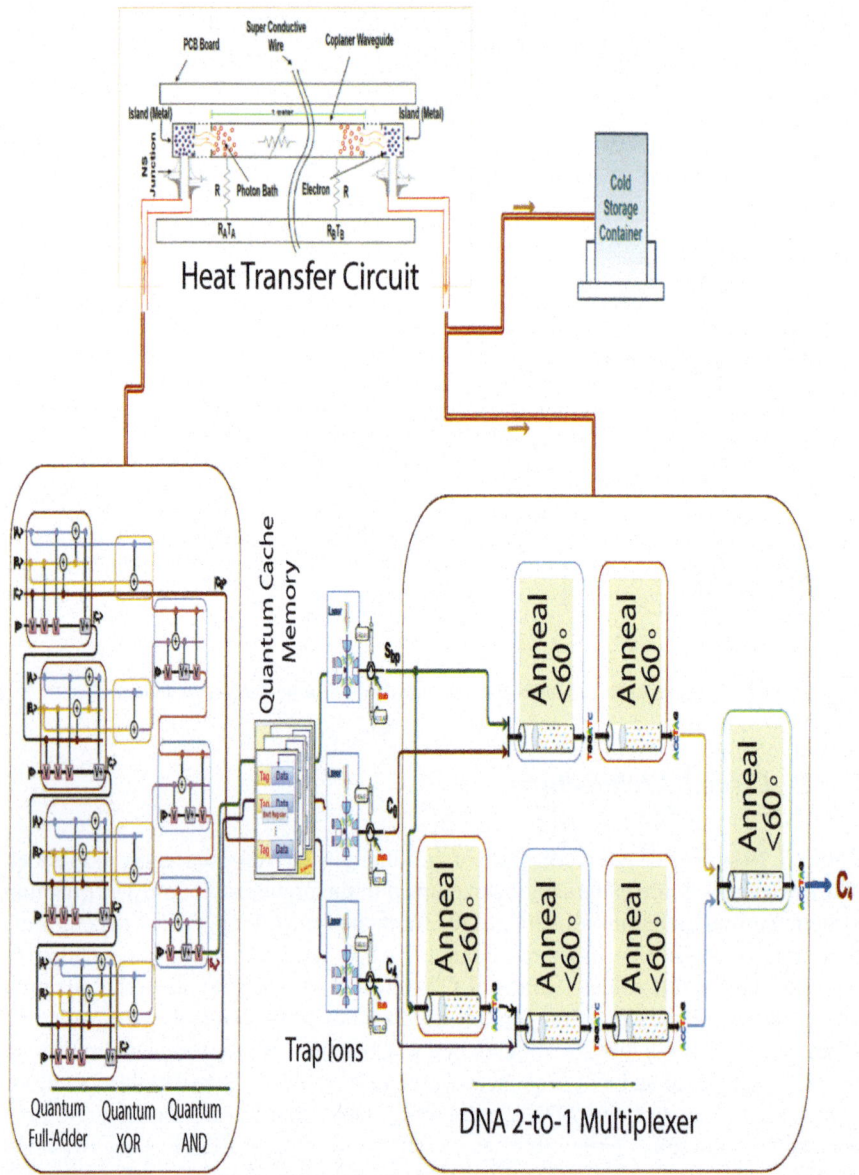

Fig. 7.19 The circuit architecture of the quantum-DNA 4-qubit carry-skip adder

DNA multiplexer operations will be discussed in Chap. 11. Besides, the working procedures of the circuit are discussed from the very beginning of this section.

7.8 Quantum-DNA Half Subtractor

The quantum-DNA half subtractor can also be used to subtract two numbers. It is equipped with two inputs and two outputs. This circuit is used to subtract two numbers with single qubit, $|A>$ and $|B>$ and generate two molecular sequences as the output states: 'D' and 'Bout.' Truth table for the quantum-DNA half subtractor is given in Table 7.3.

7.8.1 Block Diagram

The quantum-DNA circuit for a half subtractor is created by several quantum and DNA gates. To design a half subtractor, one quantum XOR operation is needed to produce the difference of the half subtractor. Additionally, to generate the borrow molecular sequence, one quantum NOT and DNA AND operations are used. The block diagram of the quantum-DNA half subtractor is illustrated in Fig. 7.20.

7.8.2 Circuit Architecture

Figure 7.21 depicts the quantum circuit of the half subtractor.

In this half subtractor circuit, the XOR operation for $|A>$ and $|B>$ inputs are done by a quantum operation and creates an output of $|D>$. For the output of $|Bout>$, at first a quantum NOT operation is performed for $|A>$ then both input qubits $|A>$ and $|B>$ pass through a quantum AND operation.

Table 7.3 Truth table for quantum-DNA half subtractor

Inputs		Outputs	
$|A>$	$|B>$	D	Bout
$|0>$	$|0>$	TGGATC	TGGATC
$|0>$	$|1>$	ACCTAG	ACCTAG
$|1>$	$|0>$	ACCTAG	TGGATC
$|1>$	$|1>$	TGGATC	TGGATC

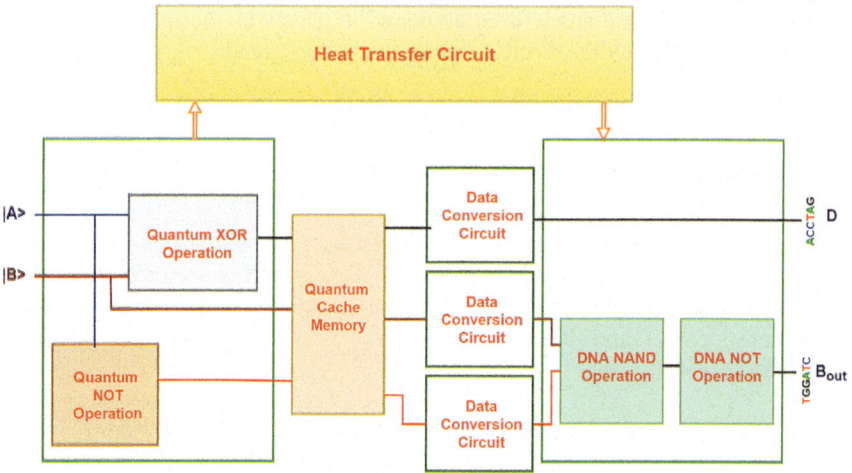

Fig. 7.20 Block diagram of quantum-DNA half subtractor

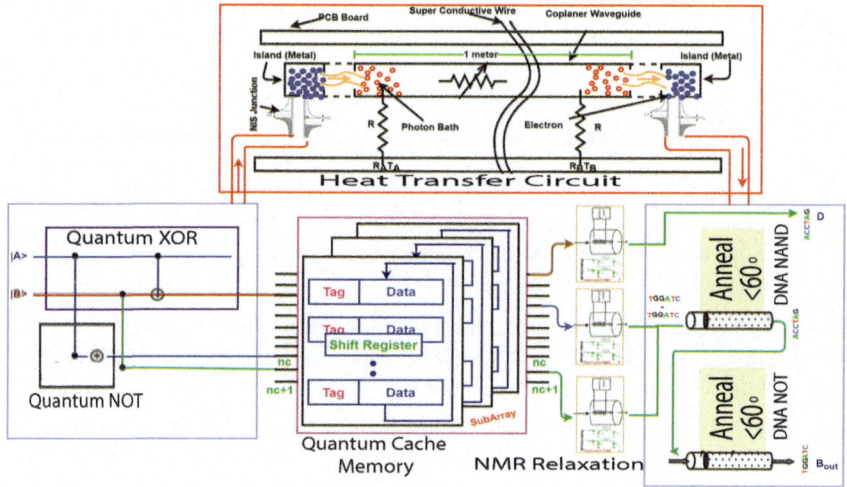

Fig. 7.21 Quantum-DNA half subtractor circuit

7.8.3 Working Principle

The quantum-DNA half adder needs input qubits, |A> and |B>. For various values of inputs A and B, consider the following cases:

(i) When both |A> and |B> are "true" |1>, the output sequences of both D and Bout are "false" **TGGATC.**

(ii) When both |A> and |B> are "false" |0>, the output qubits of both D and Bout are "false" **TGGATC**.

(iii) When |A> is "false" |0> and |B> is "true" |1>, the output qubit of D is "true" **ACCTAG** and Bout is "true" **ACCTAG**.

(iv) When |A> is "true" |1> and |B> is "false" |0>, the output qubit of D is "true" **ACCTAG** and Bout is "false" **TGGATC**.

7.9 Quantum-DNA Full Subtractor

The quantum-DNA full subtractor is a combinational circuit that performs subtraction operations among three input qubits: minuend, subtrahend, and borrow in. The difference and borrow output molecular sequences are generated by the full subtractor. The truth table for the full subtractor is given in Table 7.4.

From the truth table, it is easy to find

D = A'B'Bin + A'BBin' + AB'Bin' + ABBin
= Bin(A'B' + AB) + Bin'(AB' + A'B)
= Bin(A XNOR B) + Bin'(A XOR B)
= Bin (A XOR B)' + Bin'(A XOR B)
= Bin XOR (A XOR B)
= (A XOR B) XOR Bin
Bout = A'B'Bin + A'BBin' + A'BBin + ABBin
= Bin(AB + A'B') + A'B(Bin + Bin')
= Bin(A XNOR B) + A'B
= Bin (A XOR B)' + A'B

Table 7.4 Truth table for the quantum-DNA full subtractor

Inputs			Outputs	
|A>	|B>	|Bin>	D	Bout
|0>	|0>	|0>	TGGATC	TGGATC
|0>	|0>	|1>	ACCTAG	ACCTAG
|0>	|1>	|0>	ACCTAG	ACCTAG
|0>	|1>	|1>	TGGATC	ACCTAG
|1>	|0>	|0>	ACCTAG	TGGATC
|1>	|0>	|1>	TGGATC	TGGATC
|1>	|1>	|0>	TGGATC	TGGATC
|1>	|1>	|1>	ACCTAG	ACCTAG

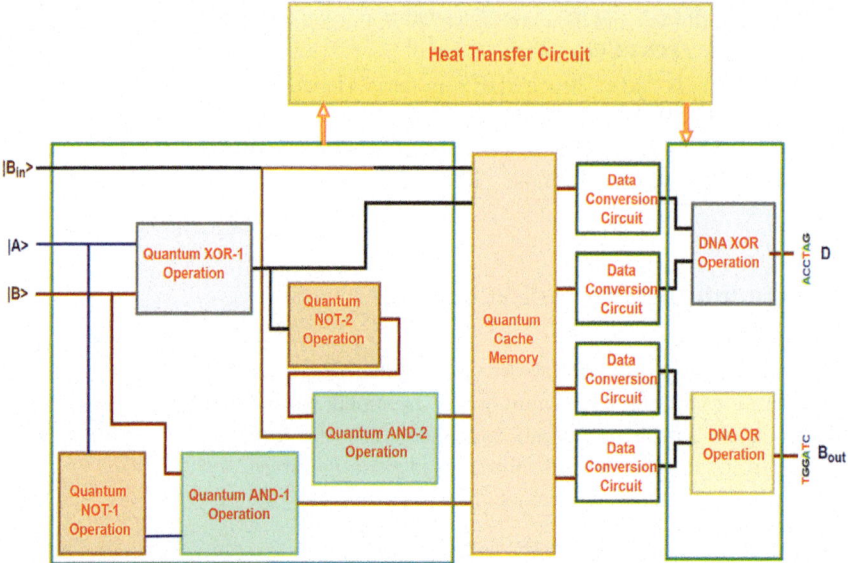

Fig. 7.22 Block diagram of quantum-DNA full subtractor

7.9.1 Block Diagram

The quantum-DNA circuit for a full subtractor needs one quantum XOR and one DNA XOR operation to produce the difference ($|D>$) of the full subtractor; and for $|Bout>$, two quantum AND operations, two quantum NOT, and one DNA OR operation are required. The block diagram of the quantum-DNA full subtractor is illustrated in Fig. 7.22.

7.9.2 Circuit Architecture

Figure 7.23 shows the quantum-DNA circuit of the full subtractor. In this full subtractor, the first XOR operation for $|A>$ and $|B>$ inputs is done by a quantum XOR operation and creates output $|A> \oplus |B>$. This output, $|A> \oplus |B>$, is used as one of the inputs of the second XOR operation, and another input qubit comes from $|Bin>$. Finally, the output $|D>$ qubit of the full subtractor is generated through this DNA XOR operation, which represents the difference of the full subtractor.

In addition, two quantum AND operations are needed to get another output qubits of $|Bout>$. At first AND operation is performed between the negation of $|A>$ input and $|B>$ input and produces $|A'>.|B>$. As for the second AND operation, one input comes from the NOT operation of $|A> \oplus |B>$ and another input is directly provided from $|Bin>$. These two input qubits then generate $(|A> \oplus |B>').|Bin>$. Finally, the

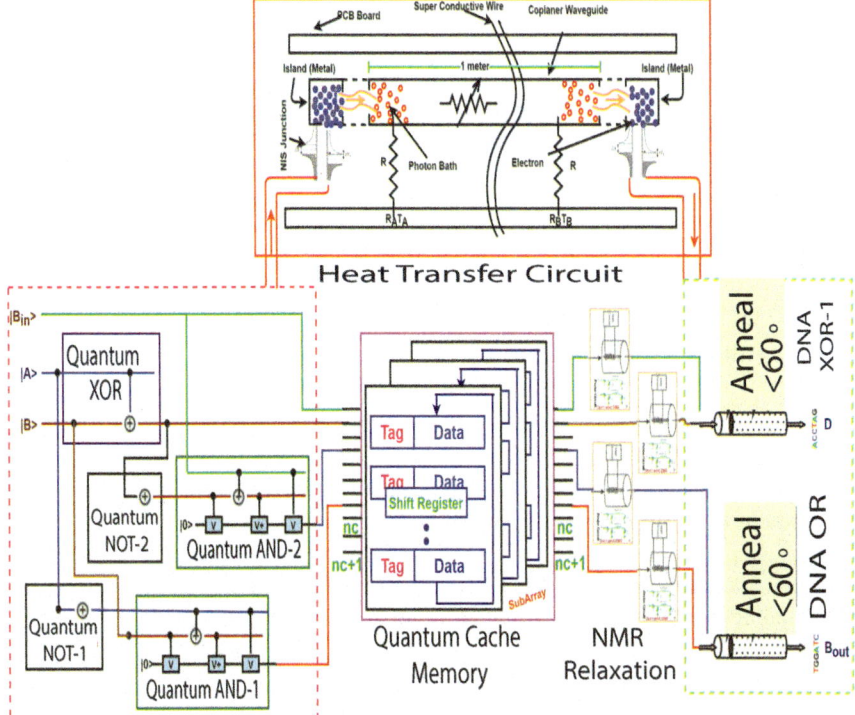

Fig. 7.23 Quantum-DNA full subtractor circuit

outputs of both quantum AND operations pass through a DNA OR operation to produce the value of |Bout> qubit.

7.9.3 Working Principle

The quantum-DNA full subtractor needs three inputs, A, B, and B_{in}; and provides two outputs D and Bout. To understand the working principle of a quantum-DNA full subtractor, consider the values of input qubits |A>, |B> and |Bin> are |1>, |0> and |1>. Now,

(i) The output of quantum XOR-1 is "true" |1>, quantum AND-1 is "false" |0> and quantum AND-2 is "false" |0>.

(ii) Finally, the value of output D is "false" **TGGATC** after performing the DNA XOR-2 operation, where the Bout is "false" **TGGATC.**

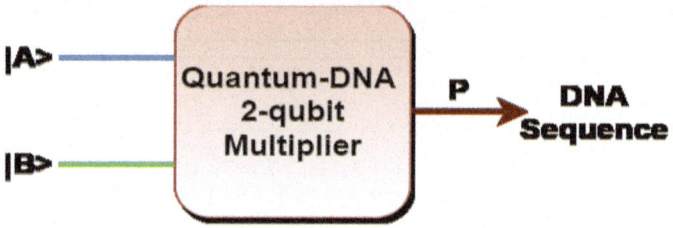

Fig. 7.24 The general working block of a quantum-DNA 2 × 2 Multiplier

7.10 Quantum-DNA Multiplier

In this section, the block diagram, working principle and design architecture of quantum-DNA multiplier are presented. Quantum biocomputing aims to model and simulate biological systems using quantum computers. Multiplexers can be used to represent the complex pathways and regulatory mechanisms within biological systems. Quantum computers can help in drug discovery by simulating interactions between molecules and drugs, and multiplexers can be used to represent different possible binding sites. Simulating protein folding, a complex process, can be facilitated by quantum circuits that use multiplexers to represent different conformations of the protein. Multiplexers are essential components in building quantum algorithms for various tasks in biology, including optimization, machine learning, and simulation.

7.10.1 Block Diagram

This is quantum-DNA cross-platform environment. A (2 × 2)-qubit quantum-DNA multiplier will be designed. The general working principle of the quantum-DNA multiplier is shown in Fig. 7.24.

Figure 7.24 depicts that, the inputs of a quantum-DNA multiplier are the qubits, and the product outputs of the inputs will be in the form of DNA nitrogen base sequences. So, this is going to be easier than before. The constructions of a quantum multiplier and a DNA multiplier have been discussed already. Now, it is needed to connect them in a way so that the properties of the quantum-DNA system are maintained and provides the expected results.

7.10.2 Circuit Architecture

Figure 7.25 shows the simplest form of the quantum-DNA (2 × 2)-multiplier, where both the multiplicand and the multiplier are 2-qubit. The circuit shows that the quantum system consists of three quantum AND operations. The output of the

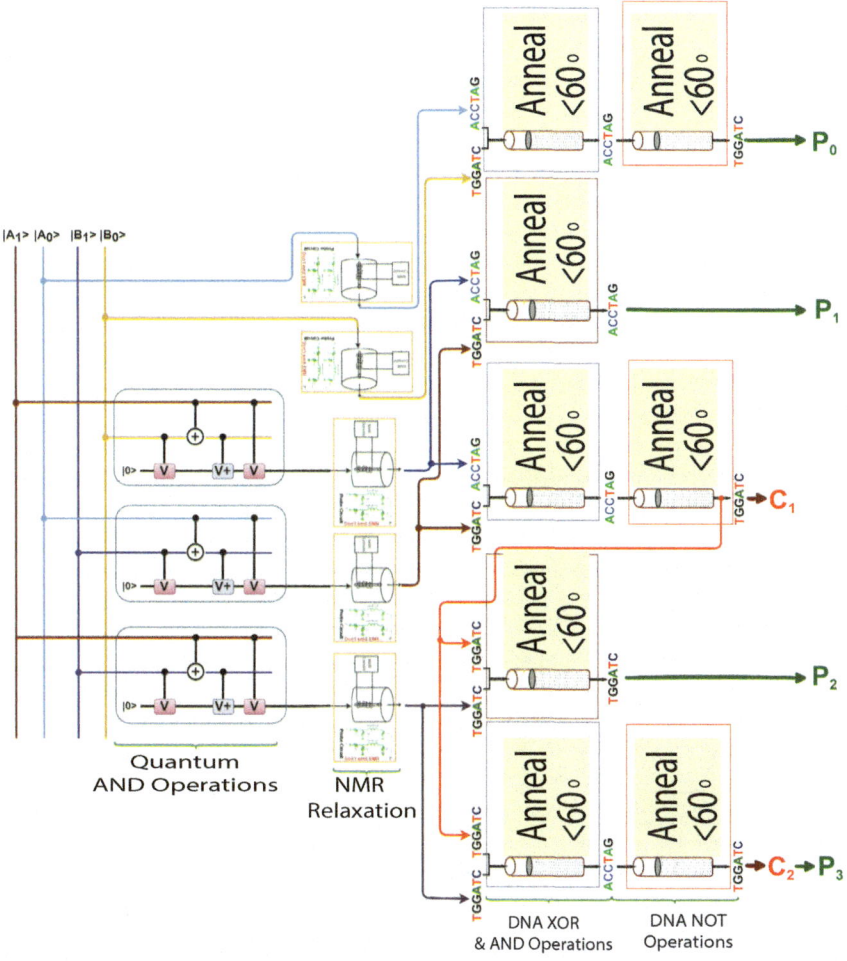

Fig. 7.25 The circuit diagram of a quantum-DNA 2 × 2 multiplier with NMR relaxation at zero Kelvin temperature

quantum AND operations' and values of $|A_0>$ and $|B_0>$ are passing through the NMR relaxation for the data conversion process. And then the converted DNA sequences are entered into the DNA system. In the DNA system, two DNA XOR, three DNA NAND, and three DNA NOT operations generate the product value of the corresponding inputs.

Quantum cache memory and a heat conductance circuit are needed to add and also use trap ion instead of NMR relaxation. Figure 7.26 depicts the quantum-DNA (2 × 2)-qubit multiplier's circuit, which comprises all of the necessary components.

The quantum system and the DNA system are both the same as in Fig. 7.26, but here trap ions are used as a data conversion unit instead of NMR relaxation. Besides, the quantum cache memory is present here to store the qubits. Also, the heat transfer circuit is presented in the figure.

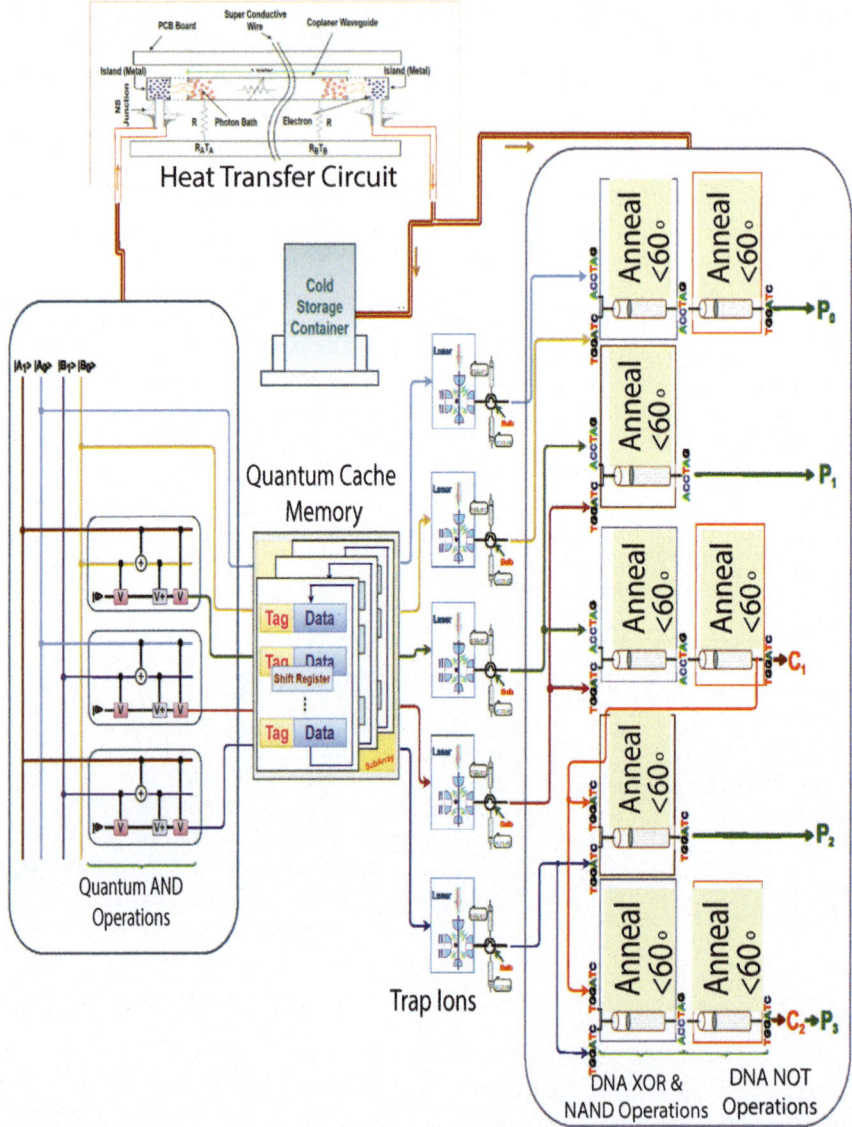

Fig. 7.26 The complete circuit diagram of a quantum-DNA (2×2)-qubit multiplier with quantum cache memory, trap ions, a heat conductance circuit, and a cooler

7.10.3 Working Principle

The working procedures are exactly the same as explained in the quantum and DNA multiplier section. The DNA sequences shown in the circuit are just taken as examples, where the values of qubits $|A>$, and $|B>$ are considered as $|01>$ and $|10>$, respectively.

Let the input qubits be $|A>$ and $|B>$. Here, $|A> = |A_1A_0>$, and $|B> = |B_1B_0>$. Suppose, $|A> = |10>$ and $|B> = |10>$.

Therefore, $|A_0> = |0>$, $|A_1> = |1>$, $|B_0> = |0>$, and $|B_1> = |1>$.

1. In the quantum system, the 1st quantum AND operation will receive input qubits $|A_1> = |1>$, and $|B_0> = |0>$, which will produce the output $|0>$, and it will be stored in the quantum cache memory.
2. The 2nd quantum AND operation will receive input qubits $|A_0> = |0>$, and $|B_1> = |0>$, which will again produce the output $|0>$, and it will be stored in the quantum cache memory.
3. The 3rd quantum AND operation will receive input qubits $|A_1> = |1>$, and $|B_1> = |1>$, which will produce the output $|1>$, and it will be stored in the quantum cache memory.
4. Now, there are five qubit values in the cache memory, such as the values of $|A_0>$ and $|B_0>$, and the three quantum AND operations output values. They will be passed through the trap ions which will be converted into their corresponding DNA sequences.
5. The qubit values of $|A_0>$ and $|B_0>$ are $|0>$. Therefore, they will be converted as TGGATC. In the DNA system, they will be working as the input of the 1st DNA AND operation (which is DNA NAND followed by the DNA NOT operation). Therefore, they will generate the first product result $P_0 = $ TGGATC.
6. Next, the DNA XOR operation will receive both inputs as TGGATC (the converted values from Step 1) and Step 2). The DNA XOR operation will produce the output sequence TGGATC, which is the second product value of the multiplication operation. Therefore, $P_1 = $ TGGATC.
7. Then, the 2nd DNA AND operation will be performed with the same input of Step 6. And it will produce the carry output sequence as TGGATC. This carry will work as an input to the next two steps.
8. The 2nd DNA XOR gets input TGGATC (from Step 7) and ACCTAG (the converted DNA sequence of the output from Step 3). So, it will generate ACCTAG as the output sequence. This is the value of P_2. Therefore, $P_2 = $ ACCTAG.
9. And finally, the last DNA AND operation will also get the inputs as the same as Step 8 (ACCTAG and TGGATC). As a result, it will generate TGGATC as the output sequence, which will be the value of P_3.

7.11   Quantum-DNA Divider

Now, in this section, a divider in the quantum-DNA environment will be discussed. The divider in both quantum and DNA computing systems has already been discussed. So, the design architectures and working principles of quantum divider and DNA divider are known to us. In the context of quantum computing, a "divider" refers to a specialized quantum circuit used to perform division operations. Specifically, in quantum biological computing, this might involve using quantum algorithms and circuits to simulate or perform computations related to biological processes that involve division or partitioning. This could include simulating molecular interactions, analyzing genetic sequences, or modeling biological systems that involve division or partitioning. A divider in the quantum biocomputing explores the potential of using quantum principles to understand and manipulate biological systems. This could involve using quantum algorithms to analyze complex biological data, model biological processes, or design new biological systems. In the context of quantum biocomputing, a "divider" could be a quantum circuit or algorithm used to simulate or analyze processes that involve division or partitioning. For example:

  i) Simulating cell division: Quantum computers could be used to simulate the complex molecular processes involved in cell division, potentially leading to a better understanding of cancer or other diseases;
 ii) Analyzing genetic sequences: Quantum algorithms could be used to analyze large genetic sequences, identifying patterns and variations that may be relevant to disease or evolution; and
iii) Modeling biological networks: Quantum simulations could be used to model complex biological networks, such as neuronal networks or metabolic pathways, by using quantum circuits to perform the necessary division or partitioning operations.

In essence, a "divider" in quantum biological computing refers to a quantum computational tool that can be used to simulate or analyze biological processes that involve division or partitioning, potentially leading to new discoveries in the fields of biology, medicine, and biotechnology.

7.11.1 Block Diagram

The general block diagram of the quantum-DNA division operation is shown in Fig. 7.27. There are two operational parts such as the quantum system and the DNA system. A quantum cache memory is used again to store the qubits for a while. Data conversion circuits will convert the output from the quantum system and that will work as input to the DNA system. Finally, the DNA system will perform the rest operations and will generate the resulting output.

Figure 7.28 shows the general architecture of the quantum-DNA (2×2)-divider operations, which can only generate the quotient of the dividing elements.

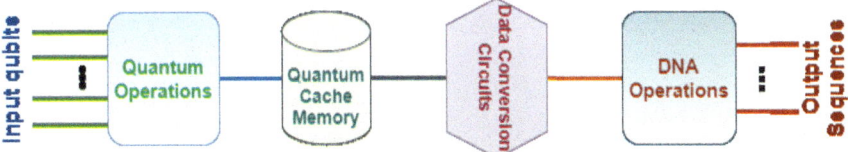

Fig. 7.27   The general view of quantum-DNA division operation

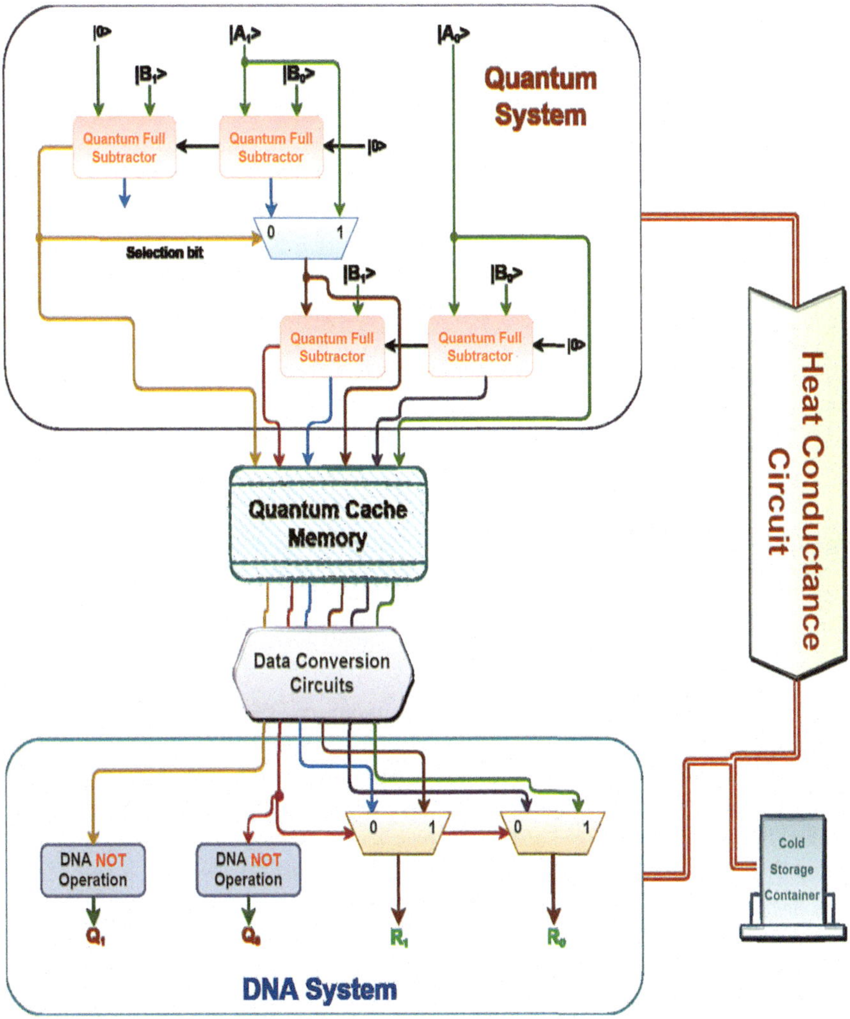

Fig. 7.28   The general circuit architecture of a quantum-DNA (2×2)-qubit divider

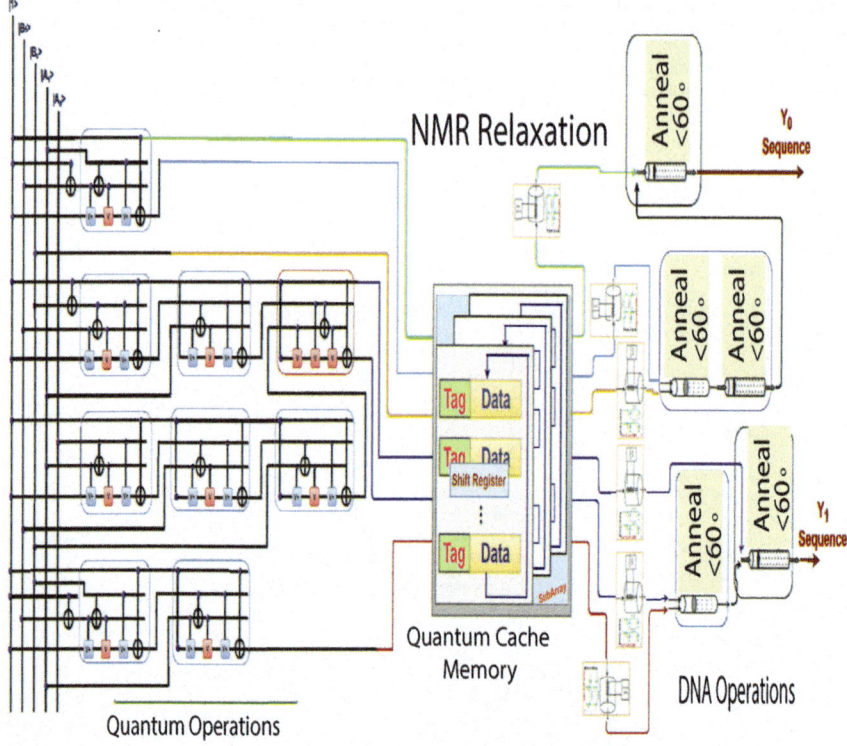

Fig. 7.29 The circuit architecture of quantum-DNA (2 × 2)-qubit divider

7.11.2   Circuit Architecture

Figure 7.29 shows a circuit diagram of the quantum-DNA divider. The quantum system in the circuit architecture includes four quantum full-subtractors and one quantum 2-to-1 multiplexer. And the DNA system consists of two DNA NOT operations and two DNA 2-to-1 multiplexers. The quotient output sequence is generated by the DNA NOT operations; and the remainders of the division operation are generated from the DNA multiplexers operations. Afterward, the data conversion circuit is added and NMR relaxation or trap ions is used to convert the qubits into the equivalent DNA sequences. The output from those DNA NOT operations will generate the quotients of the division operations and then, two DNA 2-to-1 multiplexers have to be connected accordingly.

7.11.3   Working Principle

The quantum system consists of quantum full subtractors and quantum multiplexer. On the other hand, the DNA system consists of DNA NOT operations and DNA multiplexers. The working principles of all these quantum and DNA operations are discussed in the previous chapters of this part.

The output from the quantum system will be stored in the quantum cache memory. And from the cache memory, the qubits will be passed through the data conversion units, and these qubits will be converted into their corresponding DNA nitrogen base sequence values eventually. Then, the DNA sequence inputs will enter into the DNA system. After performing the operations, the DNA system will produce the results.

However, the massive heat generated by the quantum system will be transferred to a cooling system through a heat conductance circuit. And it is known that the DNA system requires a specific amount of heat to perform any operation. Therefore, it is provided the heat from the quantum system if required.

7.12   Quantum-DNA Comparator

A comparator compares two input signals and gives output as to which input is larger or smaller or equal. In the context of quantum computing, a comparator is a circuit or algorithm that compares two numerical values, typically represented in qubits, and determines which is greater, smaller, or equal. Quantum-DNA computing combines the principles of both quantum computing and DNA computing, and comparators are fundamental for various operations within these systems. In a quantum-DNA computing system, comparators could be implemented by leveraging the properties of both quantum and DNA components. For example, DNA molecules could be used to store and manipulate quantum states, and quantum comparators could be used to analyze the results of DNA reactions. Quantum-DNA 1-qubit comparator takes two single-qubit inputs, $|A>$ and $|B>$. The values of the inputs can be $|0>$ and $|1>$. This circuit gives three outputs X, Y, and Z, where X is true when A is smaller than B, Y is true when A is equal to B, and Z is true when A is greater than B. Here, Table 7.5 shows the truth table of the 1-qubit comparator.

From the truth table, the equations for $|X>$, $|Y>$ and $|Z>$ are $\mathbf{X} = A^0.B^1$
$\mathbf{Y} = A^0.B^0 + A^1.B^1 = A \text{ XNOR } B$; and
$\mathbf{Z} = A^1.B^0$

7.12.1   Circuit Architecture

Quantum-DNA 1-qubit comparator needs two quantum AND operations, two quantum NOT operations, and one DNA XNOR operation (Fig. 7.30). Quantum AND-1

Table 7.5 Truth table of a 1-qubit comparator

Inputs		Outputs		
|A>	|B>	X	Y	Z
		A < B	A= B	A > B
|0>	|0>	TGGATC	ACCTAG	TGGATC
|0>	|1>	ACCTAG	TGGATC	TGGATC
|1>	|0>	TGGATC	TGGATC	ACCTAG
|1>	|1>	TGGATC	ACCTAG	TGGATC

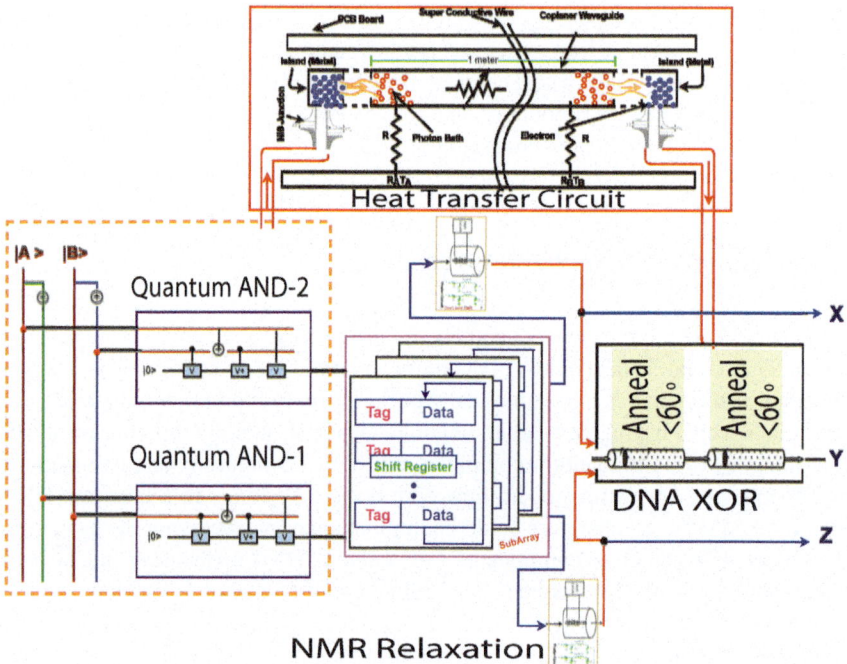

Fig. 7.30 Quantum-DNA 1-qubit comparator circuit

is connected to A′ (complement of A) and B and it performs A′. B, which is the result of Z. In the same way, the quantum AND-2 provides the output of X. This quantum qubit output converts into DNA sequence using NMR relaxation. Then, with these converted DNA sequences, a DNA XNOR operation is performed to get the value of Y.

7.12.2 Working Principle

The quantum-DNA comparator needs two input qubits |A> and |B>. For various values of inputs A and B, consider the following cases:

(i) When both |A> and |B> are "false" |0>, the output of Y would be "true" **ACCTAG** and X and Z will be "false" **TGGATC** as both AND operations will produce **TGGATC**.

(ii) When |A> is "false" |0> and |B> is "true" |1>, the output of X would be "true" **ACCTAG** and Y and Z will be "false" **TGGATC**.

(iii) When |A> is "true" |1> and |B> is "false" |0>, the output of Z would be "true" **ACCTAG** and X and Y will be "false" **TGGATC**.

(iv) When both |A> and |B> are "true" |1>, the output of Y would be "true" **ACC-TAG** and X and Z will be "false" **TGGATC** as both AND operations will produce **TGGATC**.

7.13 Summary

In this chapter, all the arithmetic operations in quantum-DNA computing are presented. Their working principles and appropriate figures have also been shown. The cross-platform of quantum computing and DNA computing is a new challenge that has been introduced here in this book for the first time in the history. The conversion circuits and heat transfer circuits are used here to improve the performance. In this chapter, the quantum-DNA arithmetic circuits are discussed. In the next chapter, another cross-platform DNA-quantum arithmetic circuits will be discussed.

Bibliography

1. T.A. Brubaker, J.C. Becker, Multiplication using logarithms implemented with read-only memory. IEEE Trans. Comput. **100**(8), 761–765 (1975)
2. V. Giovannetti, S. Lloyd, L. Maccone, Architectures for a quantum random access memory. Phys. Rev. A **78**(5), 052310 (2008)
3. M. Morris Mano, C.R. Kime, G. Guerrero, *Fundamentos de diseño lógico y de computadoras*. Number Sirsi i8420543993. Pearson Educación (2005)
4. J. Ian Munro, V. Raman, Selection from read-only memory and sorting with minimum data movement. Theor. Comput. Sci. **165**(2), 311–323 (1996)
5. Z.-R. Wang, Y.-T. Su, Y. Li, Y.-X. Zhou, T.-J. Chu, K.-C. Chang, T.-C. Chang, T.-M. Tsai, S.M. Sze, X.-S. Miao, Functionally complete Boolean logic in 1T1R resistive random access memory. IEEE Electron Dev. Lett. **38**(2), 179–182 (2016)

Chapter 8
Arithmetic Operations in DNA-Quantum Computing

8.1 Introduction

An arithmetic circuit is a set of gates with a separate set of inputs for each number that has to be processed. The gates are connected so as to carry out an arithmetic action and the outputs of the gate circuit are the qubits of the result. Arithmetic circuits are crucial in modern computing, serving as the fundamental building blocks for performing calculations within computers and other computer systems. They enable the execution of arithmetic operations like addition, subtraction, multiplication, and division, which are essential for various tasks, from basic calculations to complex algorithms. DNA-quantum is also a new idea which is introduced here for the first time in the history. It is also called quantum biocomputing. The arithmetic operations have discussed here in such a way which have never been described before. There are two cross-platform environments with quantum computing and DNA computing. One is quantum-DNA computing, and another is DNA-quantum computing. In the previous chapter, arithmetic operations in quantum-DNA computing have already been discussed. This chapter will describe another innovative unique idea for cutting-edge technology which is called DNA-quantum arithmetic operation.

The general organizations of any DNA-quantum operations, the block diagram and circuit architecture with their working principle will be discussed here. In these designs, the NMR process is used as the data conversion unit. Sometimes quadrupole trap ion has been used to convert the DNA sequences to qubits. So, the themes which will be discussed here are given below:

1. DNA-Quantum Half Adder
2. DNA-Quantum Full Adder
3. DNA-Quantum Molecular Adder
4. DNA-Quantum BCD Adder
5. DNA-Quantum Carry look ahead Adder
6. DNA-Quantum Carry-Skip Adder
7. DNA-Quantum Half Subtractor

H. M. Hasan Babu, *Quantum Biocomputing in Quantum Biology Volume I*,
https://doi.org/10.1007/978-981-97-7154-7_8

8. DNA-Quantum Full Subtractor
9. DNA-Quantum Multiplier
10. DNA-Quantum Divider
11. DNA-Quantum Comparator

The circuit constructions, working principles, and details of the above themes are discussed in this chapter.

8.2   DNA-Quantum Half Adder

A half-adder circuit, a fundamental component in digital logic, can be implemented in both classical and quantum computing. In biological quantum computing, the concept of a half-adder can be explored by simulating or mimicking its function using biological systems or molecules. In essence, the half-adder is a fundamental component of both classical and quantum computing, and its principles are relevant in the emerging field of bioquantum computing, where exploring biological systems for quantum information processing is a key area of research. The concept of a half-adder, both classical and quantum, is relevant in bioquantum computing as it provides a basic building block for arithmetic operations, which are essential for various computational tasks. Bioquantum computing explores the use of biological systems for quantum computing, potentially offering advantages in certain applications. Research in bioquantum computing may explore how biological systems can be manipulated to perform quantum logic operations, including half-adder circuits, potentially leading to novel computational paradigms. A DNA-quantum half adder is a circuit that adds two DNA molecular sequences to produce a sum and a carry-in qubits. This adder can only add two values at a time. The truth table for the DNA-quantum half adder is given in Table 8.1.

From the above truth table, we get

$$S = A' B + A B'$$
$$= A \, XOR \, B$$
$$Cout = A \, B$$

Table 8.1 Truth table of a DNA-quantum half adder

Inputs		Outputs	
A	B	|S>	|Cout>
TGGATC	TGGATC	|0>	|0>
TGGATC	ACCTAG	|1>	|0>
ACCTAG	TGGATC	|1>	|0>
ACCTAG	ACCTAG	|0>	|1>

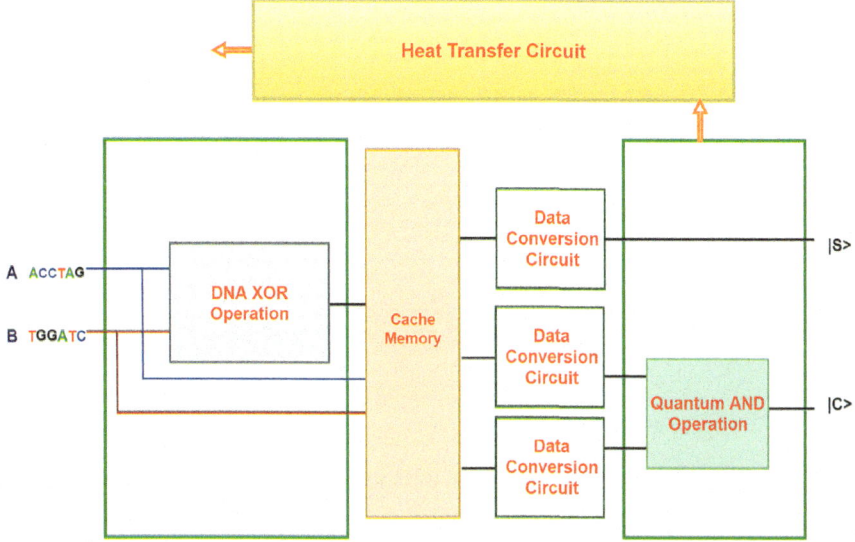

Fig. 8.1 Block diagram of DNA-quantum half adder

8.2.1 Block Diagram

The DNA-quantum circuit for a half adder is created by several DNA gates and Quantum gates. To design a half adder, one XOR gate and one AND gate are required. In this DNA-quantum circuit, one DNA XOR operation is needed to produce the SUM of the half adder. Additionally, to generate the carry qubit, one quantum AND operation is used. To balance the speed between DNA operation and quantum operation, a DNA cache memory is used. The block diagram of the DNA-quantum half adder is illustrated in Fig. 8.1.

8.2.2 Circuit Architecture

Algorithm 8.2.1 represents the overall procedures of the DNA-quantum-based half-adder operation. "HALF_SUM" is used to represent the Half_Adder sum's output, which is obtained after the DNA XOR operation. "HALF_Carry" is the carry of the half adder, which is obtained after the quantum AND operation.

Algorithm 8.2.1 DNA Half Adder Algorithm

---

1. **Begin**
2. **while** i equals to 1 to n **do**
3.   T1 <-Do_DNA_XOR(Ai, Bi)
4.   CacheArrayQuant[3] = { T1 , Ai , Bi}
5.   PerformDNAToQuant (ArrayQuant, i);
6.   HALF_SUM <- |T1>
7.   HALF_Carry <- Do_Quant_AND(|Ai>, |Bi>)
8. **end while**
9. Procedure PerformDNAToQuant (CacheArrayQuant, i)
10. |T1> <- DO_Trap_Ion(CacheArrayQuant [0]);
11. |Ai> <- DO_Trap_Ion(CacheArrayQuant [1]);
12. |Bi> <- DO_Trap_Ion(CacheArrayQuant [2]);
13. End Procedure

Figure 8.2 depicts the DNA-quantum circuit of the half adder. In this half adder circuit, the XOR operation for A and B inputs is done by a DNA operation and creates an output of S. As for the output of C, both inputs A and B pass through a quantum AND operation. In order to match the speed of DNA and quantum operations, a DNA cache memory is used. To convert the molecular sequence into a quantum qubit, the quadrupole trap ion method is applied.

8.2.3   Working Principle

The DNA-quantum half adder needs two inputs, A and B. For various values of inputs A and B, consider the following cases:

[i] When A and B are both "true" **ACCTAG**, the output of |S> is "false" |0> and |C> is "true" |1>.
[ii] When both A and B are "false" **TGGATC**, the outputs of both |S> and |C> are "false" |0>.
[iii] When A is "false" **TGGATC** and B is "true" **ACCTAG**, the output of |S> is "true" |1> and |C> is "false" |0>.
[iv] When A is "true" **ACCTAG** and B is "false" **TGGATC**, the output of |S> is "false" |0> and |C> is "true" |1>.

8.3   DNA-Quantum Full Adder

A full adder is a combinational circuit that is created to address the shortcomings of the half adder circuit. It takes three inputs together to add them together and produces two outputs. In this case, the quantum and DNA circuits are used to implement

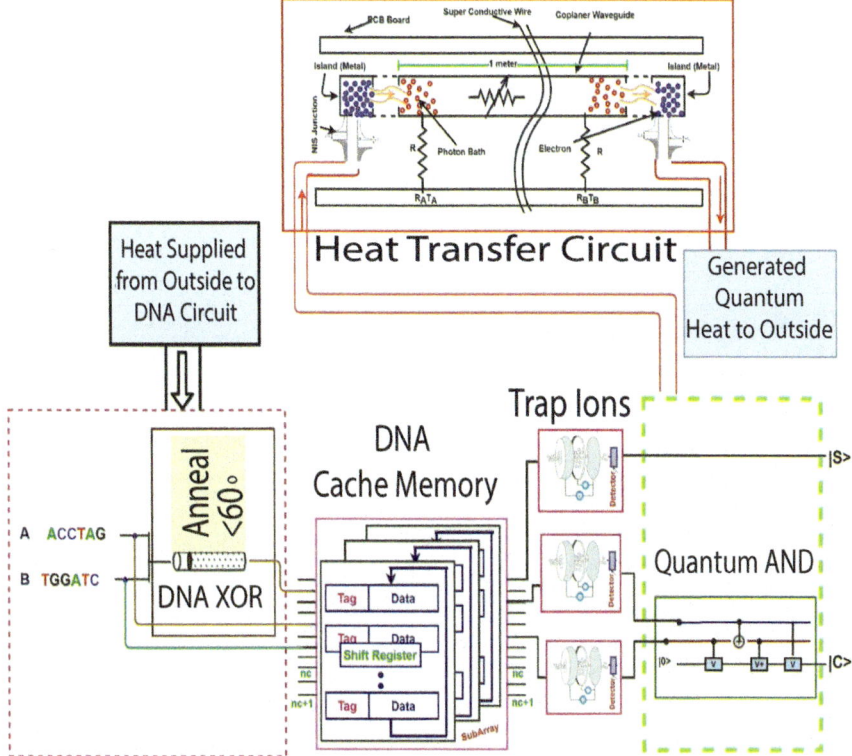

Fig. 8.2 DNA-quantum half adder circuit

the full adder. In other way, a full adder in DNA-quantum computing refers to the implementation of a full adder circuit, which adds three one-bit binary numbers, using DNA-based quantum computing techniques. This means that instead of using traditional electronic components like transistors, the logic gates and operations needed for a full adder are performed using the principles of quantum mechanics and the manipulation of DNA molecules. This field explores the potential of using DNA molecules and their properties, like entanglement and superposition, to perform computations. In a DNA-quantum full adder, the inputs (A, B, C_{in}) and outputs (Sum, C_{out}) are encoded using specific DNA sequences. DNA-based logic gates, like AND, OR, and XOR, can be constructed using techniques like DNAzyme computing and molecular recognition. DNA-quantum computing has the potential to offer advantages over traditional computing, including:

  i) Energy Efficiency: DNA-based computation might require less energy compared to electronic circuits;

 ii) Scalability: DNA molecules can be manipulated at a nanoscale, allowing for the potential to create very dense and complex circuits; and

Table 8.2 Truth table of a DNA-quantum full adder

Inputs			Outputs	
A	B	Cin	|S>	|Cout>
TGGATC	TGGATC	TGGATC	|0>	|0>
TGGATC	TGGATC	ACCTAG	|1>	|0>
TGGATC	ACCTAG	TGGATC	|1>	|0>
TGGATC	ACCTAG	ACCTAG	|0>	|1>
ACCTAG	TGGATC	TGGATC	|1>	|0>
ACCTAG	TGGATC	ACCTAG	|0>	|1>
ACCTAG	ACCTAG	TGGATC	|0>	|1>
ACCTAG	ACCTAG	ACCTAG	|1>	|1>

iii) Quantum Properties: Leveraging quantum phenomena could potentially lead to faster and more powerful computational abilities.

Table 8.2 shows the truth able for the DNA-quantum full adder.

From the above truth table, we get

$$S = A' B' Cin + A' B Cin' + A B' Cin' + A B Cin$$
$$= Cin (A' B' + A B) + Cin' (A' B + A B')$$
$$= Cin\ XOR\ (A\ XOR\ B)$$
$$Cout = A B + A Cin + B Cin (A + A')$$
$$= A B Cin + A B + A Cin + A' B Cin$$
$$= A B (1 + Cin) + A Cin + A' B Cin$$
$$= A B + A Cin + A' B Cin$$
$$= A B + A Cin (B + B') + A' B Cin$$
$$= A B Cin + A B + A B' Cin + A' B Cin$$
$$= A B (Cin + 1) + A B' Cin + A' B Cin$$
$$= A B + A B' Cin + A' B Cin$$
$$= AB + Cin (A' B + A B')$$

8.3.1 Block Diagram

The DNA-quantum circuit for a full adder is created by several DNA gates and quantum gates. To create a full adder, two XOR, two AND, and an OR operations are required. In this DNA-quantum circuit, one DNA XOR operation, two DNA AND operations (which is a combination of DNA NAND and DNA NOT), and one quantum XOR and one quantum OR operation are required to produce the SUM and the carry qubits. The block diagram of the DNA-quantum full adder is illustrated in Fig. 8.3.

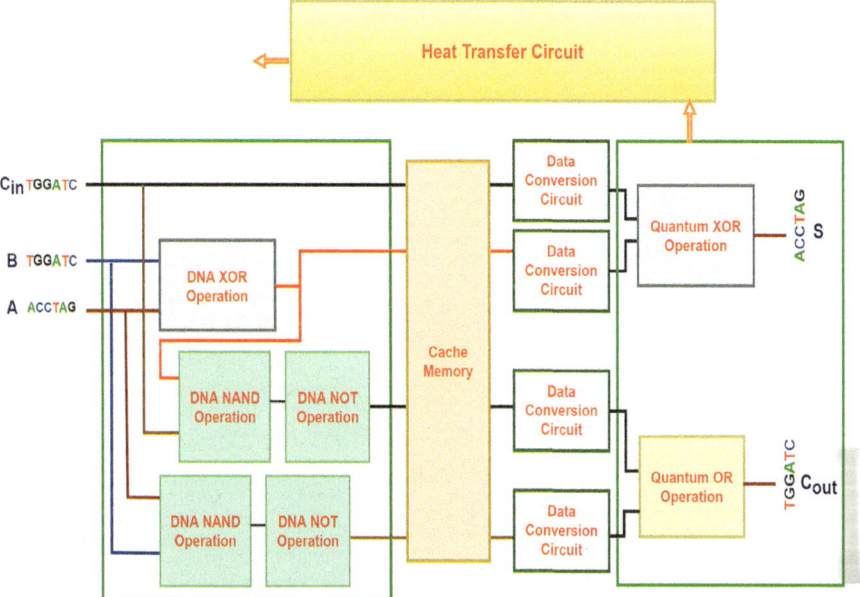

Fig. 8.3 Block diagram of a DNA-quantum full adder

8.3.2 Circuit Architecture

Figure 8.4 shows the DNA-quantum circuit of the full adder. In this full adder, the first XOR operation for A and B input sequences is done by a DNA XOR operation and creates output $A \oplus B$. This output, $A \oplus B$, is used as one of the inputs of the quantum XOR operation after going through the trap ion and other input comes from C_{in} which is converted into qubit by trap ion. Finally, the output S of the full adder is generated through this quantum XOR operation which represents the sum of the full adder.

In addition, two DNA NAND operations, two DNA NOT operations, and one XOR operation are needed to get another output of Cout. At first DNA NAND operation is performed between inputs A and B. Then, the output passes through a DNA NOT operation to produce $|A>.|B>$. As for the second DNA NAND operation, one input comes from $A \oplus B$, and another input is directly provided from Cin. The output of these two input sequences go to a NOT operation to generate $(|A> \oplus |B>).|Cin>$. Finally, the outputs of both DNA NOT operations pass through the trap ion in order to convert the molecular sequence, and after the conversion it performs a Quantum OR operation to produce the value of $|Cout>$.

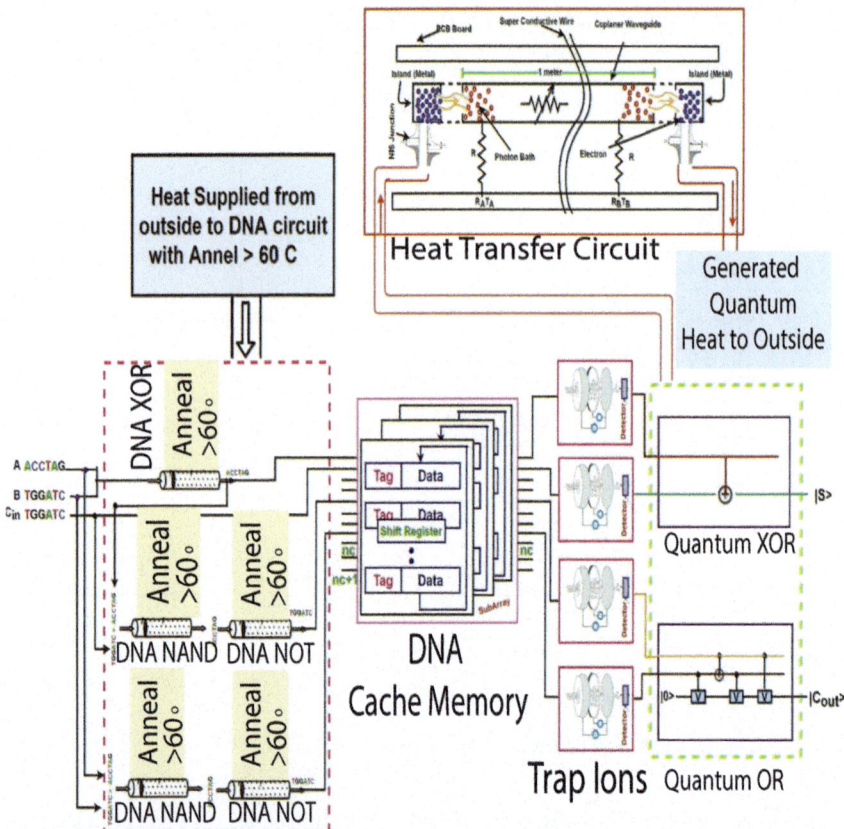

Fig. 8.4 DNA-quantum full adder circuit

8.3.3   Working Principle

The DNA-quantum full adder needs three inputs, A, B, and Cin to produce two outputs |S> and |Cout>. In order to understand the working principle of DNA-quantum full adder, consider the DNA values of inputs A, B, and Cin are **ACCTAG**, **TGGATC** and **ACCTAG**. Then,

[i] The output of DNA XOR-1 is "true" **ACCTAG,** where DNA AND-1 is "false" **TGGATC** and DNA AND-2 is "false" **TGGATC**.

[ii] Finally, the value of output |S> is "true" |1>, and the value of the |Cout> is "false" |0>.

8.4 DNA-Quantum Molecular Adder

This section describes the new concept of DNA-quantum molecular adder. In the field of bioquantum computing, a "molecular adder" refers to a system that uses the inherent quantum properties of molecules to perform arithmetic operations, specifically addition. This approach leverages the quantum behavior of biological molecules like DNA, proteins, and enzymes to achieve computations that are beyond the capabilities of classical computers. Molecular adders typically involve the manipulation and measurement of quantum states (qubits) within a molecular system. For instance, the "addition" operation might be achieved by encoding numerical values into the quantum states of a molecule and then using quantum logic gates (quantum operations) to manipulate these states to perform the addition. The inputs are DNA sequences and the outputs are qubits. DNA circuit needs extra heat to perform calculation. So, the extra heat from an external source is needed. On the other hand, the quantum circuit produces extra heat which is also needed to connect to the cooler to keep it cool.

8.4.1 Block Diagram

This circuit requires three XOR operations, three AND operations, and one OR operation in order to add the inputs of 2-molecular sequences A and B. Here, one DNA XOR, three DNA AND, two quantum XOR, and one quantum OR are required. Figure 8.5 illustrates the block diagram of the DNA-quantum 2-molecular-qubit adder.

Using the first approach, it is easy to construct the general organization of DNA-quantum 4-molecular adder which is shown in Fig. 8.6. The figure presents the 3 full-adder operations which are performed in the DNA system and the last full-adder operation is performed in the quantum system. A DNA cache memory is used to store the DNA base sequence output and the DNA sequence input for the last full-adder operation.

The molecular sequences are converted into the quantum qubits using the data conversion circuits. Besides, a heat transfer circuit is used to transfer heat from the quantum system to a cooling system. It is not right to use the generated heat from the quantum system because the DNA system operations will occur before the quantum system operations.

Figure 8.7 shows the general organization of quantum-DNA N-qubit parallel adder.

Figure 8.7 shows that $n - 1$ full-adder operations are performed in the DNA system and the last full-adder operation is performed in the quantum system. A DNA cache memory is used to store the DNA sequence output and the DNA sequence input for the last full-adder operation. The DNA base sequences are converted into qubits using the data conversion circuits. Besides, a heat transfer circuit is used to transfer heat from the quantum system to a cooling system.

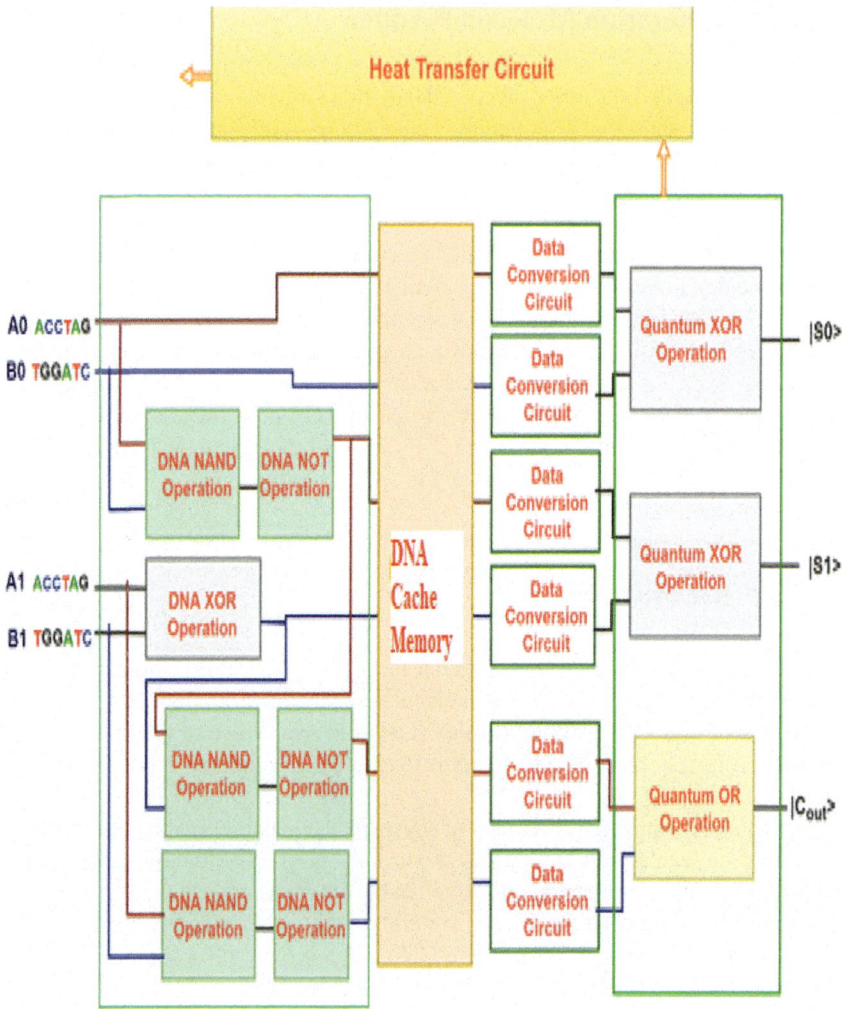

Fig. 8.5 Block diagram of 2-molecular DNA-quantum adder

8.4.2   *Circuit Architecture*

In the DNA-quantum 2-molecular adder circuit, A_0 and B_0 inputs perform a quantum XOR-1 operation to generate the output of S_0 after the data conversion using trap ion. These inputs also execute another DNA AND operation, a combination of DNA NAND operation and DNA NOT operation, whose output acts as an input for both quantum XOR-2 operation and DNA AND-2 operation, where inputs A1 and B1 pass through the DNA XOR-1 operation to produce A1 ⊕ B1. This A1 ⊕ B1 then goes to the DNA XOR-2 operation and DNA AND-2 operation as the second input.

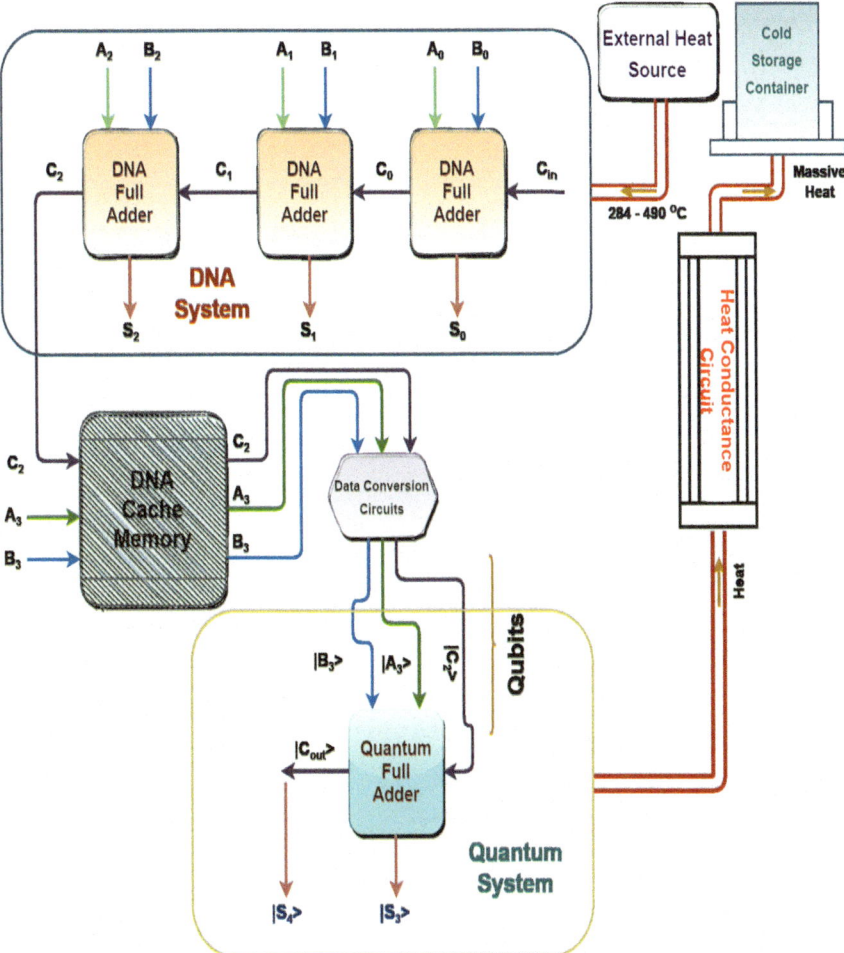

Fig. 8.6   Block diagram of DNA-quantum 4-molecular adder

The second quantum XOR then generates the output of |S1>, and the output of the DNA AND-2 operation goes to the DNA OR process in which the other input comes from the DNA AND-3 operation of A1 and B1 DNA sequences. Finally, the quantum OR operation gives the output of |Cout>. The whole architecture of the 2-Molecular DNA-quantum adder circuit is shown in Fig. 8.8.

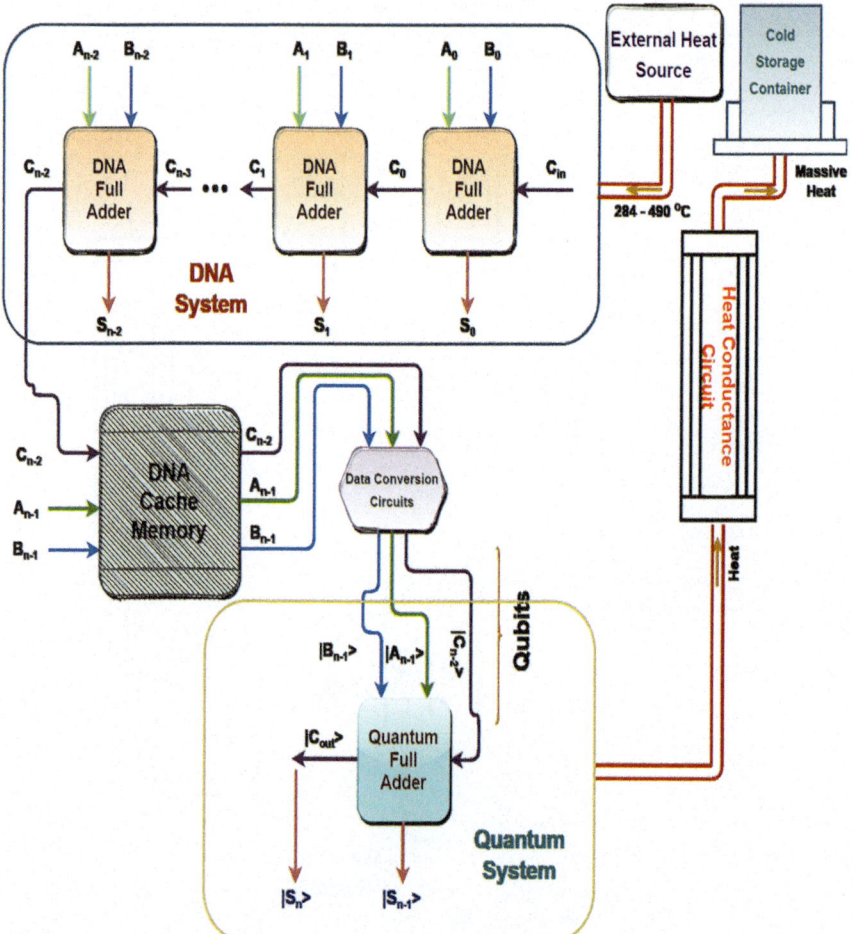

Fig. 8.7 Block diagram of DNA-quantum N-molecular adder

8.4.3 Working Principle

The DNA-quantum 2-molecular adder needs 2-molecular sequence inputs, A and B and Cin and provides three outputs |S0>, |S1> and |Cout>. In order to understand the working principle of the DNA-quantum full adder, consider the values of inputs A0, A1 and B0 and B1 which are **ACCTAG**, TGGATC, **ACCTAG**, and **ACCTAG**, respectively. Then,

[i] The output of quantum XOR-1 is |S0> which is "false" |0>.

[ii] The output of |S1> is "false" |0> **as** DNA AND-1 is "true" **ACCTAG** and DNA XOR-2 is "true" **ACCTAG**. Thus, DNA XOR-3 will be "false" **TGGATC**.

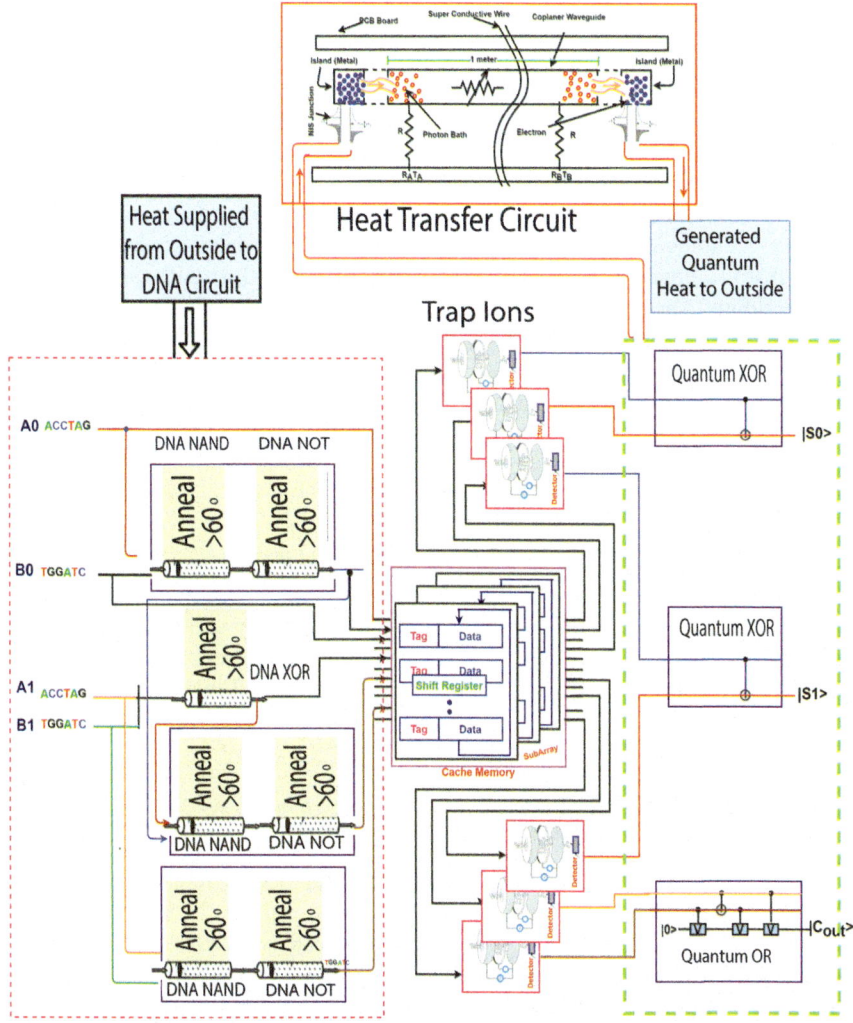

Fig. 8.8 2-Molecular DNA-quantum adder circuit

[ii] The output of |Cout> is "true" **|1>** as DNA AND-3 is "true" **ACCTAG** and DNA AND-2 is "false" **TGGATC**. Thus, DNA OR-1 will be "true" **ACCTAG**.

8.5 DNA-Quantum BCD Adder

The BCD adder is utilized in PC frameworks and adding machines that perform mathematical tasks in the decimal number framework straightforwardly. The equivalent qubits corresponding to the binary-coded form of decimal numbers are accepted by the BCD adder. A threshold of nine inputs and five outputs are required for the decimal-adder. In another way, a BCD (Binary Coded Decimal) adder is a circuit that adds two BCD numbers and produces a BCD result. It handles each decimal digit separately, ensuring the sum of each digit pair doesn't exceed 9. The term "DNA-Quantum BCD Adder" suggests exploring the implementation of a BCD adder using principles of DNA computing and quantum computing. Research in this area is exploring various approaches to design and implement BCD adders using DNA and quantum technologies. Some researchers have focused on designing reversible BCD adders that are quantum cost-efficient. Others have investigated the use of quantum algorithms to accelerate BCD addition. The DNA-Quantum BCD adder concept is a research area that explores the possibilities of combining DNA and quantum computing to implement BCD adders. It could lead to new and innovative designs for BCD adders with potential advantages in terms of speed, energy efficiency, and computational power.

8.5.1   Block Diagram

For the construction of the BCD adder, one DNA 4-molecular adder and one 4-qubit quantum adder are required along with two DNA AND operations and two DNA OR operations. The block diagram of the BCD adder is given in Fig. 8.9.

8.5.2   Circuit Architecture

Figure 8.10 shows the DNA-quantum circuit of the BCD adder. Here, the four outputs of the first 4-Molecular DNA adder directly transfer to trap ion, then the next 4-qubit quantum adder, as the inputs of A. However, in between, two DNA AND operations and two DNA OR operations are performed. The values of S1 and S3 perform AND-1 operation, whereas, AND-2 operation is operated on S2 and S3. In the next step, both the outputs of AND operation along with the Cout of the first 4-Molecular DNA adder sequentially execute two OR operations. Then, the result of the OR-2 operation acts as the inputs for the B of the last 4-qubit quantum adder.

8.5.3   Working Principle

In order to understand the working principle of DNA-quantum BCD adder, consider the values of inputs A0, A1, A2, and A3 corresponding to **ACCTAG, TGGATC, TGGATC** and **ACCTAG**, which are equal to decimal value "6"; and B0, B1, B2, and B3 corresponding to **TGGATC, ACCTAG, TGGATC** and **ACCTAG**, which are equal to decimal value '5'. Then

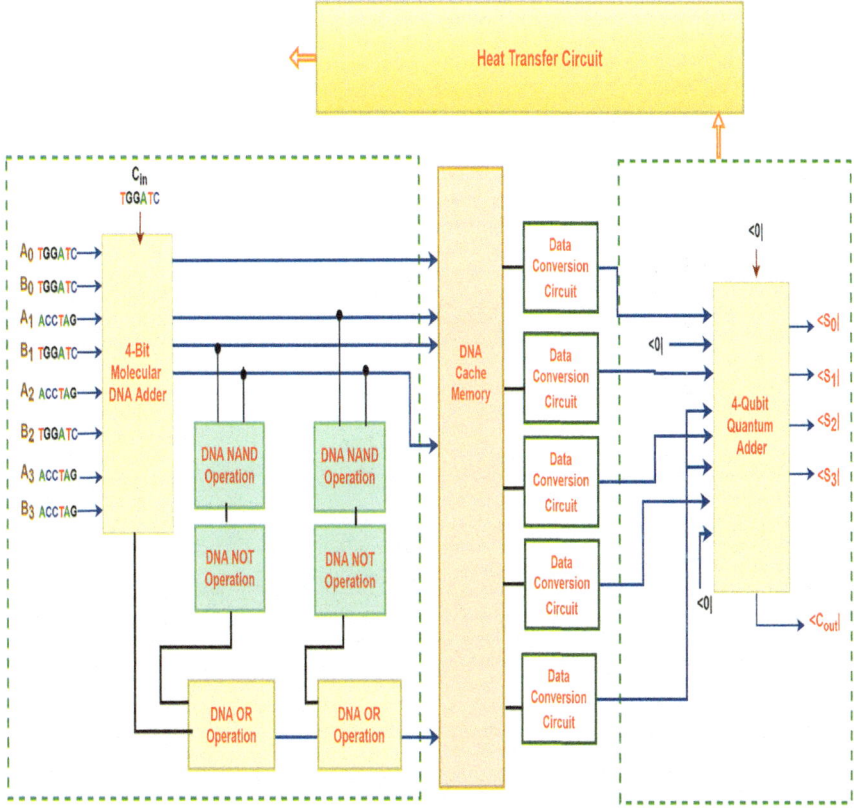

Fig. 8.9 Block diagram of DNA-quantum BCD adder

[i] The outputs of the first 4-molecular sequence adder will be S0 = **TGGATC**, S2 = **TGGATC**, S2 = **ACCTAG**, S3 = **TGGATC**, and Cout = **ACCTAG**.

[ii] Then, the result of two DNA AND operations are S2. S3 = **ACCTAG and** S1. S3 = **TGGATC**.

[iii] By continuing the above result, the final DNA OR operation generates "True" **TGGATC**.

[iv] Finally, the outputs of the last 4-qubit adder are |S0> = |**0>**, |S2> = |**1>**, |S2> = |**1>**, |S3> = |**1>**, and |Cout> = |**0>** which together form the BCD number of 11.

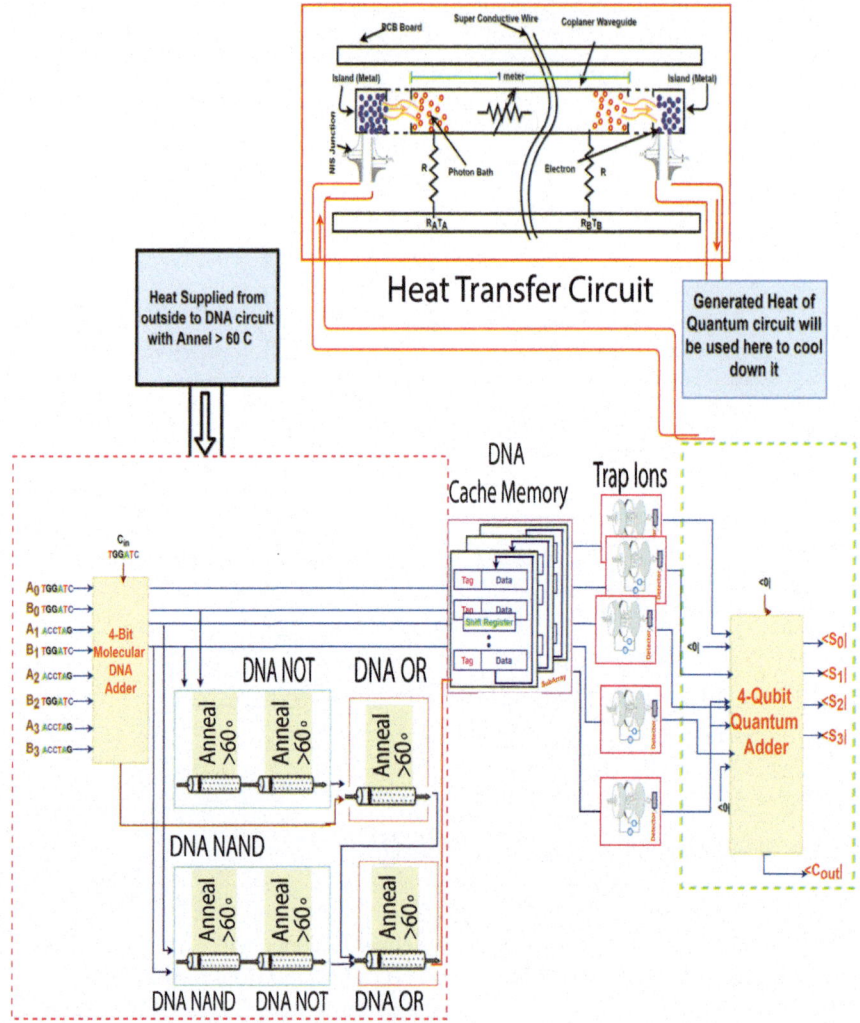

Fig. 8.10 DNA-quantum BCD adder circuit

8.6   DNA-Quantum Carry-Lookahead Adder

Figure 8.11 shows the general organization of a 4-molecular DNA-quantum carry-lookahead adder. Here, among 4-molecular input addition operations, the first three operations are performed in the DNA system, and the last addition operation is performed in the quantum system. The DNA cache memory is used to store the DNA sequences of the most significant qubit of the inputs and the carry value is obtained by the combinational DNA operations. Then, the DNA sequences are converted by data conversion unit (NMR process) into the equivalent qubits. The qubits perform

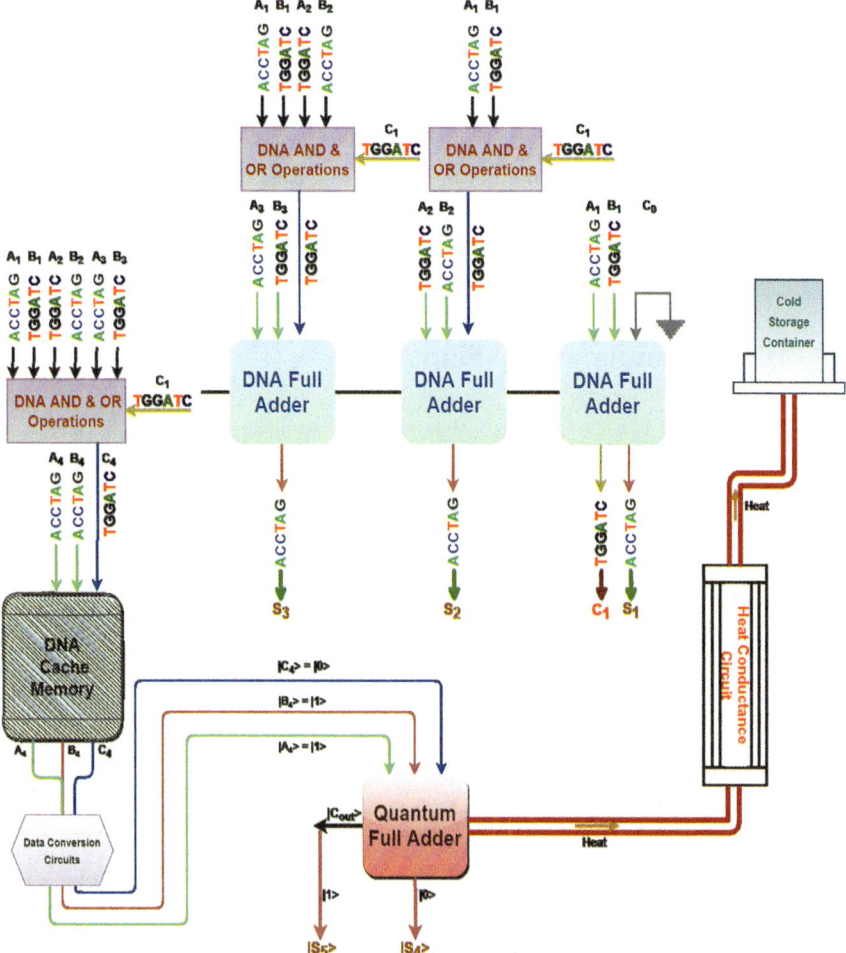

Fig. 8.11 4-Molecular DNA-quantum carry-lookahead adder

quantum full-adder operations and generate the outputs. The heat transfer circuit transfers the excessive heat from the quantum system. While carry-lookahead adders (CLAs) are primarily known for their application in traditional digital circuits, their versatility in reversible logic makes them a potential building block for quantum and DNA-based computing architectures.

8.6.1 The Circuit Architecture and Working Principle

The working procedures of the first part of the DNA-quantum system of the carry-lookahead adder are the DNA systems, which are being learned from the working procedures of the DNA carry-lookahead adder. As the general architecture shows that the first 3-Molecular DNA sequences addition and the carry input DNA sequence for the last addition operation are performed in the quantum system. Then, the carry DNA sequence along with the other two input DNA sequences (C_4, A_4, and B_4) are stored in the DNA cache memory. From the cache memory, they are passing through the NMR process to become the qubit input of the quantum full-adder. Then the quantum system will perform the quantum full-adder operation and will generate the final sum output qubits.

8.7   DNA-Quantum Carry-Skip Adder

A carry-skip adder (also known as a carry-bypass adder) is an adder implementation that improves on the delay of a ripple-carry adder with little effort compared to other adders. The improvement of the worst-case delay is achieved by using several carry-skip adders to form a block-carry-skip adder. In other way, a Carry-Skip Adder (CSA), also known as a Carry-Bypass Adder, is a type of adder that improves the performance of ripple-carry adders by skipping carries. While a direct implementation of a Carry-Skip Adder in a DNA-quantum setting might be challenging due to the complexity of bio-inspired quantum computing, the principles of carry skipping and parallel processing could be adapted. DNA-quantum computing explores the use of biological systems and quantum phenomena to perform computations, potentially leading to novel algorithms and computational models. CSAs could be explored in this context, particularly if biological components can be designed to mimic carry-skipping behavior. If a CSA-like architecture could be developed in a DNA-quantum system, it could potentially lead to faster addition operations, which could be beneficial for algorithms involving modular arithmetic or other numerical operations. Further research is needed to explore the feasibility and potential of using CSA principles within the realm of DNA-quantum computing. This would involve investigating how biological components can be engineered to efficiently perform carry-skipping operations.

The general overview to build the circuit diagram for the DNA-quantum carry-skip adder is depicted in Fig. 8.12.

From the above figure, it is easy to understand to build the DNA-quantum carry skip adder. By dividing it into two parts such as first part is in the DNA system which consists of connected DNA parallel adders, and the second part is in the quantum system which consists of a 2-to-1 quantum multiplexer that will bypasses the carry output of the block.

And a heat transfer circuit is connected to the quantum system to transfer the produced heat into the cooler.

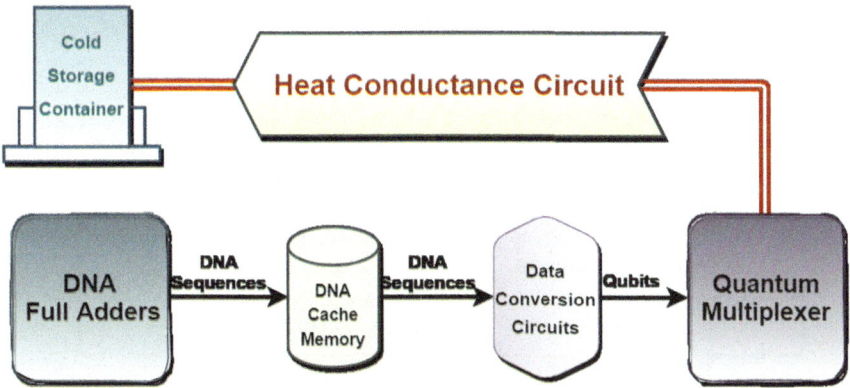

Fig. 8.12 Front-view of the DNA-quantum carry-skip adder

8.7.1 Block Diagram

Figure 8.13 shows the block diagram of a 4-bit DNA-quantum carry-skip adder, where the DNA system includes the four DNA full-adders. The first carry input to the block and carry output of the block are stored in a DNA cache memory so that they can be passed to the next operations.

The data conversion unit converts the data and delivers the qubits as the inputs to the quantum system which is a quantum 2-to-1 multiplexer. And the quantum multiplexer does the rest of the operations.

Figure 8.14 shows an inside view of the DNA-quantum 4-bit carry-skip adder. The value of P_i is determined first, and the block propagate is determined by the DNA AND operations accordingly. The value of block propagate BP, and the two carry DNA sequences such as C_0 and C_4(as the figure shows) will be stored to the DNA cache memory. The data conversion unit will convert them accordingly. The converted values will be the input of the quantum multiplexer which is generated as output of one DNA sequence. Note that, the block propagate BP is working as the select input of the quantum multiplexer. If BP is $|1>$, then the multiplexer will bypass the value of $|C_0>$ (1st carry input to the block) as output. Otherwise, the value of $|C_4>$ will be passed through the quantum multiplexer.

8.7.2 Circuit Architecture

The construction of the circuit of a DNA-quantum 4-molecular carry skip adder is done with the Table 8.3.

The design procedure is the same as explained before. The DNA system will have four DNA full-adders, the four DNA XOR operations to determine the value

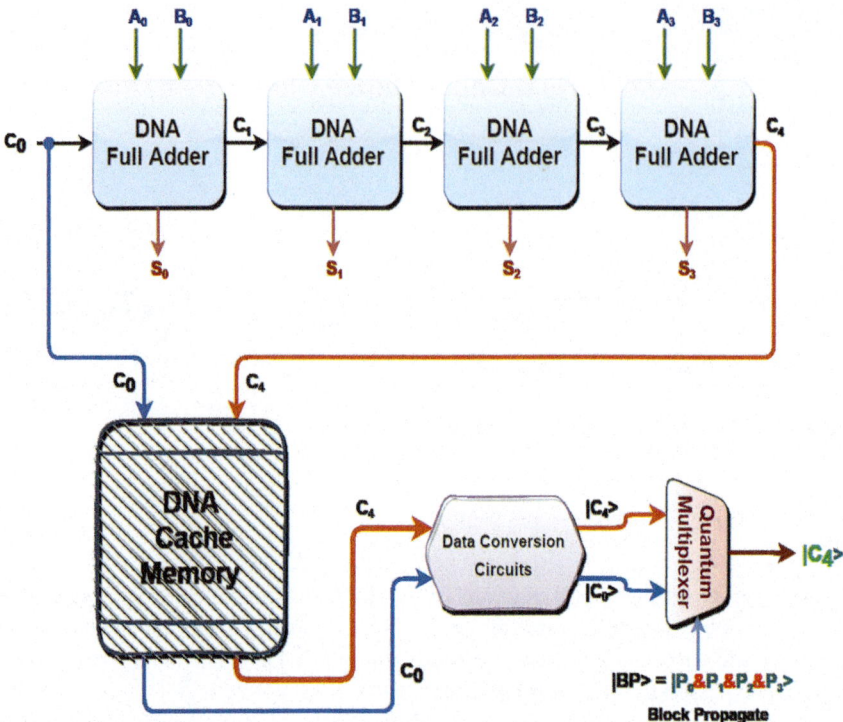

Fig. 8.13 Block diagram of the 4-bit DNA-quantum carry-skip adder

of P_i. It will also have three DNA AND operations to get the output of BP. The three-molecular sequence values (values of C_0, C_4, and BP) will be stored in the DNA cache memory. Here, NMR processes will be used to convert them into the corresponding qubits. So, the stored DNA sequences are passing through the NMR circuits and become the equivalent qubits. In the quantum system, the rest of the operations are performed according to quantum computing. Quantum multiplexer passes the output according to the select input qubit which is the value of BP. Thus, the bypassed output will be the carry input for the next cascaded block if it existed.

8.7.3 Working Principles

The working procedures of the DNA-quantum carry-skip adder include four main units such as DNA system, data storing to DNA cache memory, data conversion, and the quantum system. The operations in the DNA system and the quantum system discussed in the earlier chapters.

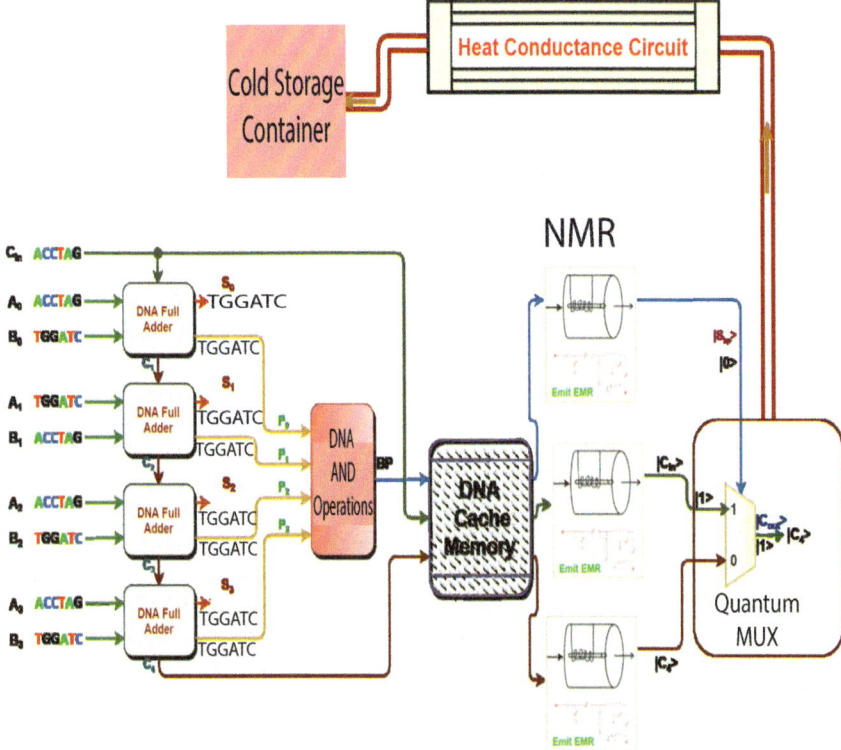

Fig. 8.14 Circuit architecture of the DNA-quantum 4-molecular carry-skip adder

8.8 DNA-Quantum Half Subtractor

A DNA-quantum half subtractor refers to a computational system that combines DNA-based molecular computing with quantum information processing to perform subtraction on two bits. It leverages DNA's ability to store and manipulate information, while using quantum principles for efficient computation. One approach is to use DNA as a physical platform for implementing quantum circuits or qubits. Another is to design DNA-based systems that can interact with quantum systems to perform computations. By leveraging the unique properties of both DNA and quantum mechanics, researchers hope to create novel computing architectures that can solve complex problems. In essence, a DNA-quantum half subtractor is a hypothetical system that combines the power of DNA-based computing with quantum mechanics to perform binary subtraction. It represents an area of ongoing research and development in the field of quantum biocomputing. The DNA-quantum half subtractor can also be used to subtract two numbers. It is equipped with two inputs and two outputs. This circuit is used to subtract two numbers with a single molecular

Table 8.3 Truth table of a DNA-quantum half subtractor

Inputs		Outputs	
A	B	$|D>$	$|Bout>$
TGGATC	TGGATC	$|0>$	$|0>$
TGGATC	ACCTAG	$|1>$	$|1>$
ACCTAG	TGGATC	$|1>$	$|0>$
ACCTAG	ACCTAG	$|0>$	$|0>$

sequence, A and B; and generates two quantum output states: 'D' and 'Bout.' The truth table for the DNA-quantum half subtractor is given in Table 8.3.

From the above truth table, we get

$D = A' B + A B'$
$= A \ XOR \ B$
$Bout = A' B$

8.8.1 Block Diagram

The DNA-quantum circuit for a half subtractor is created by a number of DNA operations and quantum operations. To design a half subtractor, one XOR operation, one NOT operation, and one AND operation are required. In this DNA-quantum circuit, one DNA XOR operation is needed to produce the difference of the half adder. Additionally, to generate the borrow qubit, one DNA NOT and quantum AND operation are used. The block diagram of the DNA-quantum half subtractor is illustrated in Fig. 8.15.

8.8.2 Circuit Architecture

Figure 8.16 depicted the DNA-quantum circuit of the half subtractor. In this half subtractor circuit, the XOR operation for A and B inputs is done by a DNA operation and creates an output of $|D>$. To generate the output of $|Bout>$, the DNA NOT operation is conducted on A. Then both inputs A' and B pass through a quantum AND operation. In order to match the speed of DNA and quantum operations, a DNA cache memory is used. To convert the molecular sequence into a quantum qubit, the trap ion method is applied.

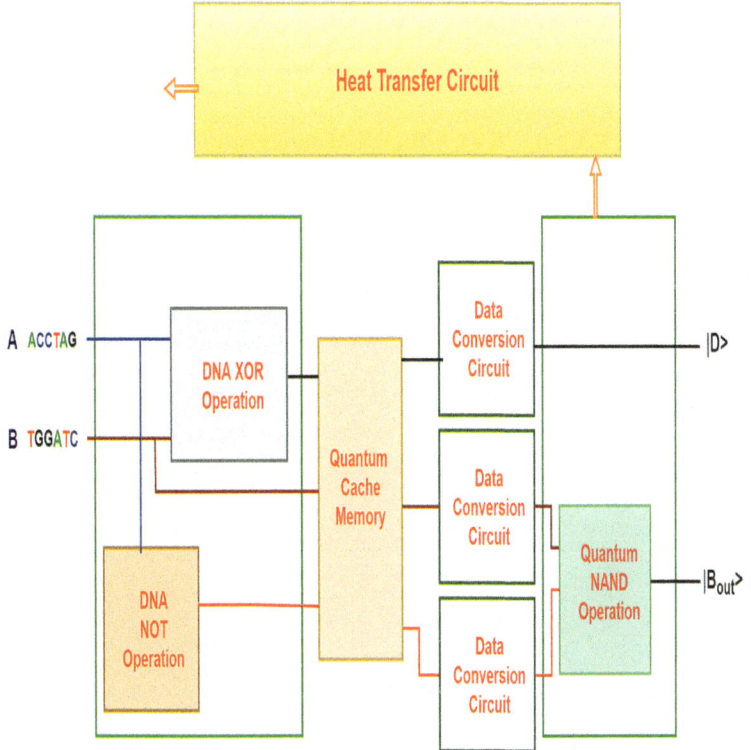

Fig. 8.15 Block diagram of DNA-quantum half subtractor

8.8.3 Working Principle

The DNA-quantum half-adder needs two inputs DNA sequences A and B. For various input values of A and B, consider the following cases:

[i] When both A and B are "true" **ACCTAG**, the outputs of both |D> and |Bout> are "false" |0>.

[ii] When both A and B are "false" **TGGATC**, the outputs of both |D> and |Bout> are "false" |0>.

[iii] When A is "false" **TGGATC** and B is "true" **ACCTAG**, the output of |D> is "true" |1> and |Bout> is "true" |1>.

[iv] When A is "true" **ACCTAG** and B is "false" **TGGATC**, the output of |D> is "true" |1> and |Bout> is "false" |0>.

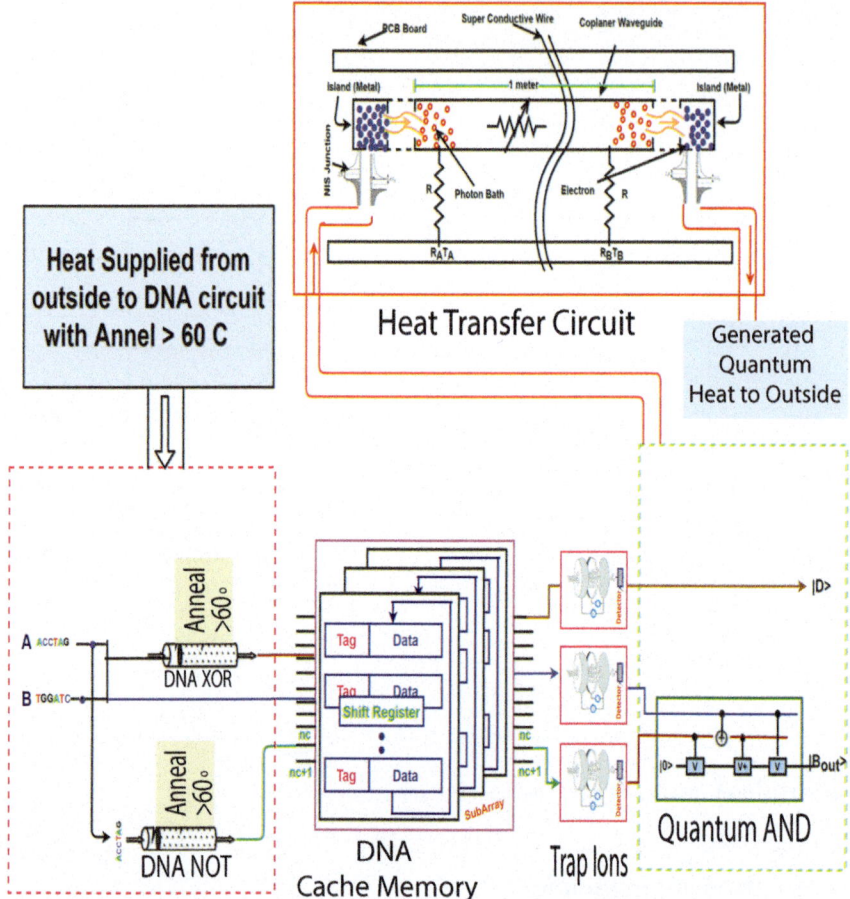

Fig. 8.16 DNA-quantum half subtractor circuit

8.9   DNA-Quantum Full Subtractor

The DNA-quantum full subtractor is a combinational circuit that performs subtraction on three input molecular sequences: minuend, subtrahend, and borrow in. The difference and borrow output qubits are generated by the full subtractor. In other way, a "DNA-Quantum Full Subtractor" refers to a subtractor circuit that combines DNA computing and quantum computing principles. It's a theoretical concept where DNA molecules are used to perform the logic operations of a full subtractor, while quantum entanglement or other quantum phenomena might be employed to enhance its functionality or performance. In essence, a DNA-quantum full subtractor is a hypothetical concept that combines the potential of DNA computing with quantum computing to create a highly efficient and potentially powerful subtractor circuit.

Table 8.4 Truth table of a DNA-quantum full subtractor

Inputs			Outputs	
A	**B**	**Cin**	**|D>**	**|Bout>**
TGGATC	TGGATC	TGGATC	|0>	|0>
TGGATC	TGGATC	ACCTAG	|1>	|1>
TGGATC	ACCTAG	TGGATC	|1>	|1>
TGGATC	ACCTAG	ACCTAG	|0>	|1>
ACCTAG	TGGATC	TGGATC	|1>	|0>
ACCTAG	TGGATC	ACCTAG	|0>	|0>
ACCTAG	ACCTAG	TGGATC	|0>	|0>
ACCTAG	ACCTAG	ACCTAG	|1>	|1>

However, it faces significant challenges in terms of design, implementation, and scalability. Table 8.4 shows the truth table of a DNA-quantum full subtractor.

From the above truth table, we get

$D = A'B'Bin + A'BBin' + AB'Bin' + ABBin$

$= Bin(A'B' + AB) + Bin'(AB' + A'B)$

$= Bin(A\ XNOR\ B) + Bin'(A\ XOR\ B)$

$= Bin (A\ XOR\ B)' + Bin'(A\ XOR\ B)$

$= Bin\ XOR\ (A\ XOR\ B)$

$= (A\ XOR\ B)\ XOR\ Bin$

$Bout = A'B'Bin + A'BBin' + A'BBin + ABBin$

$= Bin(AB + A'B') + A'B(Bin + Bin')$

$= Bin(A\ XNOR\ B) + A'B$

$= Bin (A\ XOR\ B)' + A'B$

8.9.1 Block Diagram

To design a full subtractor, two XOR, two AND, two NOT and an OR operations are required. In this DNA-quantum circuit, one DNA XOR operation and one quantum XOR are required to produce the Difference (|D>) of the full subtractor, and, as for |Bout>, two DNA AND operations, two DNA NOT, and one quantum OR operation are needed along with the quantum XOR operation. The block diagram of the DNA-quantum full subtractor is illustrated in Fig. 8.17.

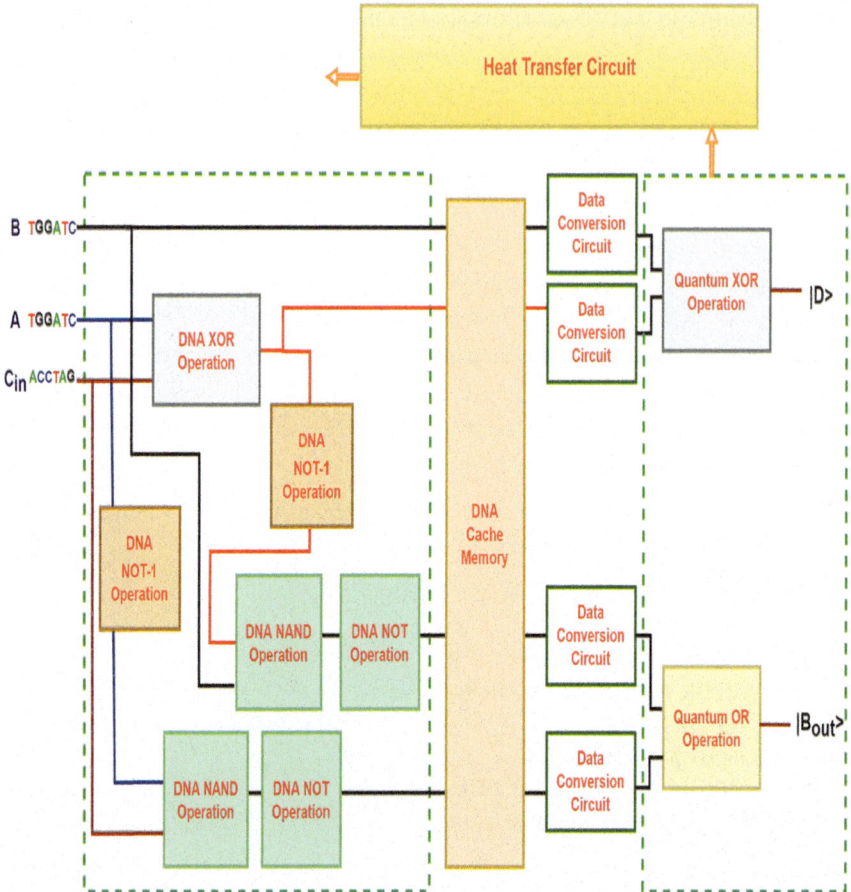

Fig. 8.17 Block diagram of DNA-quantum full subtractor

8.9.2 Circuit Architecture

Figure 8.18 shows the DNA-quantum circuit of the full subtractor. In this full sub-tractor, the first XOR operation for A and B input DNA sequences is done by a DNA XOR operation and creates the output A ⊕ B. This output, A ⊕ B, is then converted into quantum qubit using trap ion which is used as one of the inputs of the quantum XOR and other input comes from Cin which also are converted into qubit by trap ion. Finally, the output of |D> of the full subtractor is generated through this quantum XOR operation which represents the difference of the full subtractor. In addition, two DNA NAND operations, four DNA NOT operations, and one quantum

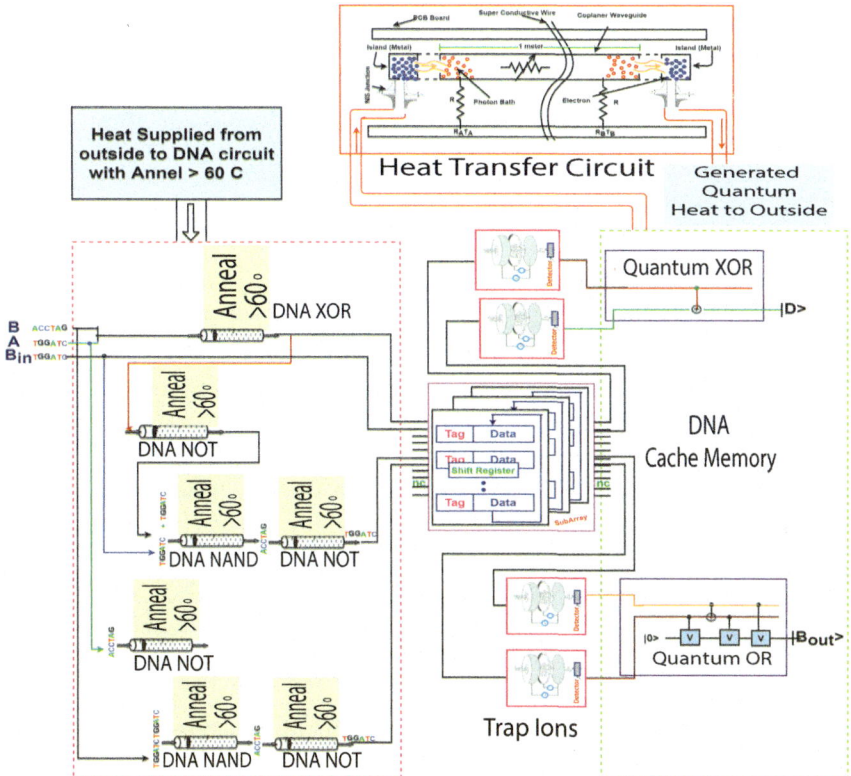

Fig. 8.18 DNA-quantum full subtractor circuit

operation are needed to get another output of Bout. First DNA NAND operation is performed between input A′ and B and then the output passes through a DNA NOT operation to produce |A′>.|B>. As for the second DNA NAND operation, one input comes from (A ⊕ B)′, and another input is directly provided from Cin. The output of these two input sequences goes to a NOT operation to generate (|A> ⊕ |B>)′.|Cin>. Finally, the outputs of both DNA NOT operations pass through the trap ion in order to convert the molecular sequence and after the conversion it performs a quantum OR operation to produce the value of |Bout>.

8.9.3 Working Principle

The DNA-quantum full subtractor needs three inputs, A, B and Bin, and provides two outputs D and Bout. In order to understand the working principle of DNA-quantum full subtractor, consider the values of input DNA sequences A, B and Bin are **ACCTAG**, **TGGATC**, and **ACCTAG**. Then,

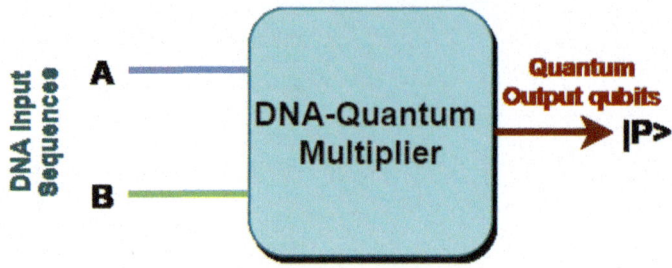

Fig. 8.19   General working block of a DNA-quantum 2 × 2 multiplier

[i] The output of DNA XOR-1 is "true" **ACCTAG**, when DNA AND-1 is "false" **TGGATC** and DNA AND-2 is "false" **TGGATC**.

[ii] Finally, the value of output |D> is "false" |0> after performing the quantum XOR operation; and |Bout> is "false" |0> after performing the quantum OR operation.

8.10   DNA-Quantum Multiplier

In the DNA-quantum multiplier, the DNA system will be implemented first and then the quantum system. A "DNA-Quantum Multiplier" is a conceptual approach that combines principles of DNA computing and quantum computing to perform multiplication operations. While not a fully developed technology, it explores the possibility of using DNA strands and their interactions as a foundation for quantum computations, including arithmetic operations like multiplication. In summary, the concept of a DNA-quantum multiplier is a promising area of research that explores the potential of combining DNA computing and quantum computing for arithmetic operations. While still in its early stages, it could offer significant advantages in terms of parallel processing and speed for specific calculations, but faces challenges related to scalability, coherence, and complexity. Figure 8.19 shows the block diagram of the multiplier.

Figure 8.19 depicts that the inputs of a DNA-quantum multiplier are the DNA-based sequences, and the product outputs of the inputs will be in the form of qubits. The constructions of a DNA multiplier and a quantum multiplier have already been discussed. Now, the connectivity in between the both multipliers will be discussed.

8.10.1 Circuit Architecture

Figure 8.20 shows the simplest form of the DNA-quantum 2×2 multiplier, where both the multiplicand and the multiplier are 2-molecular sequences.

The circuit shows that the DNA system has three DNA AND operations (which are actually three DNA NAND operations, each followed by one DNA NOT operation). The output of the DNA AND operations and values of A_0 and B_0 are stored in the DNA cache memory and from there, they are passed through the NMR process for the data conversion. Afterwards, the converted qubits are entered into the quantum system.

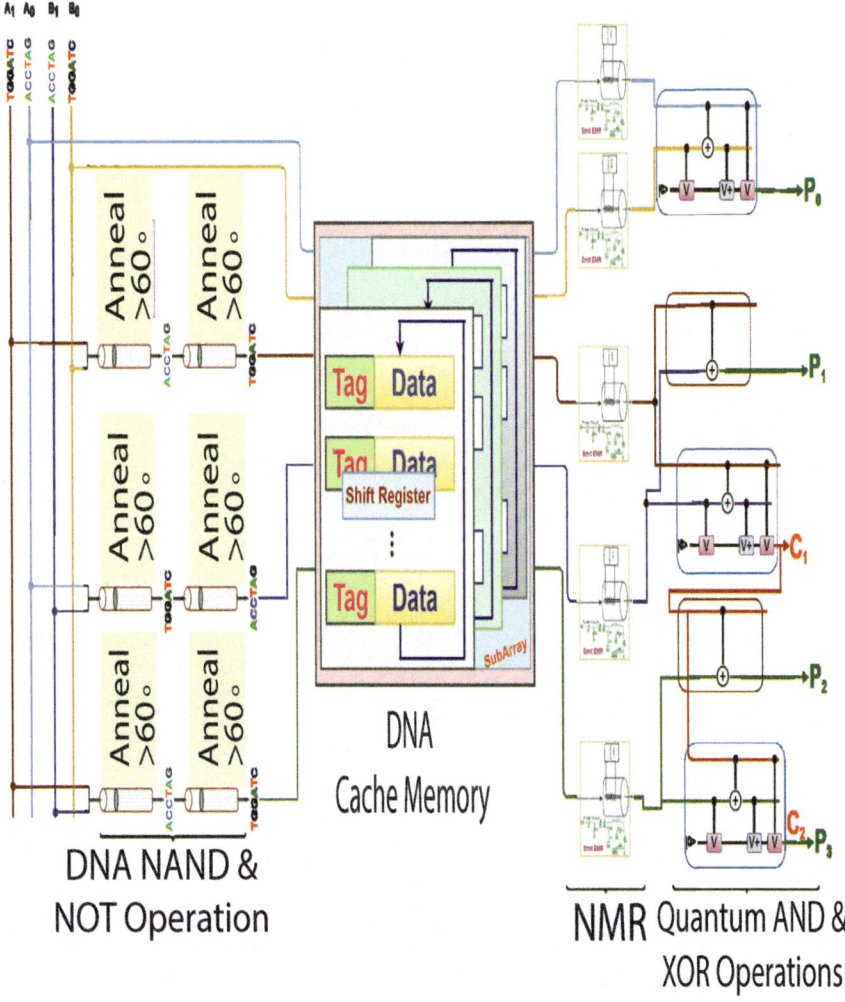

Fig. 8.20 Circuit diagram of a quantum-DNA 2×2 multiplier with the DNA cache memory and NMR process as data conversion unit

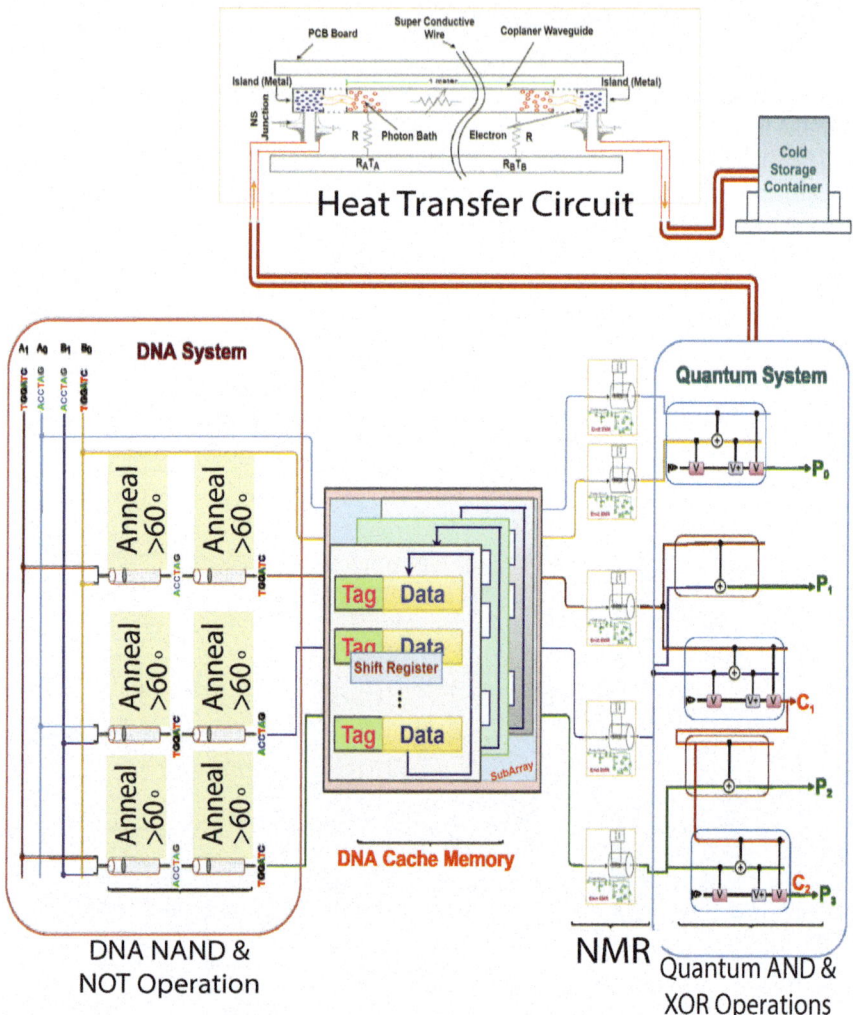

Fig. 8.21 Complete circuit diagram of a DNA-quantum 2×2 multiplier with DNA cache memory, NMR processes, a heat conductance circuit, and a cooler

In the quantum system, two quantum XOR and three quantum AND operations are performed to generate the product value for the corresponding inputs.

A heat conductance circuit is added with the circuit, where Fig. 8.21 depicts the DNA-quantum 2×2 multiplier circuit, which comprises all of the necessary components including the cooler.

8.10.2 Working Principle

Let the inputs be A and B. Here, $A = A_1A_0$, and $B = B_1B_0$. Suppose, $A = 10$ and $B = 10$.

Therefore, A_0 = TGGATC, A_1 = ACCTAG, B_0 = TGGATC, and B_1 = ACCTAG.

1. In the DNA system, the 1st DNA AND operation will receive inputs A_1 = ACCTAG, and B_0 = TGGATC, which will produce the output TGGATC, and it will be stored in the DNA cache memory.
2. The 2nd DNA AND operation will receive inputs A_0 = TGGATC, and B_1 = ACCTAG, which will again produce the output TGGATC, and it will be stored in the DNA cache memory.
3. The 3rd DNA AND operation will receive inputs A_1 = ACCTAG, and B_1 = ACCTAG, which will produce the output ACCTAG, and it will be stored in the DNA cache memory.
4. Now, there are five DNA sequence values which are stored in the DNA cache memory such as the values of A_0 and B_0, and the three DNA AND output's values. They will be passed through the NMR processes, and they will be converted into their corresponding quantum bits.
5. The values of A_0 and B_0 are TGGATC. Therefore, they will be converted as qubit $|0>$. In the quantum system, they will be working as the input of the 1st quantum AND operation. So, these qubits will generate the first product result $|P_0> = |0>$.
6. Next, the quantum XOR operation will receive both inputs as $|0>$ (the converted value from Step 1 and Step 2) and it will produce the output sequence $|0>$, which is the second product value of the multiplication operation. Therefore, $|P_1> = |0>$.
7. Then, the 2nd quantum AND operation will be performed with the same input of Step 6, which will generate the carry output qubit as $|0>$. This carry will work as an input to the next two steps.
8. The 2nd quantum XOR gets inputs $|0>$ (from Step 7) and $|1>$ (the converted qubit of the output from Step 3). So, it will generate $|1>$ as the output qubit. This is the value of $|P_2>$. Therefore, the third product value of the multiplication operation is $|P_2> = |1>$.
9. And finally, the last quantum AND operation will also get the inputs as the same as Step 8. As a result, it will generate $|0>$ as the output qubit, which will be the value of $|P_3>$.

So, after decoding the outputs, P = 0100, which is the expected product result.

8.11 DNA-Quantum Divider

The quantum and DNA divisions have already been discussed. Now this section will describe how to construct a divider in the DNA-quantum environment. A "divider"

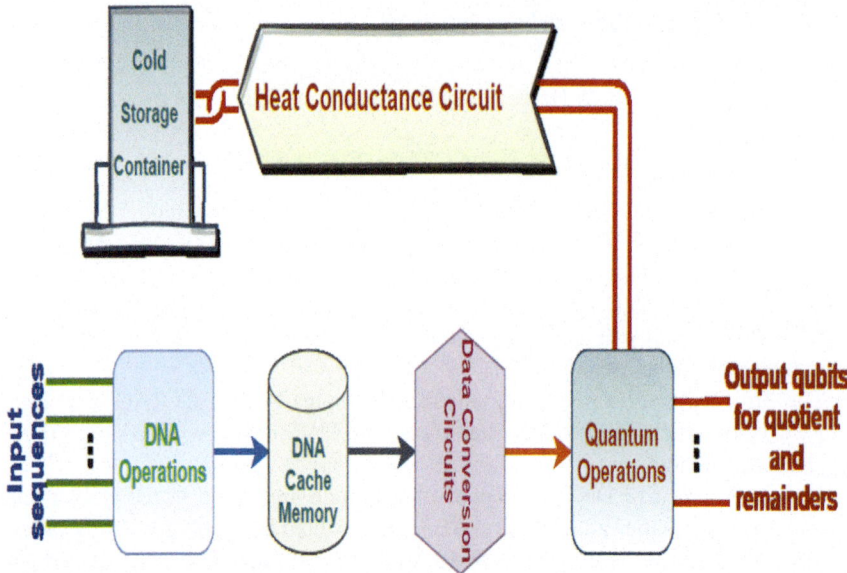

Fig. 8.22   General view of DNA-quantum division operation

in DNA-quantum computing refers to a computational circuit or algorithm that uses DNA molecules and quantum principles to perform division operations. This concept builds upon both traditional divider design and the potential of DNA-quantum computing to create novel computational approaches. DNA-quantum computing combines the use of DNA molecules for information storage and processing with quantum computing principles. In this context, a divider could be implemented by using DNA molecules to encode the operands (numbers to be divided) and then using quantum gates or algorithms to perform the division operation.

The block diagram of the DNA-quantum division operation is shown in Fig. 8.22. There is nothing surprising here. There will be two operational parts such as the DNA system and the quantum system. A DNA cache memory is used again to store the DNA sequences for a while. Data conversion circuits will convert the output from the DNA system to qubits and that will work as input to the quantum system. Finally, the quantum system will perform rest of the operations and will generate the resulted output.

8.11.1   Circuit Architecture

The design architecture of the DNA system has four DNA full-subtractors and one DNA 2-to-1 multiplexer. And the quantum system consists of two quantum NOT

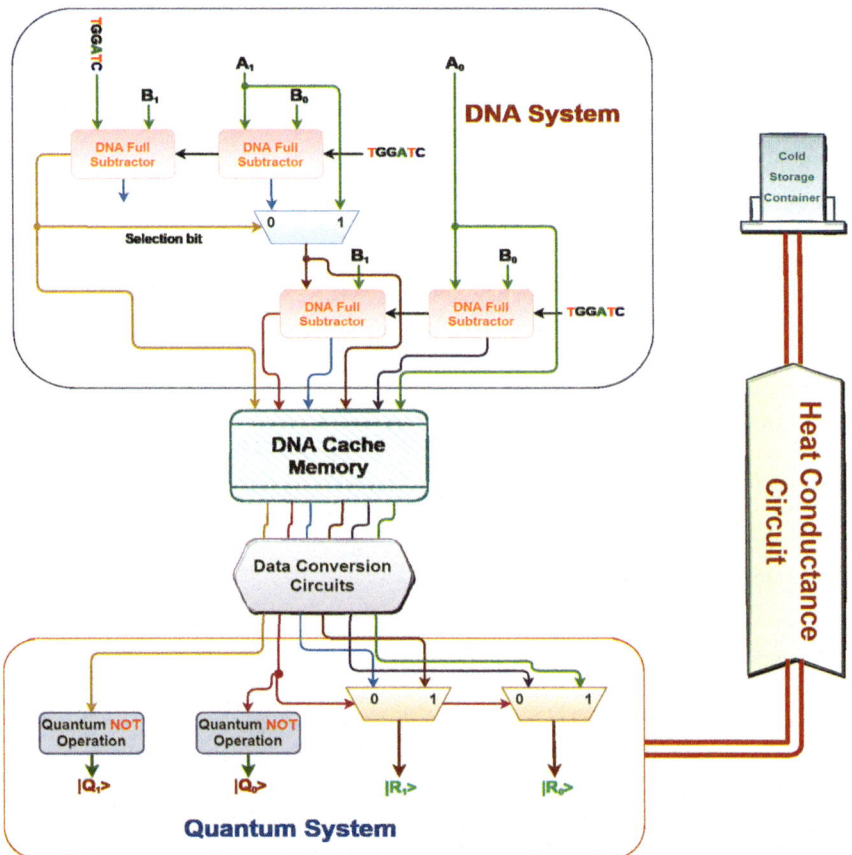

Fig. 8.23 General architecture of DNA-quantum 2 × 2 molecular divider

operations and two quantum 2-to-1 multiplexers. Figure 8.23 shows the general design architecture of a DNA-quantum 2 × 2 molecular divider.

The quotient output sequence is generated by the quantum NOT operations and the remainders of the division operation are generated from the quantum multiplexers operations. All of them are needed to connect the components accordingly. The next component is DNA cache memory, and the design architecture of the DNA cache memory, and its working mechanism have already described. Then the data conversion circuits, either NMR processes or trap ions to convert the DNA sequences to the equivalent qubits need to be used. And in the quantum system, it needs to connect the two quantum NOT operations with the converted input qubits as it is shown in the figure. The output from those quantum NOT operations will generate the quotients of the division operations. And then two quantum 2-to-1 multiplexers have to be connected accordingly.

8.11.2   Working Principle

In the earlier chapters, the working procedures of DNA divider and quantum divider have been discussed in details. So, the DNA system performs their operations as like before. Then the output from the DNA system will be stored in the DNA cache memory. And from the cache memory, the DNA sequences will be passed through the data conversion units and they will be converted into their corresponding qubit values eventually. Then the qubit inputs will enter into the quantum system. After performing the operations, the quantum system will produce the results as qubits. The quantum divider has also been discussed in detail in earlier chapters.

However, the massive heat generated by the quantum system will be transferred to a cooling system through a heat conductance circuit. The DNA system requires a specific amount of heat to perform any operation, but it is not possible to provide that heat from the quantum system, because the operations of the DNA system will be executed first. Therefore, the heat supply is needed from outside to the DNA system.

8.12   DNA-Quantum Comparator

A comparator that compares two input signals and gives output as which the input is larger or smaller or equal. In the context of DNA-based quantum computing, a comparator is a mechanism or algorithm that determines the similarity or difference between two DNA sequences. This is typically achieved by comparing the nucleotide sequences (A, C, G, T) at each position and quantifying the differences. Quantum computers can be used to speed up this comparison process by leveraging superposition and entanglement to explore multiple possibilities simultaneously. In essence, a comparator in DNA-quantum computing is a quantum algorithm designed to efficiently compare DNA sequences by leveraging the unique properties of quantum mechanics to explore the possibilities faster than classical methods. DNA-quantum 1-qubit comparator takes two single DNA input sequences, A and B. The values of the inputs can be **TGGATC** (0) and **ACCTAG** (1). This circuit gives three outputs $|X>$, $|Y>$, and $|Z>$, where $|X>$ is true, when A is smaller than B, $|Y>$ is true when A is equal to B and $|Z>$ is true when A is greater than B. Here, Table 8.5 shows the truth table of the 1-molecular comparator.

From the truth table, the equations for $|X>$, $|Y>$ and $|Z>$ are as follows:

$$|\mathbf{X>} = A^0.B^1$$
$$|\mathbf{Y>} = A^0.B^0 + A^1.B^1 \ = A \text{ XNOR } B; \text{ and}$$
$$|\mathbf{Z>} = A^1.B^0$$

Table 8.5 Truth table of 1-molecular comparator

Inputs		Outputs		
A	B	$\lvert X >$	$\lvert Y >$	$\lvert X >$
		$A < B$	$A = B$	$A > B$
TGGATC	TGGATC	$\lvert 0 >$	$\lvert 1 >$	$\lvert 0 >$
TGGATC	ACCTAG	$\lvert 1 >$	$\lvert 0 >$	$\lvert 0 >$
ACCTAG	TGGATC	$\lvert 0 >$	$\lvert 0 >$	$\lvert 1 >$
ACCTAG	ACCTAG	$\lvert 0 >$	$\lvert 1 >$	$\lvert 0 >$

8.12.1 Circuit Architecture

DNA-quantum 1-molecular comparator needs two DNA AND operations, two DNA NOT operations and one quantum XNOR operation (Fig. 8.24). DNA AND-1 is connected to A′ and B and it performs A′. B, which is the result of Z. In the same way, DNA AND-2 provides the output of X. These DNA sequence outputs are converted to quantum qubits using trap ion. Then, these converted qubits perform a quantum XNOR operation to get the value of Y.

8.12.2 Working Principle

The DNA-quantum comparator needs two DNA sequence inputs, A and B. For various input values of A and B, consider the following cases:

[i] When both A and B are "false" **TGGATC**, the output of $\lvert Y >$ will be "true" $\lvert 1 >$; and $\lvert X >$ and $\lvert Z >$ will be "false" $\lvert 0 >$; as both AND operations will produce $\lvert 0 >$.

[ii] When A is "false" **TGGATC** and B is "true" **ACCTAG**, the output of $\lvert X >$ will be "true" $\lvert 1 >$; and $\lvert Y >$ and $\lvert Z >$ will be "false" $\lvert 0 >$.

[iii] When A is "true" **TGGATC** and B is "false" **TGGATC**, the output of $\lvert Z >$ will be "true" $\lvert 1 >$; and $\lvert X >$ and $\lvert Y >$ will be "false" $\lvert 0 >$.

[iv] When both A and B are "true" **ACCTAG**, the output of $\lvert Y >$ will be "true" $\lvert 1 >$; and $\lvert X >$ and $\lvert Z >$ will be "false" $\lvert 0 >$ as both AND operations will produce $\lvert 0 >$.

8.13 Summary

The arithmetic operations in the quantum-DNA computing and DNA-quantum computing with the new cutting-edge technologies provide the benefits of execution speed

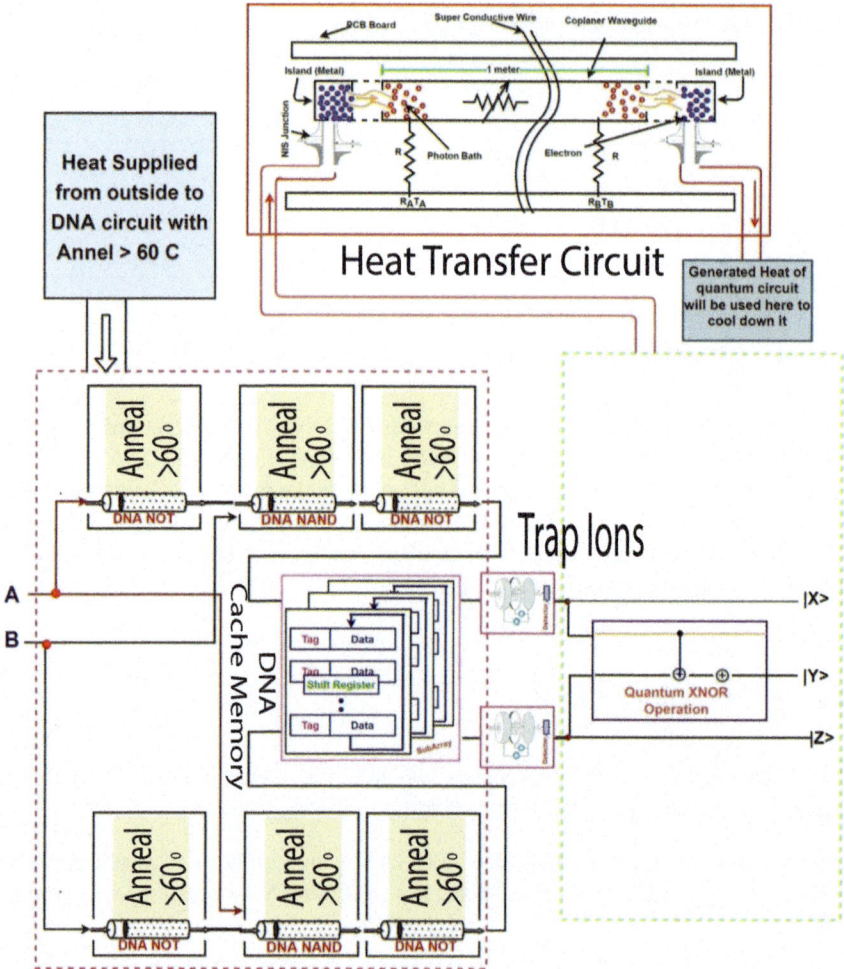

Fig. 8.24 DNA-quantum 1-molecular comparator circuit

and a large storage capability at a single computing system. The arithmetic operations can be implemented in the cross-platform environment with the same logic they belong to but requires more system resources. The required heat to perform arithmetic operations in the DNA system can be taken from the quantum system (the excessive heat is generated from the quantum system) in the quantum-DNA environment, where quantum operations are performed before the DNA operations. On the other hand, in DNA-quantum computing environment, the extra heat is needed to be supplied to the DNA part, where the quantum part is connected to the cooler to reduce the excessive heat from the quantum system.

Bibliography

1. J. Anders, D.K.L. Oi, E. Kashefi, D.E. Browne, E. Andersson, Ancilla-driven universal quantum computation. Phys. Rev. A **82**(2), 020301 (2010)
2. D. Auguin, V. Catherinot, T.E. Malliavin, J.L. Pons, M.A. Delsuc, Superposition of chemical shifts in NMR spectra can be overcome to determine automatically the structure of a protein. Spectroscopy **17**(2–3), 559–568 (2003)
3. F. Hobo, M. Takahashi, H. Maeda, S 33 NMR cryogenic probe for taurine detection. Rev. Sci. Instrum. **80**(3), 036106 (2009)
4. J.A. Jones, Quantum computing and nuclear magnetic resonance. PhysChemComm **4**(11), 49–56 (2001)
5. V.D. Kodibagkar, M.S. Conradi, Remote tuning of NMR probe circuits. J. Magn. Reson. **144**(1), 53–57 (2000)
6. Y. Takahashi, S. Tani, Power of uninitialized qubits in shallow quantum circuits. Theor. Comput. Sci. **851**, 129–153 (2021)

Part III
Combinational Circuits in Quantum Biocomputing

Overview

This part is about the design, architecture, and development of combinational circuits in quantum biocomputing which is a new and unique cutting-edge technology that will bring the changes in future tech-world. Quantum combinational circuits are circuits that utilize quantum mechanics to perform logical operations, similar to how classical combinational circuits use bits to perform calculations. They are a fundamental part of quantum computing, where qubits are manipulated through quantum gates to achieve a specific computational outcome. In summary, quantum combinational circuits are a powerful tool for exploring the potential of quantum computing. While challenges remain in their development, they offer the promise of solving problems that are currently beyond the reach of classical computers. On the other hand, in quantum-DNA combinational circuits or combinational circuits in quantum biocomputing, combining DNA-based logic gates with quantum phenomena, like superposition and entanglement, allows for the design of more powerful and complex quantum-DNA circuits. In essence, the concept of quantum-DNA combinational circuits or combinational circuits in quantum biocomputing provides a framework for building a new computational quantum systems that can utilize DNA's unique properties for performing various logic operations. This research area is pushing the boundaries of computing by exploring how biology can be harnessed to achieve new quantum computational capabilities. A combinational circuit consists of logic gates whose outputs at any instant of time are determined directly from the present combination of inputs without regard to previous input. Multiplexer, demultiplexer, encoder, and decoder are the four combinational circuits that will be discussed in this part of the book. The core part of these chapters is about the quantum and DNA circuit designs of various combinational circuits. Quantum computing is a new way of computing that is faster than any other super computer in the world, and DNA computing is a new branch of computing that replaces traditional electronic computing with DNA, biochemistry, and molecular biology hardware. Another new thing which is called quantum-DNA computing and DNA-quantum computing, a hybrid of DNA computing and quantum computing can bring change in the circuit's performance by taking the lead of the future Tech-World, and it is called the computations in quantum biology. In quantum-DNA computing, basic quantum gates and DNA gates are used. In addition, how the data of quantum circuits is transferred to the DNA circuits is also explained in details with the heat transfer technique. DNA-quantum computing is also a unique and new technology where two different technologies come together. The overall structure is the same as the quantum-DNA computing but opposite manner. In DNA-quantum computing, basic DNA gates and quantum gates are used just like quantum-DNA computing. In addition, how the data of DNA sequences are transferred to the quantum circuits is also explained in details with the heat transfer technique. So, this part will describe the combinational circuits in quantum computing, quantum-DNA computing, and DNA-quantum computing.

Chapter 9
Quantum Combinational Circuits

9.1 Introduction

Quantum computing is the development of computer technology based on the principles of quantum theory, which explains the behavior of energy and matter at the atomic and subatomic levels. Quantum computing is a new method of computation that promises to perform computational tasks and algorithms that are currently too difficult to perform on existing computing paradigms. Computers can only incorporate the information in bits that have a value of 1 or 0, limiting their ability. Quantum computing employs quantum bits, also known as qubits. Qubits are the fundamental unit of quantum computing for storing information. Qubits have three states: $|0>$ and $|1>$, as well as a third state called "superposition," which allows them to represent a one and a zero at the same time, or multiple states at the same time.

9.2 Design Architectures of Quantum Combinational Circuits

A combinational circuit is a quantum logic circuit in which the output depends on the combination of input qubits at that point in time with total disregard for the past state of the input qubits. The quantum logic gate is the building block of quantum combinational circuits. In other words, quantum combinational circuits are circuits that utilize qubits and quantum gates to perform computational tasks, similar to how classical combinational circuits use bits and logic gates. These circuits are a core component of quantum computing, enabling the manipulation and processing of quantum information. Four combinational circuits will be described here in quantum computing logic.

© The Author(s), under exclusive license to Springer Nature Singapore Pte Ltd. 2025   225
H. M. Hasan Babu, *Quantum Biocomputing in Quantum Biology Volume I*,
https://doi.org/10.1007/978-981-97-7154-7_9

Table 9.1 Quantum 2-to-1 multiplexer truth table

Inputs	Outputs
$|S>$	$|Y>$
$|0>$	$|I_0>$
$|1>$	$|I_1>$

Table 9.2 Quantum 4-to-1 multiplexer truth table

Inputs		Outputs
$|S1>$	$|S0>$	$|Y>$
$|0>$	$|0>$	$|I_0>$
$|0>$	$|1>$	$|I_1>$
$|1>$	$|0>$	$|I_2>$
$|1>$	$|1>$	$|I_3>$

9.2.1 Quantum Multiplexer

A multiplexer is a combinational circuit with 2^n input lines and a single output line, indicating that it is a multi-input and single-output combinational circuit. A quantum multiplexer is a circuit in quantum computing that selects one of multiple input quantum states and directs it to a single output. The primary purpose of a quantum multiplexer is to select which input quantum state is passed to the output. When the control qubit is in a specific state (e.g., $|0>$), the multiplexer connects one input to the output. When the control qubit is in a different state (e.g., $|1>$), it connects another input to the output. Quantum multiplexers are essential for implementing various quantum algorithms and circuits. For example, they can be used in quantum state preparation, quantum error correction, and other quantum information processing tasks.

9.2.1.1 Block Diagram

There are only two inputs, I_0 and I_1, one selection line, S, and one output, Y in a 2-to-1 multiplexer. Table 9.1 shows the truth table of a 2-to-1 multiplexer, and Table 9.2 shows the truth table of a 4-to-1 multiplexer.

From the truth table, we get

$$Y = S_0'.I_0 + S_0.I_1$$

Figure 9.1 presents the block diagram of a 2-to-1 multiplexer. For this circuit, one quantum NOT operation, two quantum AND operations, and one quantum OR operation are needed.

From the truth table, we get

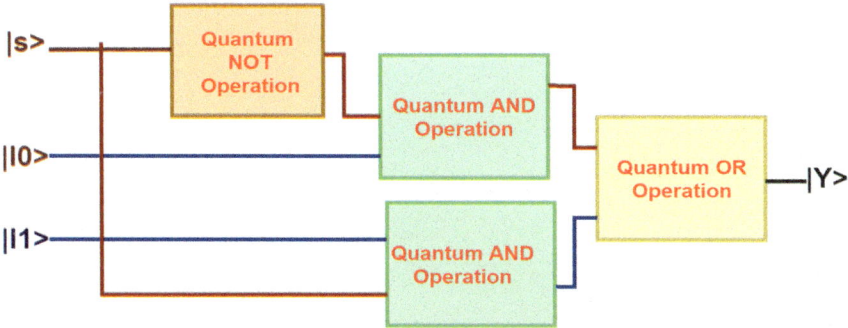

Fig. 9.1 Block diagram of a quantum 2-to-1 multiplexer

$$Y = S1'\ S0'\ I0 + S1'\ S0\ I1 + S1\ S0'\ I2 + S1\ S0\ I3$$

Figure 9.2 displays the block diagram of a 4-to-1 quantum multiplexer. In this circuit, two quantum NOT operations, eight quantum AND operations, and three quantum OR operations are needed.

9.2.1.2 Circuit Architecture

Figure 9.3 depicts the quantum circuit of the 2-to-1 multiplexer. Firstly, each input $|I0>$ and $|I1>$ separately connect with two quantum AND operations. The input of the selection line, $|S>$ is connected directly to the AND-1 operation, whereas, another input is connected to the AND-2 operation after completing the NOT operation. Finally, both the outputs of quantum AND operations are passed through quantum OR operation generating the value of $|Y>$.

Figure 9.4 gives the view of the 4-to-1 quantum multiplexer circuit. Firstly, each of input qubits $|I0>$, $|I1>$, $|I2>$, and $|I3>$ separately connect with four quantum AND operations. The other input qubits of these AND operations come from the output of another four quantum AND operations which are executed between two selection lines, $|S0>$ and $|S1>$. Finally, the outputs of AND-2, AND-4, AND-6, and AND-8 conduct three quantum OR operations to produce the value of $|Y>$.

9.2.1.3 Working Principle

Quantum 2-to-1 Multiplexer: The Quantum 2-to-1 multiplexer needs one selection line S0 and two input qubits, I0 and I1. When the input qubits I0-I1 are "true", $|1>$, consider the following cases:

[i] The output Y = I0, when the input qubit, S, is "false" $|0>$.
[ii] The output Y = I1, when the input qubit, S, is "true" $|1>$.

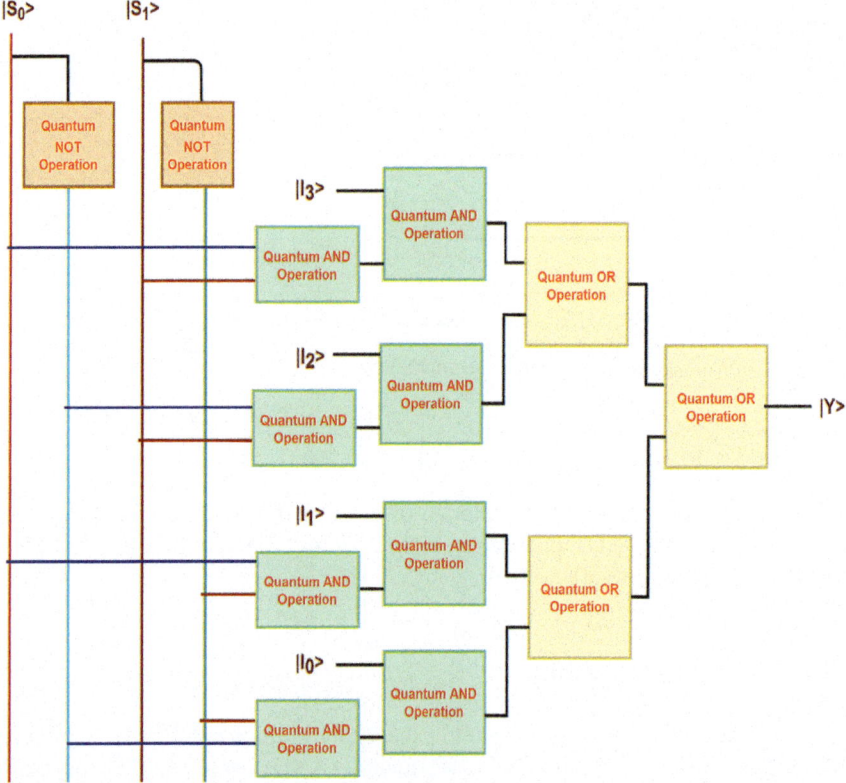

Fig. 9.2 Block diagram of a quantum 4-to-1 multiplexer

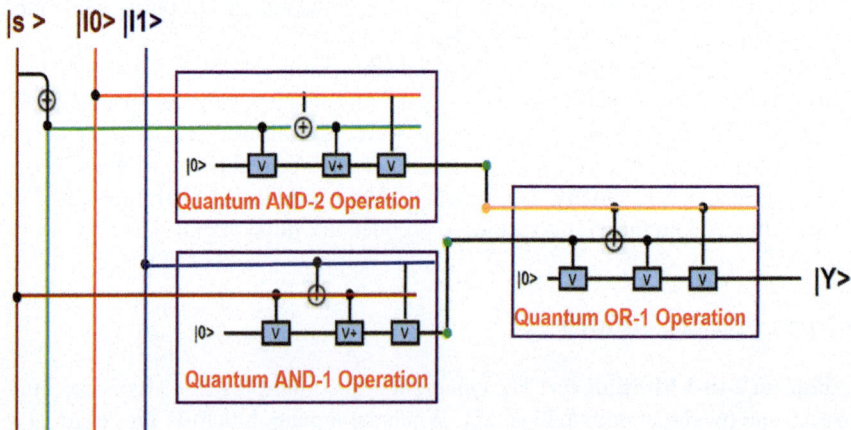

Fig. 9.3 Quantum 2-to-1 multiplexer circuit

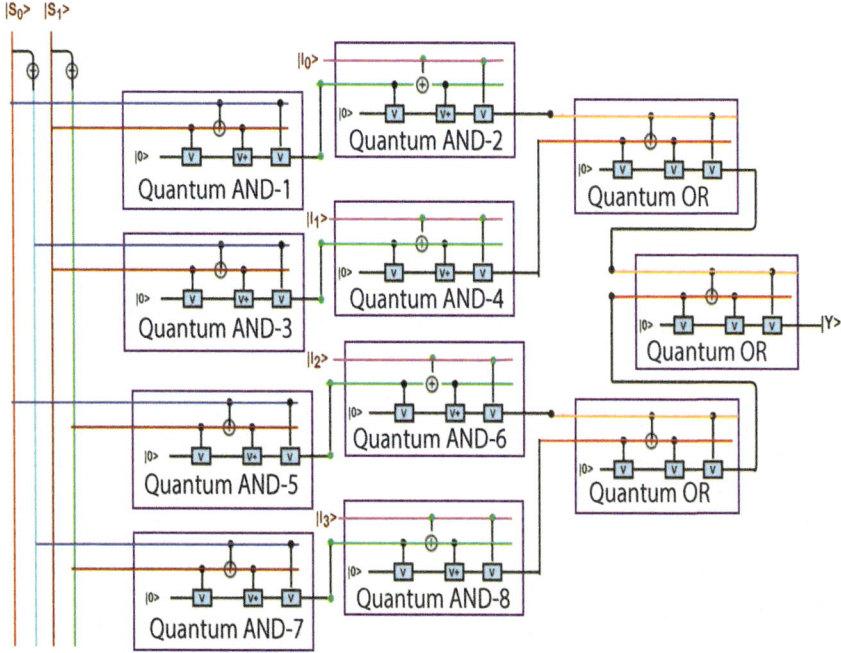

Fig. 9.4 Quantum 4-to-1 multiplexer circuit

Quantum 4-to-1 Multiplexer: The Quantum 4-to-1 multiplexer needs two selection lines S0 and S1, and four inputs, I0-I3. If all inputs I0-I3 are "true", |1>, then consider the following cases:

[i] The output Y = I0, when the input qubits S1 and S0 are "false" |0>.

[ii] The output Y = I1, when the input qubit S1 is "false" |0> and input S0 is "true" |1>.

[iii] The output Y = I2, when the input qubit S0 is "false" |0> and input qubit S1 is "true" |1>.

[iv] The output Y = I3, when the inputs qubits S0 and S1 are "true" |1>.

9.2.2 Quantum Demultiplexer

A demultiplexer is a combinational quantum circuit with a single input line and 2^n output lines, indicating that it is a single-input and multi-output combinational quantum circuit. A quantum demultiplexer is a quantum-mechanical device that distributes a single quantum state or signal to multiple output lines. In essence, a quantum demultiplexer takes a single quantum input and directs it to one of several potential output channels, determined by a set of control qubits. A quantum demultiplexer uses a set of control qubits to select which output line the quantum input should be routed to. The state of these control qubits dictates which of the multiple output lines is activated. Quantum demultiplexers are crucial for quantum communication networks, allowing for the distribution of quantum information to multiple receivers. They also find use in quantum computing, where they're essential for implementing quantum algorithms and control flow.

9.2.2.1   Block Diagram

There are only two input qubits, I0 and I1, one selection line, S, and one output, Y in a 2-to-1 quantum multiplexer. The truth table of the 1-to-2 quantum demultiplexer is given in Table 9.3 and the truth table of the 1-to-4 quantum demultiplexer is given in Table 9.4.

From the truth table, we get

I0 = S'.D
I1 = S.D

Figure 9.5 shows the block diagram of a 1-to-2 demultiplexer. For this circuit, one quantum NOT operation and two quantum AND operations are needed.

From the truth table, we get

Y0 = S1' S0' I
y1 = S1' S0 I
y2 = S1 S0' I
y3 = S1 S0 I

Table 9.3 Truth table of a 1-to-2 demultiplexer	Inputs	Outputs	
	$|S>$	$|I_1>$	$|I_0>$
	$|0>$	$|0>$	$|D>$
	$|1>$	$|D>$	$|0>$

Table 9.4   Truth table of 1-to-4 demultiplexer

Inputs		Outputs			
$|S1>$	$|S0>$	$|Y0>$	$|Y1>$	$|Y2>$	$|Y3>$
$|0>$	$|0>$	$|I>$	$|0>$	$|0>$	$|0>$
$|0>$	$|1>$	$|0>$	$|I>$	$|0>$	$|0>$
$|1>$	$|0>$	$|0>$	$|0>$	$|I>$	$|0>$
$|1>$	$|1>$	$|0>$	$|0>$	$|0>$	$|I>$

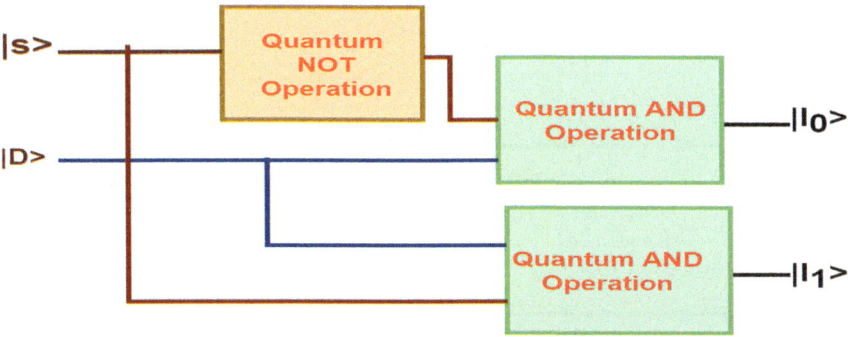

Fig. 9.5 Block diagram of quantum 1-to-2 demultiplexer

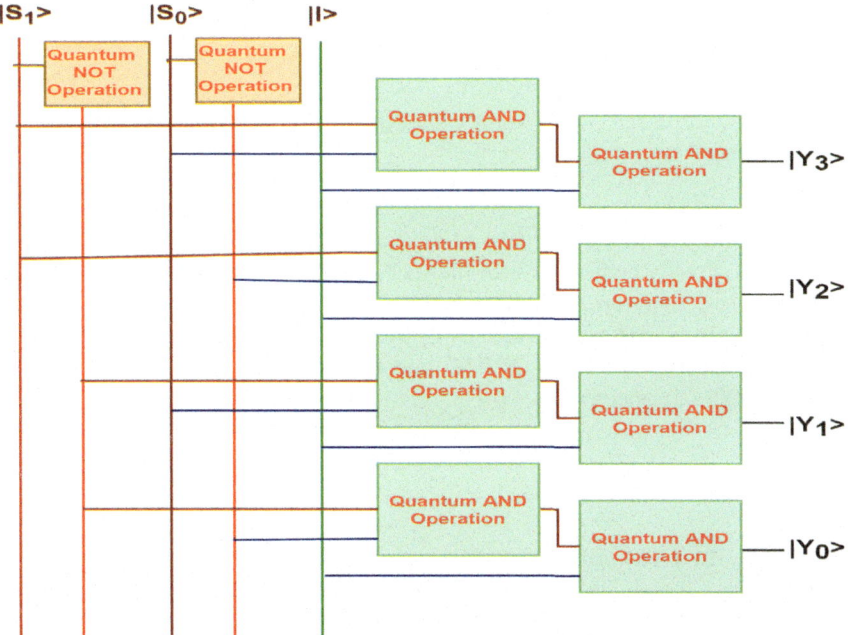

Fig. 9.6 Block diagram of quantum 1-to-4 demultiplexer

Figure 9.6 displays the block diagram of a 1-to-4 demultiplexer. In this circuit, two quantum NOT and eight quantum AND operations are needed.

9.2.2.2 Circuit Architecture

Figure 9.7 depicts the quantum circuit of the 1-to-2 demultiplexer. Firstly, input $|D>$ connects with two quantum AND operations directly, which is the same as selection input, $|S>$. Based on the values of $|S>$ and $|D>$, only one quantum AND operation is needed to generate the output of $|1>$.

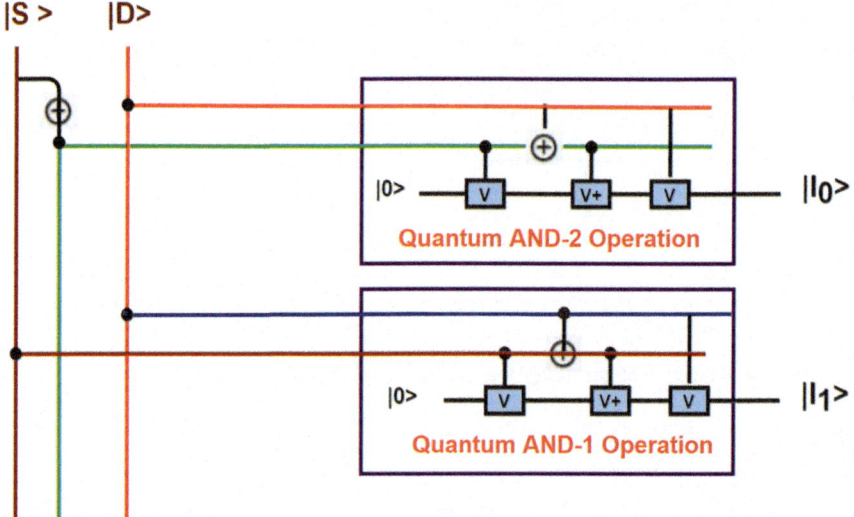

Fig. 9.7 Quantum 1-to-2 demultiplexer circuit

Figure 9.8 gives the view of the 1-to-4 quantum demultiplexer circuit. Firstly, the input |I> is directly connected with the four quantum AND operations. The other inputs of these AND operations come from the outputs of another four quantum AND operations which are executed by two selection lines, |S0> and |S1>. Finally, the outputs of AND-2, AND-4, AND-6, and AND-8 are the qubits of |Y0>, |Y1>, |Y2>, and |Y3>, respectively.

9.2.2.3   Working Principle

Quantum 1-to-2 Demultiplexer: The quantum 1-to-2 demultiplexer needs one selection line S0 and one input, D. For DNA 1-to-2 demultiplexer, when input D is "true", |1>, consider the following cases:

> **[i]** The output I0 = S'.D is "true" |1> when the input S is "false" |0>. I1 will be "false" |0>.
> **[ii]** The output I1 = S.D is "true" |1> when the input S is "true" |1>. I0 will be "false" |0>.

Quantum 1-to-4 Demultiplexer: The 1-to-4 quantum demultiplexer needs two selection lines S0 and S1, and one input, I. For 1-to-4 quantum demultiplexer, when the input I is "true", |1>, consider the following cases:

> **[i]** The output Y0 = S0'.S1'. I is "true" |1> when the inputs S1 and S0 are "false" |0>. The output lines Y0 will be "true" |1> and Y1 to Y3 will be "false" |0>.

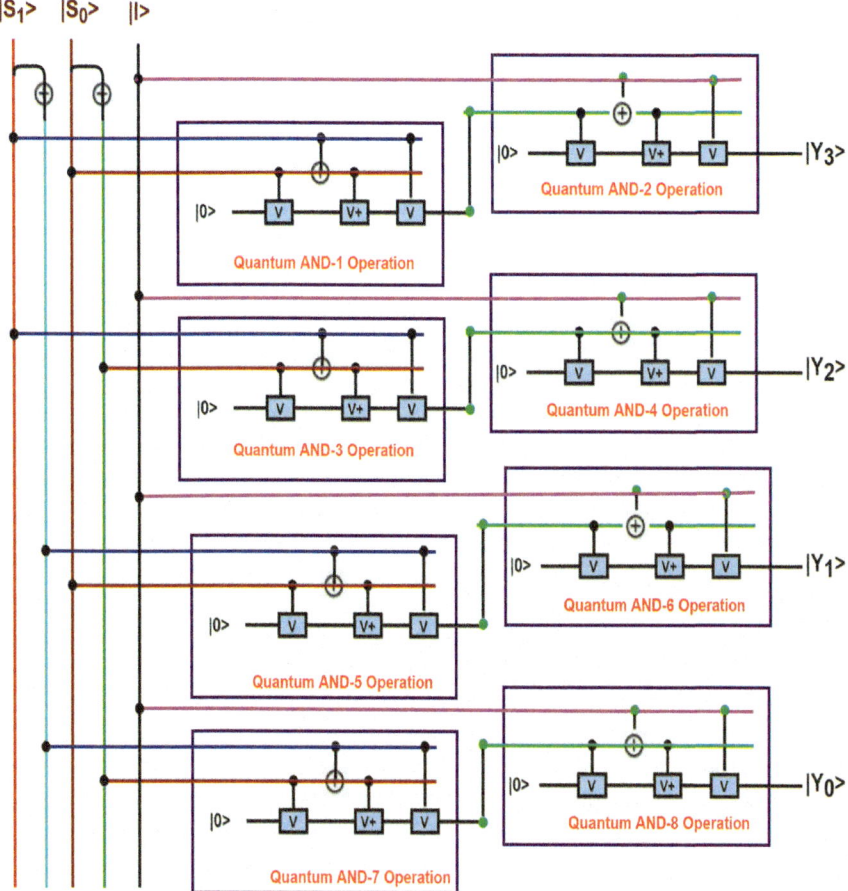

Fig. 9.8 Quantum 1-to-4 demultiplexer circuit

[ii] The output Y1 = S0.S1'.I is "true" |1> when the input S1 is "false" |0> and input S0 is "true" |1>. The output lines Y1 will be "true" |1> and Y0, Y2 and Y3 will be "false" |0>.

[iii] The output Y2 = S0'.S1.I is "true" |1> when the input S0 is "false" |0> and input S1 is "true" |1>. The output lines Y2 will be "true" |1> and Y0, Y1 and Y3 will be "false" |0>.

[iv] The output Y3 = S0.S1.I is "true" |1> when the inputs S0 and S1 are "true" |1>. The output lines Y3 will be "true" |1> and Y0 to Y2 will be "false" |0>.

Table 9.5 Truth table of a 2-to-1 quantum encoder

Inputs		Outputs
|Y1>	|Y0>	|A>
|0>	|1>	|0>
|1>	|0>	|1>

9.2.3 Quantum Encoder

A quantum device or a quantum process that translates data from one format to another is known as a quantum encoder. Quantum encoder is a quantum combinational circuit that takes 2^n inputs and generates n outputs. In quantum computing, an encoder is a crucial component that transforms classical data into a form suitable for quantum computation, typically by mapping it to the state of qubits. The primary goal of an encoder is to prepare the initial state of qubits based on the input data, making it accessible to quantum algorithms. Encoding plays a vital role in quantum machine learning, where classical data is mapped to quantum states for processing by quantum algorithms. In quantum error correction (QEC), encoding helps protect quantum information from noise and decoherence by encoding logical qubits into physical qubits. Many quantum algorithms, like Shor's algorithm, assume a certain encoding method, with some limitations on the complexity of the encoding process. In essence, encoders are essential for bridging the gap between classical data and the quantum realm, enabling quantum computers to process and manipulate information in ways that are not possible with classical computers.

9.2.3.1 Block Diagram

There are only two inputs, Y0 and Y1, and one output, A in a 2-to-1 quantum encoder. The truth table of the 2-to-1 quantum encoder is shown in Table 9.5 and the truth table of 4-to-2 quantum encoder is shown in Table 9.6.

From the truth table, we get

$$A = Y0'.Y1$$

Figure 9.9 presents the block diagram of a 2-to-1 encoder. For this circuit, one quantum NOT operation and one quantum OR operation are needed (Table 9.5).

From the truth table, we get

$$A_1 = Y_3 + Y_2$$
$$A_0 = Y_3 + Y_1$$

Figure 9.10 displays the block diagram of a 4-to-2 quantum encoder. In this circuit, two quantum OR operations are needed.

Table 9.6 Truth table of a 4-to-2 quantum encoder

Inputs				Outputs	
|Y0>	|Y1>	|Y2>	|Y3>	|A1>	|A0>
|I>	|0>	|0>	|0>	|0>	|0>
|0>	|I>	|0>	|0>	|0>	|1>
|0>	|0>	|I>	|0>	|1>	|0>
|0>	|0>	|0>	|I>	|1>	|1>

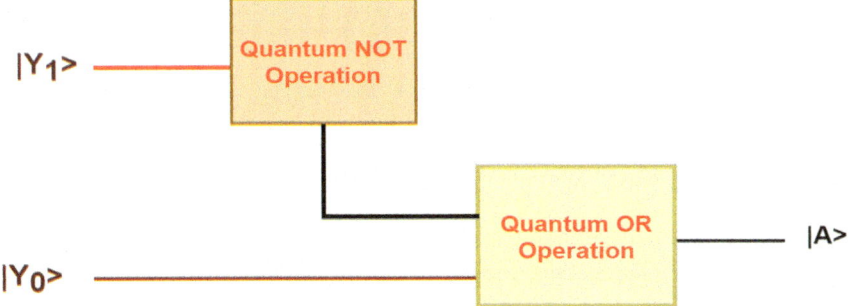

Fig. 9.9 Block diagram of a quantum 2-to-1 encoder

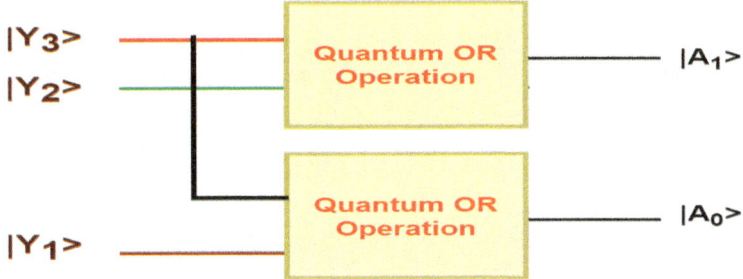

Fig. 9.10 Block diagram of a quantum 4-to-2 encoder

9.2.3.2 Circuit Architecture

Figure 9.11 depicts the quantum circuit of the 2-to-1 encoder. Firstly, the input \|Y0> performs a quantum NOT operation. After that, the output of this operation along with \|Y1> input transfers to a quantum OR operation to get the result of \|A>.

Figure 9.12 gives the view of the 4-to-2 quantum encoder circuit. In 4-to-2 quantum encoder circuit, input qubits \|Y2> and \|Y3>, and \|Y1> and \|Y3> parallelly perform two quantum AND operations to generate the values of \|A1> and \|A0>.

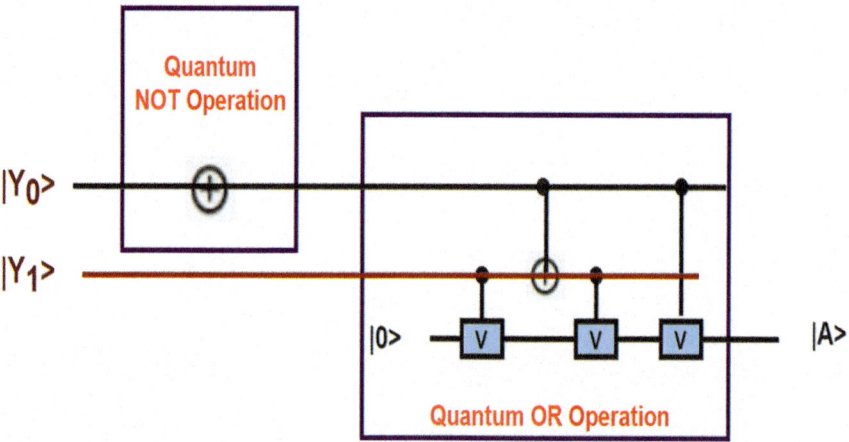

Fig. 9.11   Quantum 2-to-1 encoder circuit

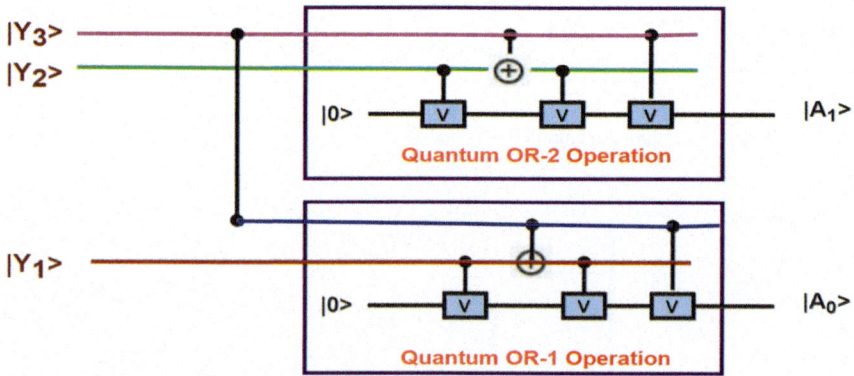

Fig. 9.12   Quantum 4-to-2 encoder circuit

9.2.3.3   Working Principle

Quantum 2-to-1 Encoder: The quantum 2-to-1 encoder needs two inputs Y0 and Y1, and one output A0. Consider the following cases:

  [i] The output A0 = Y0 is "true" |1>, when the input Y1 is "false" |0> and the input Y0 is "true" |1>.
  [ii] The output A0 = Y1 is "true" |1>, when the input Y0 is "false" |0> and the input Y1 is "true" |1>.

Quantum 4-to-2 Encoder: The quantum 4-to-2 encoder needs four inputs Y1, Y3 and Y2, Y3 and two outputs A0 and A1. Consider the following cases:

[i] The output A0 = Y1 + Y3 is "true" |1> when the input Y1 or the input Y3 is "true" |1>. The output A1 = Y2 + Y3 is "true" |1>, when the input Y2 or input Y3 is "true" |1>.

[ii] The output A0 = Y1 + Y3 is "false" |0>, when the both inputs Y1 and Y3 are "false" |0>. The output A1 = Y2 + Y3 is "true" |1>, when the input Y2 or the input Y3 is "true" |1>.

[iii] The output A0 = Y1 + Y3 is "true" |1>, when the input Y1 or the input Y3 is "true" |1>. The output A1 = Y2 + Y3 is "false" |0>, when the both inputs Y1 and Y3 are "false" |0>.

[iv] The output A0 = Y1 + Y3 is "false" |0>, when the both inputs Y1 and Y3 are "false" |0>. The output A1 = Y2 + Y3 is "false" |0>, when the both inputs Y1 and Y3 are "false" |0>.

9.2.4 Quantum Decoder

Decoders are combinational quantum circuits that convert quantum information into 2^N output lines. The quantum data is transmitted in the form of N input lines. In quantum computing, a decoder is a critical component of quantum error correction (QEC). It analyzes syndrome data, which is information gathered during QEC about errors that have affected the system, to determine the specific errors and apply corrections. This process is essential for fault-tolerant quantum computing, ensuring the accuracy of calculations by mitigating the impact of noise. Quantum error correction schemes involve measuring the state of qubits to detect errors. This measurement generates syndrome data, which provides information about the nature and location of errors. The decoder uses the syndrome data to interpret the errors and determine the specific corrections needed to restore the logical state of the qubits. The decoder applies corrections to the logical qubit, effectively neutralizing the effect of the detected errors. Decoding algorithms need to be fast enough to keep up with the rate at which errors occur in a quantum computer, which can be very rapid. Efficient decoding algorithms are crucial for scaling up quantum computers to handle more complex computations. Surface codes are a type of quantum error-correcting code that uses a 2D lattice of qubits. Decoders for surface codes, such as the collision clustering algorithm, are designed to efficiently detect and correct errors in these codes. Artificial neural networks can also be used as decoders, offering potential for scalability and speed. Decoders need to be accurate in identifying and correcting errors, minimizing the chance of introducing new errors. Decoders should use resources efficiently to allow for scaling up to larger quantum computers.

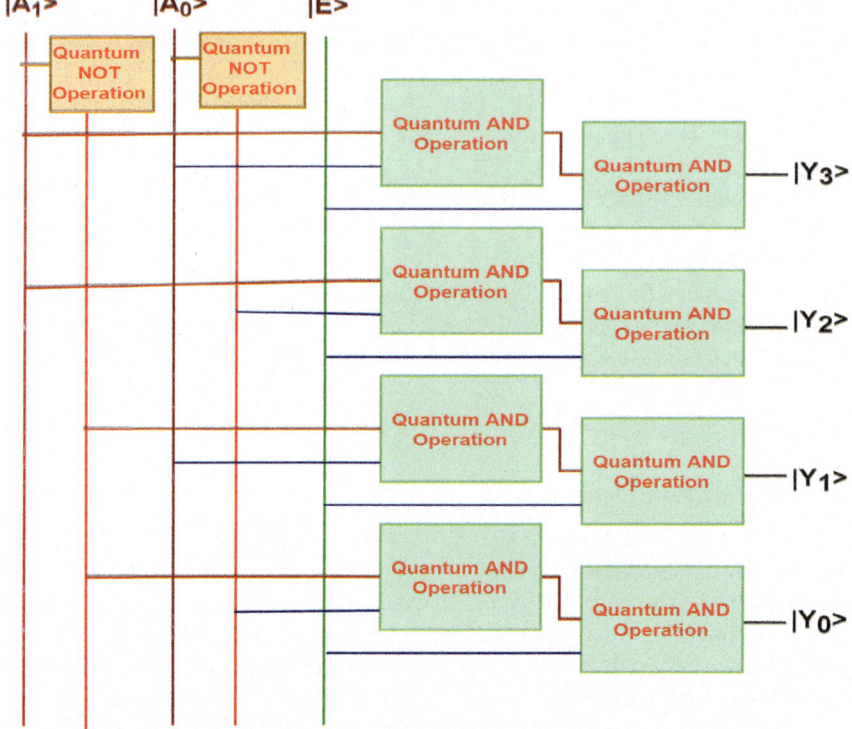

Fig. 9.13 Block diagram of a quantum 2-to-4 decoder

Table 9.7 Truth table of a quantum 2-to-4 decoder

Enable	Inputs		Outputs			
$|E>$	$|S1>$	$|S0>$	$|Y0>$	$|Y1>$	$|Y2>$	$|Y3>$
$|0>$	X	X	$|0>$	$|0>$	$|0>$	$|0>$
$|1>$	$|0>$	$|0>$	$|1>$	$|0>$	$|0>$	$|0>$
$|1>$	$|0>$	$|1>$	$|0>$	$|1>$	$|0>$	$|0>$
$|1>$	$|1>$	$|0>$	$|0>$	$|0>$	$|1>$	$|0>$
$|1>$	$|1>$	$|1>$	$|0>$	$|0>$	$|0>$	$|1>$

9.2.4.1   Block Diagram

Figure 9.13 presents the block diagram of the 2-to-4 quantum decoder. In this circuit, eight quantum AND operations and two quantum NOT operations are needed. Then the truth table of a 2-to-4 quantum decoder is shown in Table 9.7.

From the truth table, we get

Y3 = E.A1.A0
Y2 = E.A1.A0'
Y1 = E.A1'.A0
Y0 = E.A1'.A0'

9.2.4.2   Circuit Architecture

Figure 9.14 presents the view of the 2-to-4 quantum decoder circuit. Firstly, the qubit input |E> connects with four quantum AND operations. Then the other inputs of these AND operations come from the output of another four quantum AND operations, which are executed by two selection lines, |A0> and |A1>. Finally, the outputs of AND-2, AND-4, AND-6, and AND-8 are |Y3>, |Y2>, |Y1>, and |Y0>.

9.2.4.3   Working Principle

The quantum 2-to-4 decoder needs two inputs A1 and A0, and one enable input E, which activates the circuit. For a quantum 2-to-4 decoder, suppose E is "true", |1>. Now, consider the following cases:

> **[i]** The output D0 = A0'. A1' is "true" |1>, when the inputs A1 and A0 are "false" |0>. The output line D0 will be "true" |1> and D1 to D3 will be "false" |0>.
> **[ii]** The output D1 = A0.A1' is "true" |1>, when the input A1 is "false" |0> and input A0 is "true" |1>. The output line D1 will be "true" |1> and D0, D2 and D3 will be "false" |0>.
> **[iii]** The output D2 = A0'. A1 is "true" |1>, when the input A0 is "false" |0> and input A1 is "true" |1>. The output line D2 will be "true" |1> and D0, D1 and D3 will be "false" |0>.
> **[iv]** The output D3 = A0.A1 is "true" |1>, when the inputs A0 and A1 are "true" |1>. The output line D3 will be "true" |1> and D0 to D2 will be "false" |0>.

9.3   Summary

This chapter covers the block diagram, circuit structure, and functioning principle of combinational circuits in quantum computing. The required figures and description have also been provided. A quantum logic system underpins all of

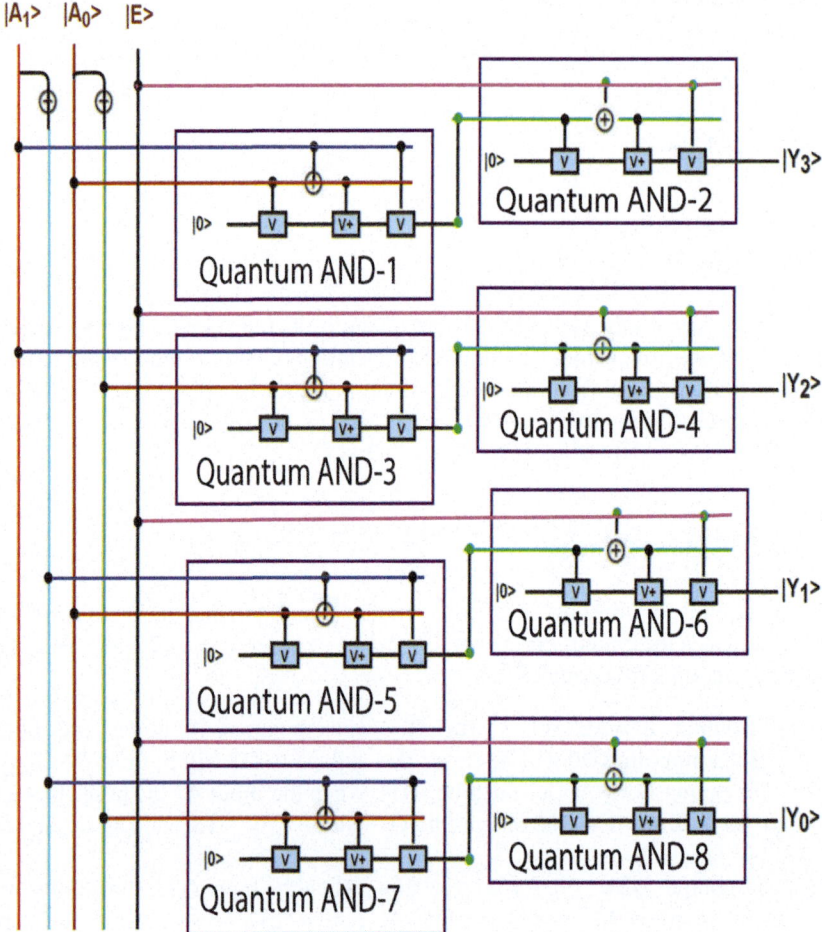

Fig. 9.14 Quantum 2-to-4 decoder circuit

the circuits. These circuits are built with quantum gates, and the explanations are quite straightforward. Here, the multiplexer, demultiplexer, encoder, and decoder in quantum systems are explained in depth. Quantum computing is an emerging field in modern technology. It has a deep impact on the development of the computing world for nanotechnology. Combinational circuits in quantum computing will bring a revolutionary change in the upcoming future of the cutting-edge technologies.

Bibliography

1. J. Anders, D.K.L. Oi, E. Kashefi, D.E. Browne, E. Andersson, Ancilla-driven universal quantum computation. Phys. Rev. A **82**(2), 020301 (2010)
2. D. Auguin, V. Catherinot, T.E. Malliavin, J.L. Pons, M.A. Delsuc, Superposition of chemical shifts in NMR spectra can be overcome to determine automatically the structure of a protein. Spectroscopy **17**(2–3), 559–568 (2003)
3. F. Hobo, M. Takahashi, H. Maeda, S 33 NMR cryogenic probe for taurine detection. Rev. Sci. Instrum. **80**(3), 036106 (2009)
4. J.A. Jones, Quantum computing and nuclear magnetic resonance. PhysChemComm **4**(11), 49–56 (2001)
5. V.D. Kodibagkar, M.S. Conradi, Remote tuning of NMR probe circuits. J. Magn. Resonance **144**(1), 53–57 (2000)
6. Y. Takahashi, S. Tani, Power of uninitialized qubits in shallow quantum circuits. Theor. Comput. Sci. **851**, 129–153 (2021)

Chapter 10
Quantum-DNA Combinational Circuits

10.1 Introduction

Quantum biocomputing or quantum-DNA computing with combinational circuits refers to the use of quantum computing principles and techniques to design and implement circuits that perform specific logical operations, similar to how classical computers use combinational circuits. This involves creating circuits that manipulate qubits, the quantum equivalent of bits, to perform complex calculations and transformations, often inspired by biological systems or processes. In essence, quantum biocomputing with combinational circuits represents a way to leverage quantum mechanics to perform complex computations, drawing inspiration from biological systems and processes. As discussed in the previous chapters, biomolecules and biomolecular processes are meant to apply computational algorithms in DNA computing. Quantum computing involves performing calculations at a scale where quantum mechanical effects are significant. Both of these new computing paradigms have been considered as potential successors to solid-state computers. Both of these qualities could be captured by combining quantum computing and DNA computing. DNA computers could self-assemble quantum logic circuits from DNA strand-attached gates. Furthermore, quantum computers might be built directly from the physical properties of the DNA molecules. This chapter will describe combinational circuits in quantum biology which are known as quantum biocomputing or quantum-DNA computing or quantum biological computing.

10.2 Design Architecture of Quantum-DNA Combinational Circuits

This chapter will present the details of Quantum-DNA multiplexer, Quantum-DNA demultiplexer, Quantum-DNA encoder, and Quantum-DNA decoder.

Table 10.1 Truth table of a 2-to-1 quantum-DNA multiplexer

Inputs	Outputs
$|S>$	Y
$|0>$	I_0
$|1>$	I_1

10.2.1 Quantum-DNA Multiplexer

Quantum-DNA multiplexer is the combination of quantum circuits and DNA circuits. The first part is the quantum part and the last part is the DNA part of the whole circuit. In the context of quantum-DNA computing, a multiplexer (often abbreviated as mux) plays a role similar to its function in traditional digital electronics, but with a quantum-DNA twist. It acts as a selector, choosing between multiple inputs and directing the selected data (or, in this case, quantum information or DNA strands) to a single output. This is crucial for building larger, more complex quantum-DNA computing systems by allowing for the efficient selection and manipulation of quantum resources or DNA strands. In quantum-DNA computing, the "inputs" could be qubits or DNA strands representing different states of information. The multiplexer would then choose which of these inputs to pass on to the next step in the computation. The multiplexer acts as a control switch, allowing the quantum-DNA system to dynamically choose which data to process at any given time. This is crucial for building complex quantum-DNA algorithms and circuits. A multiplexer might be used to select a specific qubit from a larger pool of qubits for a particular quantum operation. It could select a particular DNA strand from a set of DNA strands for use in a molecular computation. Multiplexers allow for

  i) Building larger quantum-DNA computers from smaller, interconnected units;
 ii) Optimizing the use of limited quantum resources or DNA strands; and
iii) Adjusting the flow of information within the quantum-DNA system.

10.2.1.1 Block Diagram

Two types of quantum-DNA multiplexer are described here: One is the 2-to-1 multiplexer and another is the 4-to-1 multiplexer in quantum-DNA computing. In 2-to-1 multiplexer, there are only two inputs, $|I0>$ and $|I1>$, one selection line, $|S>$, and one output, Y. Table 10.1 presents the truth table of a 2-to-1 quantum-DNA multiplexer.

Figure 10.1 portrays the block diagram of a quantum-DNA 2-to-1 multiplexer. In this circuit, one quantum NOT operation, two quantum AND operations, and one DNA OR operation are needed.

Table 10.2 presents the truth table of a 4-to-1 quantum-DNA multiplexer. Here inputs are S1 and S2, and the output is Y.

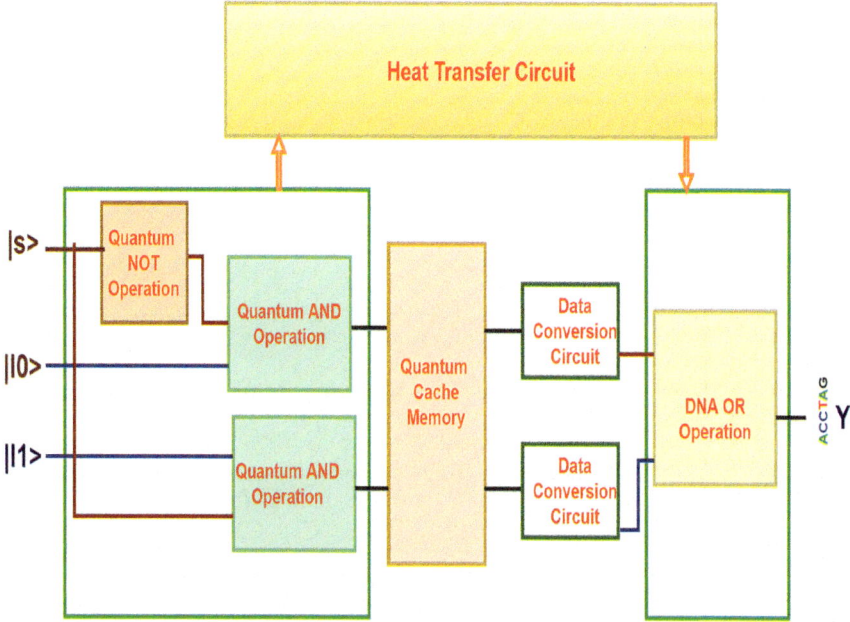

Fig. 10.1 Block diagram of a quantum-DNA 2-to-1 multiplexer

Table 10.2 Truth table of a 4-to-1 quantum-DNA multiplexer

Inputs		Outputs
$|S1\rangle$	$|S0\rangle$	Y
$|0\rangle$	$|0\rangle$	I_0
$|0\rangle$	$|1\rangle$	I_1
$|1\rangle$	$|0\rangle$	I_2
$|1\rangle$	$|1\rangle$	I_3

Figure 10.2 displays the block diagram of a quantum-DNA 4-to-1 multiplexer. In this circuit, two quantum NOT operations, eight quantum AND operations, and three DNA OR operations are needed.

10.2.1.2 Circuit Architecture

Figure 10.3 depicts the quantum circuit of the 2-to-1 multiplexer. Firstly, each of inputs $|I0\rangle$ and $|I1\rangle$ separately connect with two quantum AND operations. The input of the selection line, $|S\rangle$ is connected directly to the AND-1 operation, whereas another input is connected to the AND-2 operation after completing the NOT opera-

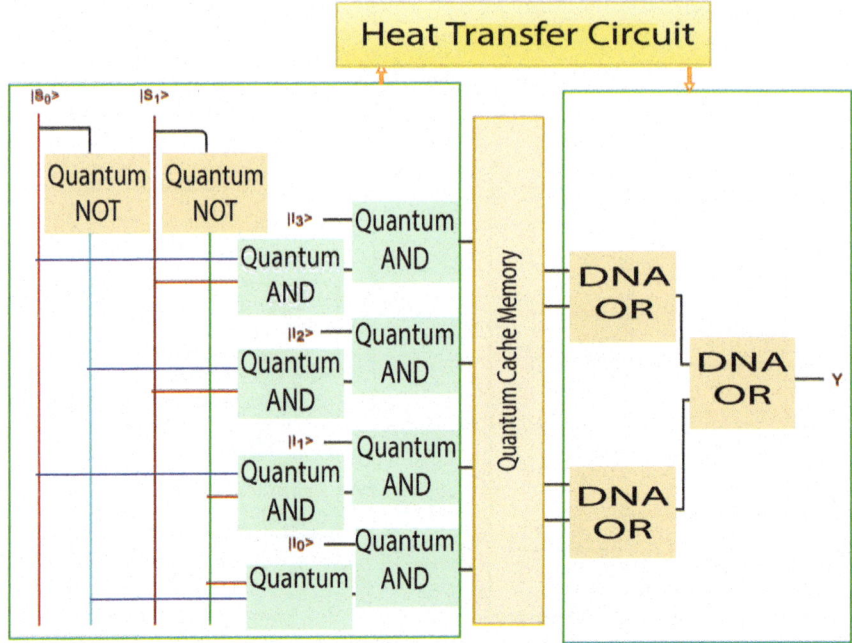

Fig. 10.2 Block diagram of a quantum-DNA 4-to-1 multiplexer

tion. Finally, both the outputs of quantum AND operation are passed through quantum OR operation generating the value of |Y>.

Figure 10.4 gives the view of the 4-to-1 multiplexer quantum circuit. Firstly, each of qubit inputs |I0>, |I1>, |I2>, and |I3> separately connect with four quantum AND operations. Then, the other inputs of these AND operations come from the output of another four quantum AND operations which are executed through two selection lines, |S0> and |S1>. Finally, the outputs of AND-2, AND-4, AND-6, and AND-8 conduct three quantum OR operations to produce the value of |Y>.

10.2.1.3 Working Principle

Quantum-DNA 2-to-1 Multiplexer: The quantum-DNA 2-to-1 multiplexer needs one selection line S0 and two inputs, I0 and I1. When qubit inputs I0-I1 are "true", |1>, consider the following cases:

   [i] The output Y = I0 when the input qubit, S, is "false" |0>.
   [ii] The output Y = I1 when the input qubit, S, is "true" |1>.

Quantum-DNA 4-to-1 Multiplexer: The quantum-DNA 4-to-1 multiplexer needs two selection lines S0 and S1, and four inputs, I0-I3. When all inputs I0-I3 are "true", |1>, consider the following cases:

   [i] The output Y = I0 when the input qubits S1 and S0 are "false" |0>.

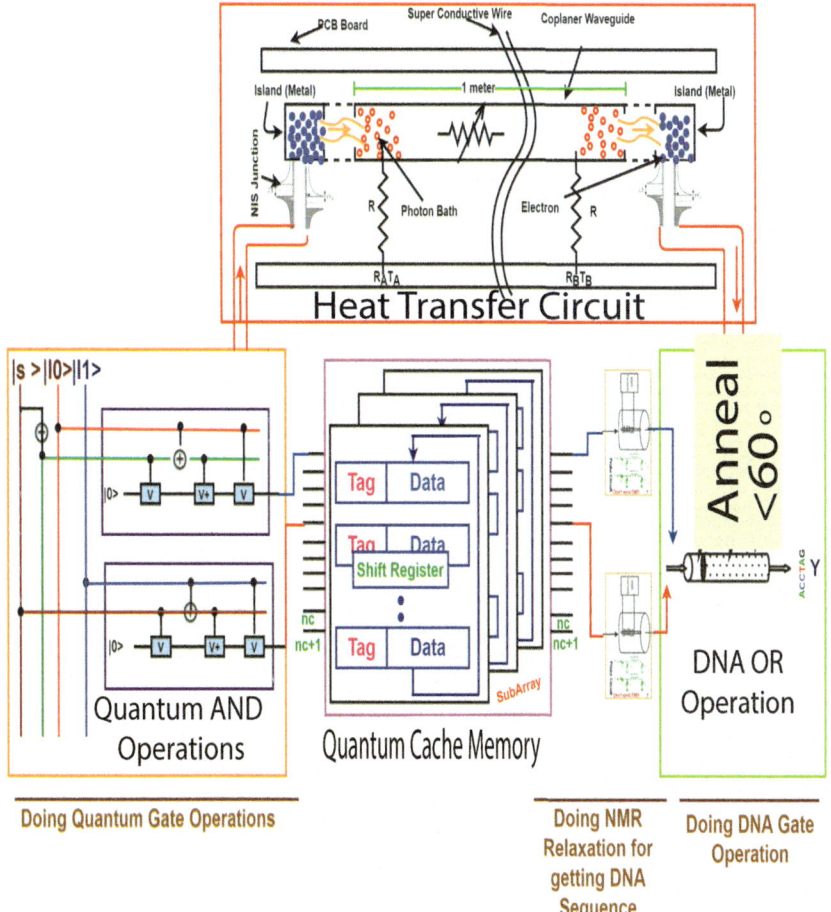

Fig. 10.3 Quantum-DNA 2-to-1 multiplexer circuit

[ii] The output Y = I1 when the input qubit S1 is "false" |0> and input qubit S0 is "true" |1>.
[iii] The output Y = I2 when the input qubit S0 is "false" |0> and input qubit S1 is "true" |1>.
[iv] The output Y = I3 when the input qubits S0 and S1 are "true" |1>.

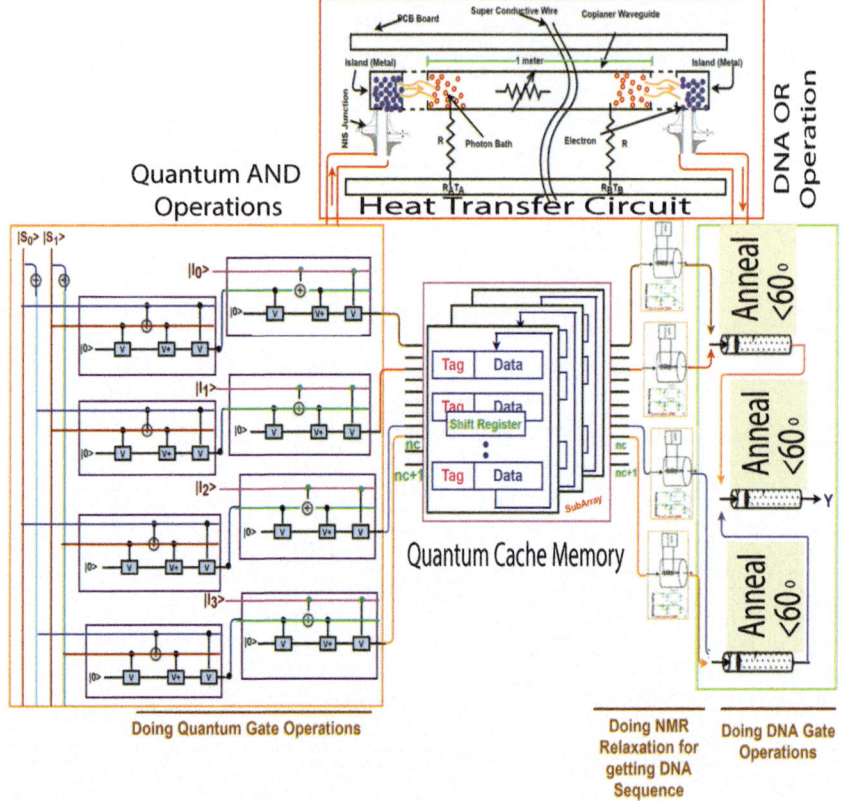

Fig. 10.4 Quantum-DNA 4-to-1 multiplexer circuit

10.2.2 Quantum-DNA Demultiplexer

A quantum-DNA demultiplexer is a combinational circuit which is the combination of the quantum demultiplexer circuit and DNA demultiplexer circuit. In the context of quantum-DNA computing, a demultiplexer is a logic circuit, similar to its classical counterpart, that selects one of multiple output signals based on a control input. The term "quantum-DNA computing" refers to the use of quantum mechanics and DNA's molecular structure to perform computation. Demultiplexers play a role in routing quantum signals and manipulating quantum states within these systems.

10.2.2.1   Block Diagram

In the 1-to-2 quantum-DNA demultiplexer, there is only one input D, one selection line S, and two outputs, I0 and I1 as shown in truth Table 10.3 and the truth table of 1-to-4 quantum-DNA demultiplexer is given in Table 10.4 where one input is I, two selection lines S1 and S2, and four outputs are Y0, Y1, Y2, and Y3.

Table 10.3 Truth table of a 1-to-2 quantum-DNA demultiplexer

Inputs	Outputs	
$|S>$	I_0	I_1
$|0>$	**TGGATC**	**D**
$|1>$	**D**	**TGGATC**

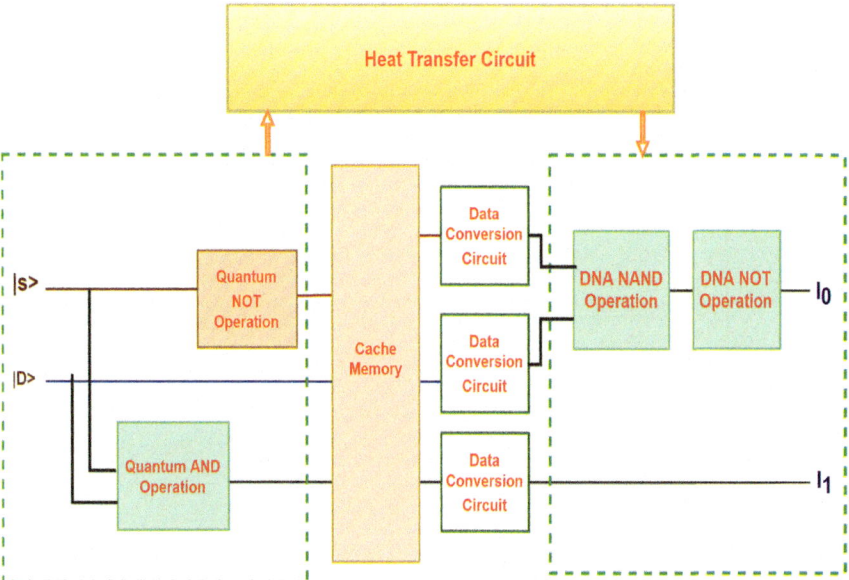

Fig. 10.5 Block diagram of a quantum-DNA 1-to-2 demultiplexer

Table 10.4 Truth table of a 1-to-4 quantum-DNA demultiplexer

Inputs		Outputs			
$|S1>$	$|S0>$	**Y0**	**Y1**	**Y2**	**Y3**
$|0>$	$|0>$	**I**	**TGGATC**	**TGGATC**	**TGGATC**
$|0>$	$|1>$	**TGGATC**	**I**	**TGGATC**	**TGGATC**
$|1>$	$|0>$	**TGGATC**	**TGGATC**	**I**	**TGGATC**
$|1>$	$|1>$	**TGGATC**	**TGGATC**	**TGGATC**	**I**

From the truth table, we get

I0 = S′.D
I1 = S.D

Figure 10.5 presents the block diagram of a 1-to-2 quantum-DNA demultiplexer. In this circuit, one quantum NOT operation, one quantum AND operation, and one DNA AND operation are needed.

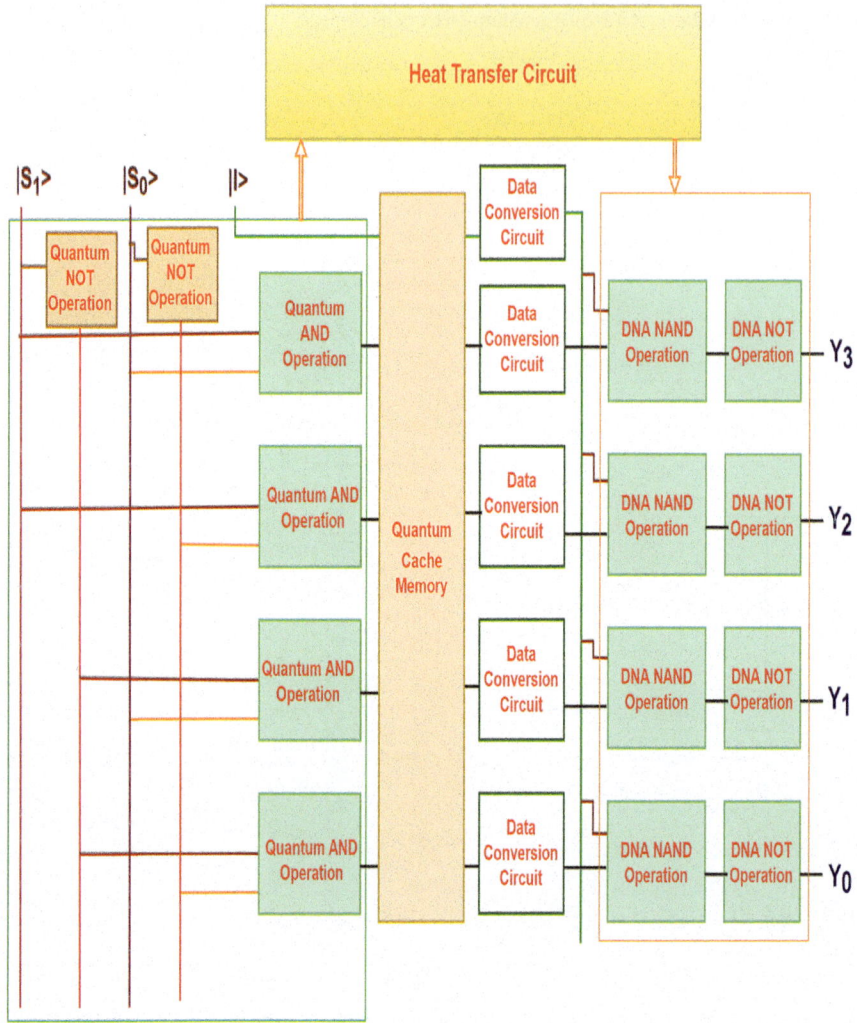

Fig. 10.6 Block diagram of a quantum-DNA 1-to-4 demultiplexer

From the truth table, we get

Y0 = S1′ S0′ I
y1 = S1′ S0 I
y2 = S1 S0′ I
y3 = S1 S0 I

The block diagram of a quantum-DNA 1-to-4 demultiplexer is shown in Fig. 10.6, where two quantum NOT operations, four quantum AND operations, and four DNA AND operations are needed.

10.2.2.2 Circuit Architecture

Figure 10.7 depicts the quantum circuit of the 1-to-2 demultiplexer. Firstly, the input qubit |D> is connected with the quantum AND operation directly as the same as the selection input, |S> and after passing through the NMR relaxation, it produces output I_1. Based on the value of |S> (after NOT operation) and |D>, only one DNA AND operation provides the output I_0.

Figure 10.8 gives the view of the 1-to-4 quantum demultiplexer circuit. Firstly, the input |I> is directly connected with four quantum AND operations. The other input qubits of these AND operations come from the output of another four quantum AND operations which are executed through two selection lines, |S0> and |S1>. Finally, the outputs of AND-1, AND-2, AND-3, and AND-4 generate the qubits of |Y0>, |Y1>, |Y2>, and |Y3>.

10.2.2.3 Working Principle

Quantum-DNA 1-to-2 Demultiplexer: The quantum-DNA 1-to-2 demultiplexer needs one selection line |S0> and one input, D. For DNA 1-to-2 demultiplexer, when the qubit input |D> is "true", |1>, consider the following cases:

[i] The output I0 = S'.D is "true" **ACCTAG** when the qubit input |S> is "false" |0> and I1 will be also "false" **TGGATC**.
[ii] The output I1 = S.D is "true" **ACCTAG** when the input |S> is "true" |1> and I0 will be "false" **TGGATC**.

Quantum-DNA 1-to-4 Demultiplexer: The quantum-DNA 1-to-4 demultiplexer needs two selection lines S0 and S1, and one input, I. For quantum-DNA 1-to-4 demultiplexer, when input |I> is "true" |1>, consider the following cases:

[i] The output Y0 = S0'.S1'.I is "true" **ACCTAG** when the qubit inputs |S1> and |S0> are "false" |0>. The output line Y0 will be "true" **ACCTAG** and Y1 to Y3 will be "false" **TGGATC**.
[ii] The output Y1= S0.S1'.I is "true" **ACCTAG**, when the input |S1> is "false" |0> and the input |S0> is "true" |1> The output line Y1 will be "true" **ACCTAG** and Y0, Y2 and Y3 will be "false" **TGGATC**.
[iii] The output Y2 = S0'.S1.I is "true" **ACCTAG**, when the input |S0> is "false" |0> and the input |S1> is "true" |1>. The output line Y2 will be "true" **ACCTAG** and Y0, Y1 and Y3 will be "false" **TGGATC**.
[iv] The output Y3 = S0.S1.I is "true" **ACCTAG**, when the inputs |S0> and |S1> are "true" |1>. The output line Y3 will be "true" **ACCTAG** and Y0 to Y2 will be "false" **TGGATC**.

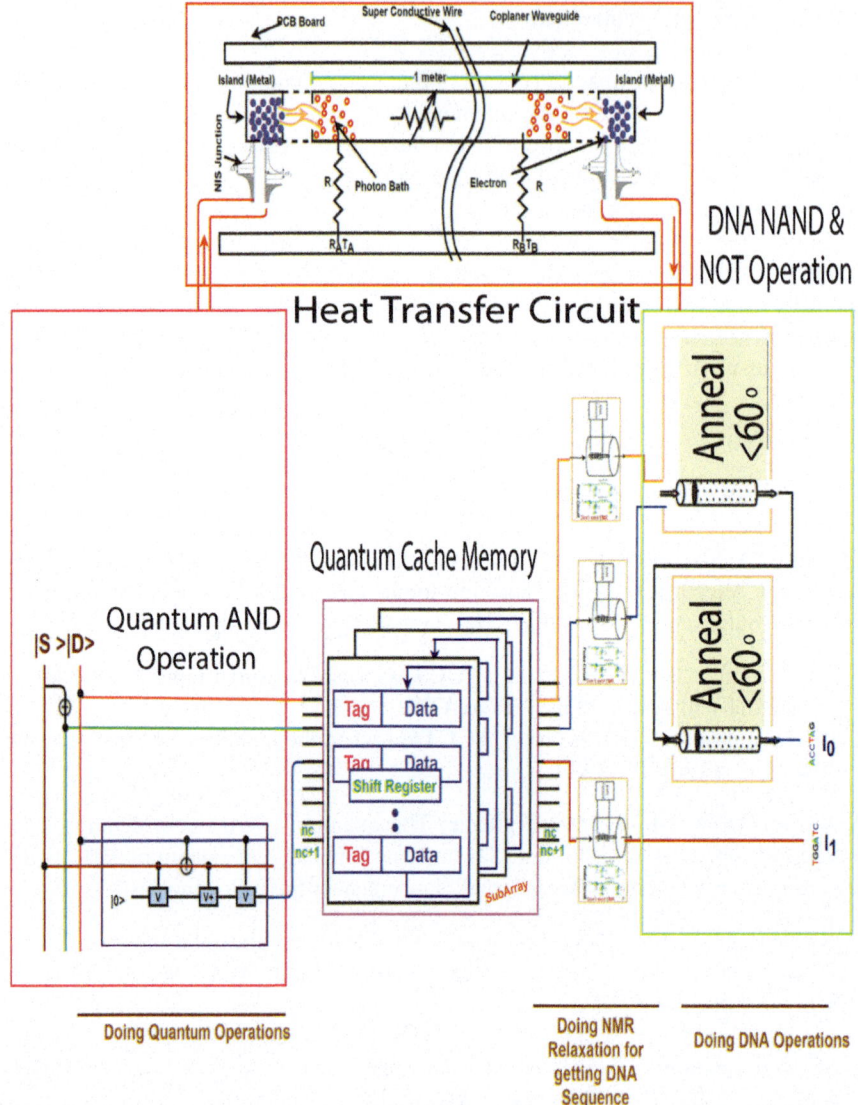

Fig. 10.7 Quantum-DNA 1-to-2 demultiplexer circuit

10.2.2.4 Quantum-DNA Encoder

Quantum encoding involves mapping classical information (e.g., the sequence of nucleotides in a DNA strand) to a quantum state, which is a superposition of multiple states. This allows for the use of quantum algorithms to process and analyze this biological data in a way that is not possible with classical computers.

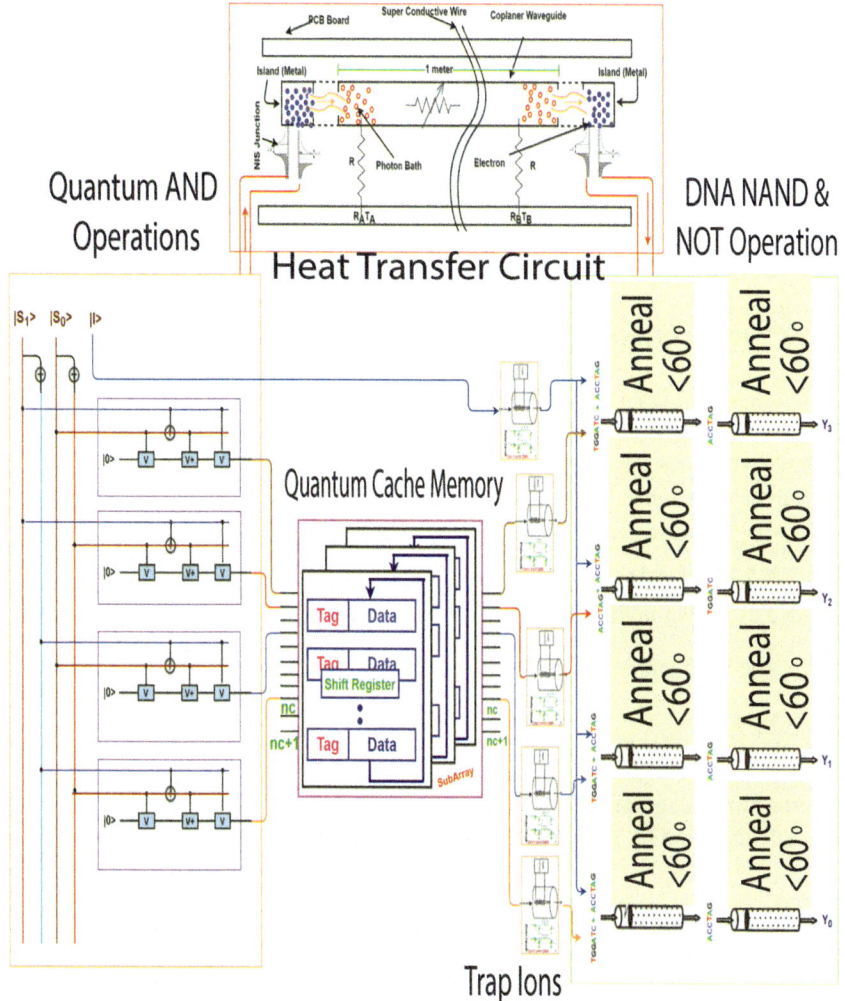

Fig. 10.8 Quantum-DNA 1-to-4 demultiplexer circuit

Biological data, like DNA or protein sequences, is complex and often high-dimensional. Quantum encoding allows for a more efficient and compact representation of this data in the quantum domain. Once biological data is encoded into a quantum state, it can be manipulated by quantum algorithms that can perform tasks like searching for patterns, simulating molecular interactions, or predicting protein structures. Quantum algorithms can potentially offer exponential speedups or outperform classical algorithms for certain biological problems, especially those involving large datasets or complex interactions. In essence, quantum encoding is a fundamental step in harnessing the power of quantum computing to address biological problems, by translating classical biological data into a format that can be processed by quantum algorithms. In essence, quantum encoding is a fundamental step in harnessing the power of quantum computing to address biological problems, by translating classical biological data into a format that can be processed by quantum algorithms. On the other way, quantum-DNA encoder is a combinational circuit in which the first part of the circuit is a quantum circuit and the last part of the circuit is a DNA circuit. Both parts will combinedly present the quantum-DNA encoder circuit.

10.2.2.5   Block Diagram

In 2-to-1 quantum-DNA encoder, there are only two inputs, |Y0> and |Y1>, and one output, A. The truth table of a 2-to-1 quantum-DNA encoder is given in Table 10.5 and Table 10.6 shows the truth table of a 4-to-2 quantum-DNA encoder.

From the truth table, it can be written as follows:

$A = Y0'.Y1$

For this circuit, one quantum NOT operation and one DNA OR operation are needed as shown in Fig. 10.9.

From the truth table, we get

$A_1 = Y_3 + Y_2$
$A_0 = Y_3 + Y_1$

As the block diagram demonstrates in Fig. 10.10, 4-to-2 Quantum-DNA encoders need one quantum OR operation and one DNA OR operation.

Table 10.5 Truth table of a 2-to-1 quantum-DNA encoder

Inputs		Outputs
|Y1>	|Y0>	A
|0>	|1>	TGGATC
|1>	|0>	ACCTAG

Table 10.6 Truth table of a 4-to-2 quantum-DNA encoder

Inputs				Outputs	
|Y0>	|Y1>	|Y2>	|Y3>	A1	A0
|1>	|0>	|0>	|0>	TGGATC	TGGATC
|0>	|1>	|0>	|0>	TGGATC	ACCTAG
|0>	|0>	|1>	|0>	ACCTAG	TGGATC
|0>	|0>	|0>	|1>	ACCTAG	ACCTAG

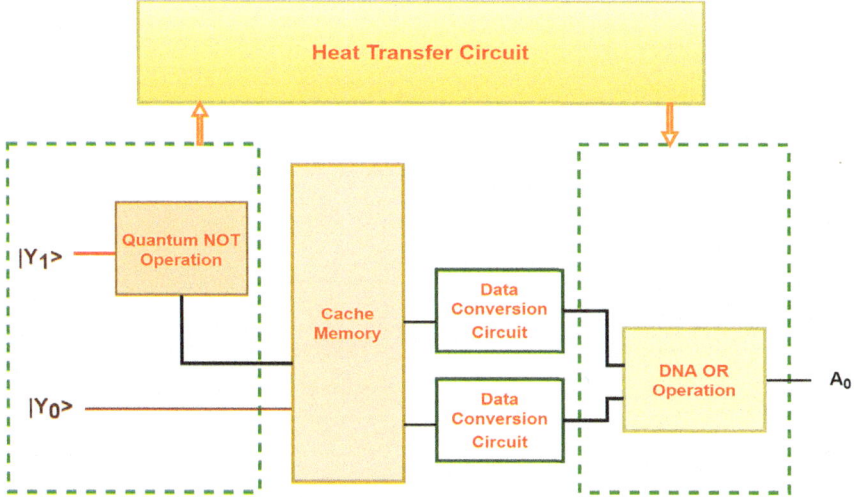

Fig. 10.9 Block diagram of quantum-DNA 2-to-1 encoder

10.2.2.6 Circuit Architecture

Figure 10.11 depicts the quantum circuit of the 2-to-1 encoder. Firstly, input |Y0> performs a quantum NOT operation. After that, the output of this operation along with |Y1> input transfers to a DNA OR operation to get the result of A_0.

Figure 10.12 gives the view of a 4-to-2 encoder quantum circuit. In the 4-to-2 encoder quantum circuit, input |Y2> and input |Y3>, and inputs |Y1> and |Y3> parallelly perform two quantum AND operations to generate the values of A1 and A0.

10.2.2.7 Working Principle

Quantum-DNA 2-to-1 Encoder: The quantum-DNA 2-to-1 encoder needs two inputs Y0 and Y1, and one output A0. Consider the following cases:

[i] The output A0 = Y0 is "true" **ACCTAG**, when the input |Y1> is "false" |0> and |Y0> is "true" |1>.

[ii] The output A0 = Y1 is "true" **ACCTAG**, when the input |Y0> is "false" |0> and |Y1> is "true" |1>.

Quantum-DNA 4-to-2 Encoder: The quantum-DNA 4-to-2 encoder needs four inputs Y1, Y3 and Y2, Y3 and two outputs A1 and A0. Consider the following cases:

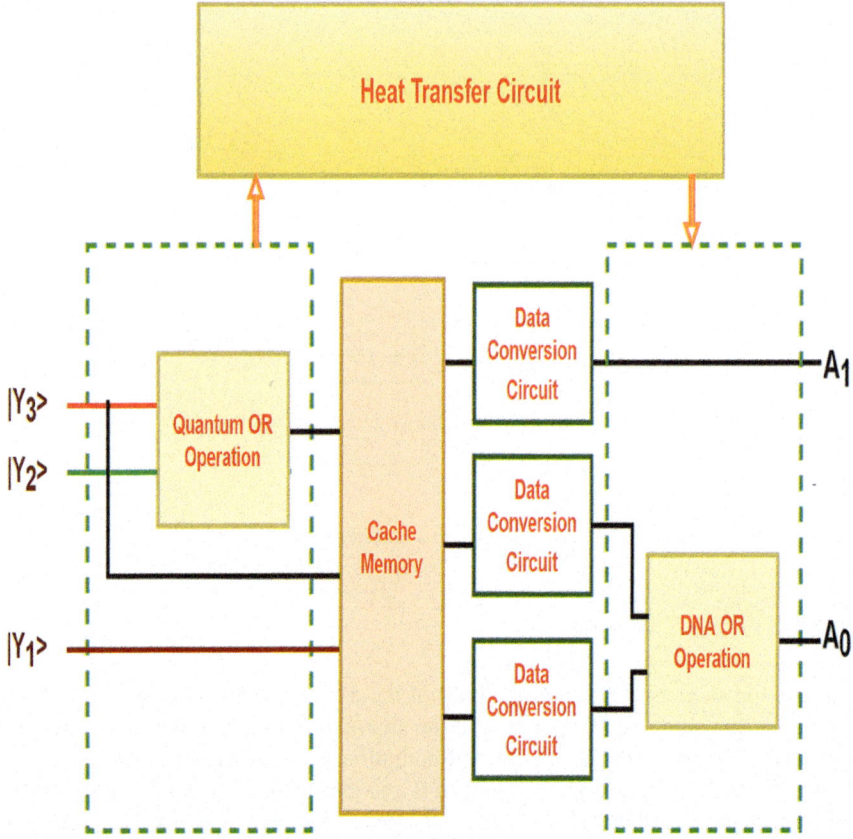

Fig. 10.10 Block diagram of a quantum-DNA 4-to-2 encoder

[i] The output A0 = Y1 + Y3 is "true" **ACCTAG**, when the input |Y1> or input |Y3> is "true" |1>. The output A1 = Y2 + Y3 is "true" **ACCTAG**, when the input |Y2> or input |Y3> is "true" |1>.

[ii] The output A0 = Y1 + Y3 is "false" **TGGATC**, when both inputs |Y1> and |Y3> are "false" |0>. The output A1 = Y2 + Y3 is "true" **ACCTAG**, when the input |Y2> or input |Y3> is "true" |1>.

[iii] The output A0 = Y1 + Y3 is "true" **ACCTAG**, when the input |Y1> or input |Y3> is "true" |1>. The output A1 = Y2 + Y3 is "false" **TGGATC**, when the both inputs |Y1> and |Y3> are "false" |0>.

[iv] The output A0 = Y1 + Y3 is "false" **TGGATC**, when the both inputs |Y1> and |Y3> are "false" |0>. The output A1 = Y2 + Y3 is "false" **TGGATC**, when the both inputs |Y1> and |Y3> are "false" |0>.

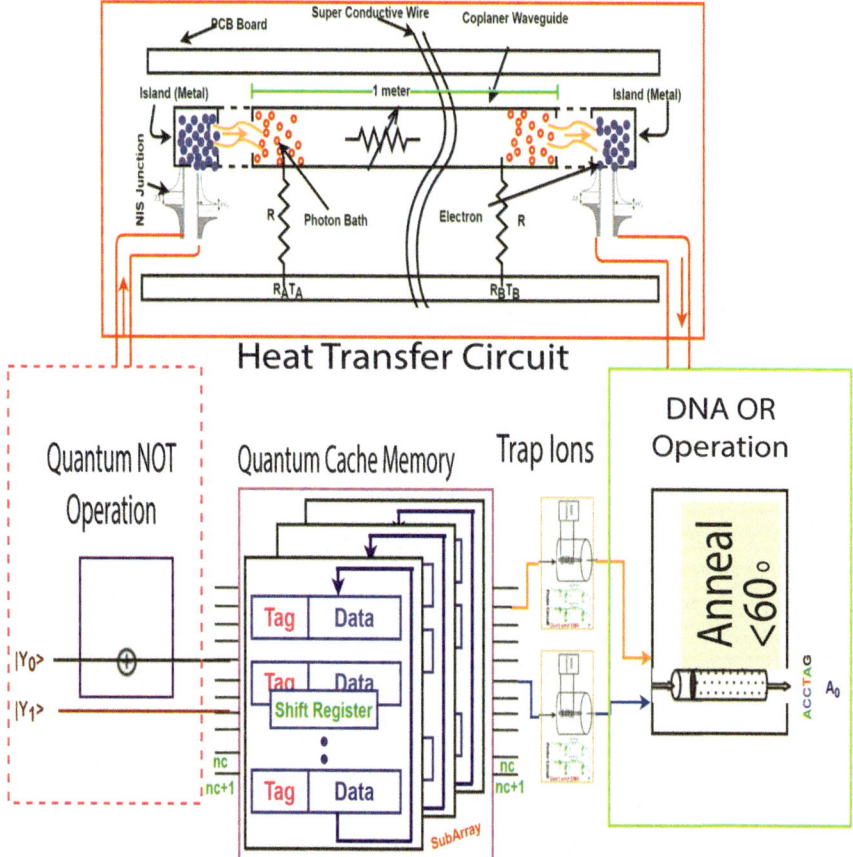

Fig. 10.11 Quantum-DNA 2-to-1 encoder circuit

10.2.3 Quantum-DNA Decoder

The decoder is a combinational quantum circuit that converts quantum information into 2^N output lines. The quantum data is transmitted in the form of N input lines. In the context of quantum biocomputing, a decoder plays a crucial role in quantum error correction (QEC) by interpreting syndrome data to identify and correct errors that may have occurred during computation. This process is essential for ensuring the reliability of quantum-DNA computations and realizing fault-tolerant quantum computers. By effectively correcting errors, decoders contribute to the overall reliability of quantum-DNA computations, which is crucial for trustworthy results in areas like drug discovery or materials science. Research and development in decoding algorithms based on quantum-DNA systems are ongoing, with advancements like AI-based decoders showing promise for improving speed and accuracy.

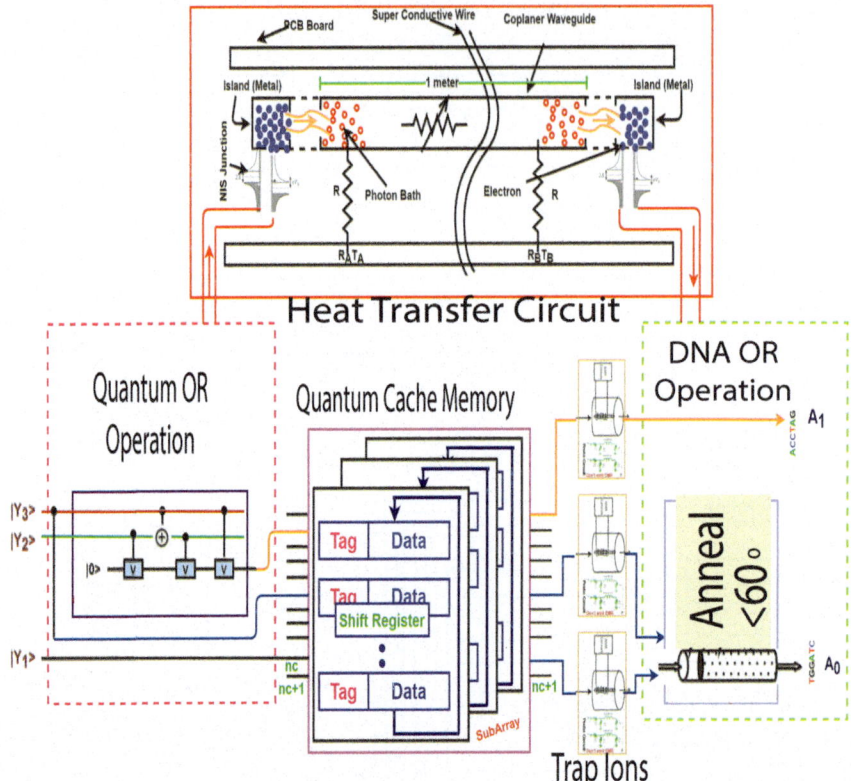

Fig. 10.12   Quantum-DNA 4-to-2 encoder circuit

10.2.3.1   Block Diagram

Figure 10.13 presents the block diagram of a quantum-DNA 2-to-4 decoder. In this circuit, two quantum NOT, four quantum AND, and four DNA AND operations are needed. The truth table of a quantum-DNA 2-to-4 decoder is given in Table 10.7.

Table 10.7   Truth table of a quantum-DNA 2-to-4 decoder

Enable	Inputs		Outputs			
$|E>$	$|S1>$	$|S0>$	Y0	Y1	Y2	Y3>
$|0>$	X	X	TGGATC	TGGATC	TGGATC	TGGATC
$|1>$	$|0>$	$|0>$	I	TGGATC	TGGATC	TGGATC
$|1>$	$|0>$	$|1>$	TGGATC	I	TGGATC	TGGATC
$|1>$	$|1>$	$|0>$	TGGATC	TGGATC	I	TGGATC
$|1>$	$|1>$	$|1>$	TGGATC	TGGATC	TGGATC	I

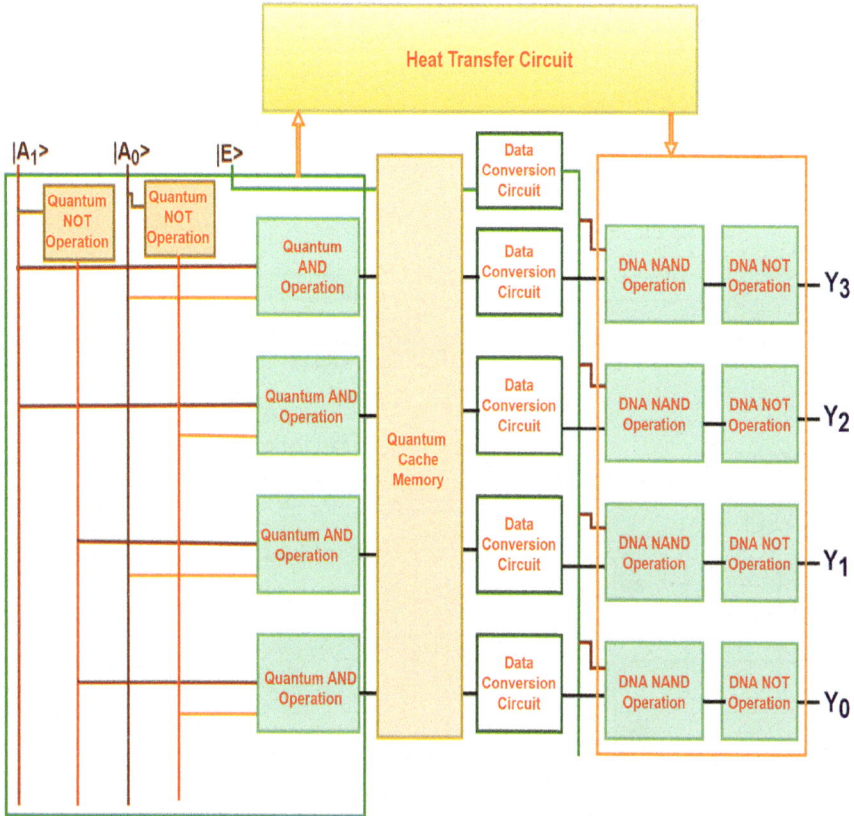

Fig. 10.13 Block diagram of a quantum-DNA 2-to-4 decoder

From the truth table, it can be written as follows:

Y3 = E.A1.A0
Y2 = E.A1.A0′
Y1 = E.A1′.A0
Y0 = E.A1′.A0′

10.2.3.2   Circuit Architecture

Figure 10.14 gives the view of the 2-to-4 quantum-DNA decoder circuit. Firstly, each of inputs $|A_1 >$ and $|A_0 >$ separately is connected with four quantum AND operations. The enable signal $|E>$ also goes to the DNA part directly as input. Then, the outputs of the four quantum AND operations go through the quantum cache memory and data conversion circuit to DNA part. All these then pass through the

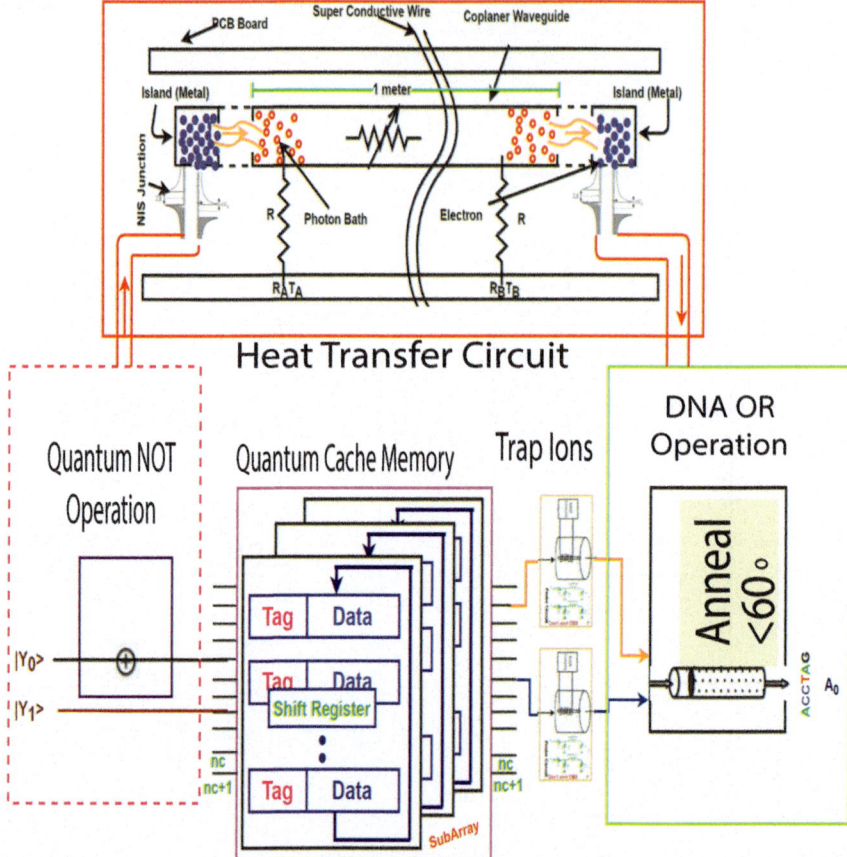

Fig. 10.14 Quantum-DNA 2-to-4 decoder circuit

four DNA AND operations and produce the four output DNA sequences Y0, Y1, Y2, and Y3.

10.2.3.3 Working Principle

The quantum-DNA 2-to-4 decoder needs two inputs |A1> and |A0>, and one enable input |E>, which activates the circuit. For quantum 2-to-4 decoder, suppose |E> is "true", |1>, then consider the following cases:

[i] The output D0 = A0′. A1′ is "true" **ACCTAG**, when the inputs |A1> and |A0> are "false" |0>. The output line D0 will be "true" **ACCTAG**; and D1 to D3 will be "false" **TGGATC**.

[ii] The output D1 = A0.A1' is "true" **ACCTAG**, when the input |A1> is "false" |0> and input |A0> is "true" |1>. The output line D1 will be "true" **ACCTAG** and D0, D2 and D3 will be "false" **TGGATC**.

[iii] The output D2 = A0'. A1 is "true" **ACCTAG**, when the input |A0> is "false" |0> and input |A1> is "true" |1>. The output line D2 will be "true" **ACCTAG** and D0, D1 and D3 will be "false" **TGGATC**.

[iv] The output D3 = A0.A1 is "true" **ACCTAG**, when the inputs |A0> and |A1> are "true" |1>. The output line D3 will be "true" **ACCTAG** and D0 to D2 will be "false" **TGGATC**.

10.3 Summary

In this chapter, a new concept of quantum-DNA computing with the possible architectural ideas for constructing various combinational circuits are presented. This chapter states specific design architectures for high-performance quantum-DNA combinational circuits. Some combinational circuits are also designed for 2-qubit and 4-qubit, together with their design techniques and functioning principles. Quantum computers generate a lot of heat, which causes disorder among qubits. On the other hand, DNA computers need a lot of heat to execute the reactions. As a result, there is a scope to utilize the amount of heat which can be needed or can be generated by these quantum-DNA combinational circuits. The heat transfer circuit transfers the needed heat from quantum part to DNA part. All of these themes have been discussed here in details.

Bibliography

1. L.M. Adleman, Molecular computation of solutions to combinatorial problems. Science **266**(5187), 1021–1024 (1994)
2. K.J. Breslauer, R. Frank, H. Blöcker, L.A. Marky, Predicting DNA duplex stability from the base sequence. Proc. Natil. Acad. Sci. **83**(11), 3746–3750 (1986)
3. N. Isailovic, Y. Patel, M. Whitney, J. Kubiatowicz, Interconnection networks for scalable quantum computers, in *33rd International Symposium on Computer Architecture (ISCA'06)* (IEEE, 2006), pp. 366–377
4. L.B. Levitin, T. Toffoli, Z. Walton, Operation time of quantum gates. arXiv preprint quant-ph/0210076 (2002)
5. M.M. Mano, *Digital Logic and Computer Design* (Pearson Education India, 2017)
6. D.D. Thaker, T.S. Metodi, A.W. Cross, I.L. Chuang, F.T. Chong, Quantum memory hierarchies: efficient designs to match available parallelism in quantum computing, in *33rd International Symposium on Computer Architecture (ISCA'06)* (IEEE, 2006), pp. 378–390
7. J. Watada, DNA computing and its application, in *Computational Intelligence: A Compendium* (Springer, 2008), pp. 1065–1089
8. X. Zheng, J. Yang, C. Zhou, C. Zhang, Q. Zhang, X. Wei, Allosteric DNAzyme-based DNA logic circuit: operations and dynamic analysis. Nucleic Acids Res. **47**(3), 1097–1109 (2019)

Chapter 11
DNA-Quantum Combinational Circuits

11.1 Introduction

As discussed in the previous chapter, the new and unique computing technology which is called quantum biocomputing or quantum-DNA computing, is the combination of quantum computing and DNA computing. Another new combination is DNA-quantum computing, which is the same as quantum-DNA computing but in an opposite manner. Both of these new computing paradigms have been considered as potential and challenging sector for the researchers. This book introduces these two technologies globally for the first time. This chapter will discuss the combinational circuits, such as multiplexer, demultiplexer, encoder, and decoder in DNA-quantum computing. DNA-quantum combinational circuits or biological quantum combinational circuits refer to the use of quantum principles and biological systems, like DNA, to create computational circuits that perform logical operations. These circuits leverage the unique properties of quantum mechanics, such as superposition and entanglement, to potentially enhance computational capabilities beyond classical circuits. These circuits could potentially offer advantages in speed, efficiency, and complexity compared to classical circuits, making them promising for applications in fields like biology, medicine, and artificial intelligence. Developing and implementing these circuits faces significant challenges, including the need to understand and control quantum phenomena in biological systems and the challenges associated with manipulating and reading information from biological molecules.

11.2 Design Architectures of DNA-Quantum Combinational Circuits

A combinational circuit combines numerous gates or elements together to form a single circuit, such as encoders, decoders, multiplexers, and demultiplexer, etc. are the

H. M. Hasan Babu, *Quantum Biocomputing in Quantum Biology Volume I*,
https://doi.org/10.1007/978-981-97-7154-7_11

examples of combinational circuits. Some of the features of combinational circuits are as follows:

1. A combinational circuit's output is wholly defined by the levels present at the input terminals at any given time.
2. The combinational circuit does not have any memory. The previous state of the input has no influence on the current state of the circuit.
3. A combinational circuit can have an n numbers of inputs and an m numbers of outputs.

DNA-quantum combinational circuits have two parts such as DNA combinational circuits and quantum combinational circuits. DNA combinational circuits use the principles of molecular biology and DNA strand displacement to create circuits that perform logical operations, similar to electronic circuits, but with DNA as the building material. These circuits leverage the specific base pair recognition in DNA to design logic gates and perform computations. In essence, DNA combinational circuits provide a way to create computational logic using DNA molecules, opening up possibilities for new applications in biotechnology, medicine, and other fields. On the other hand, quantum combinational circuits utilize qubits and quantum gates to perform computations, similar to how classical combinational circuits use bits and logic gates. These circuits are designed to manipulate qubits through a series of quantum operations, aiming to achieve computational advantages over their classical counterparts. Quantum combinational circuits are being explored for various applications, including cryptography, drug discovery, and materials science. Combinational circuits in terms of DNA-quantum computing are discussed in this section.

11.2.1 DNA-Quantum Multiplexer

In the DNA-quantum computing platform, the multiplexer works in opposite direction of quantum-DNA multiplexer. In the context of DNA-based quantum computing, a multiplexer is a circuit that selects one of multiple input signals and forwards it to a single output, similar to its role in electronic circuits. DNA-quantum computing utilizes the principles of quantum mechanics to perform calculations using DNA molecules as qubits, and multiplexers can be implemented in this system to control the flow of information between these qubits. This emerging field explores using DNA as a medium for quantum information processing. DNA strands can act as qubits, and their interactions can be manipulated to perform computations.

11.2.1.1 Block Diagram

In 2-to-1 DNA-quantum multiplexer, there are only two inputs I0 and I1, one selection line S, and one output Y. The truth table of a 2-to-1 DNA-quantum multiplexer is

Table 11.1 Truth table of a 2-to-1 DNA-quantum multiplexer

Inputs	Outputs
S	$|Y>$
TGGATC	$|I_0>$
ACCTAG	$|I_1>$

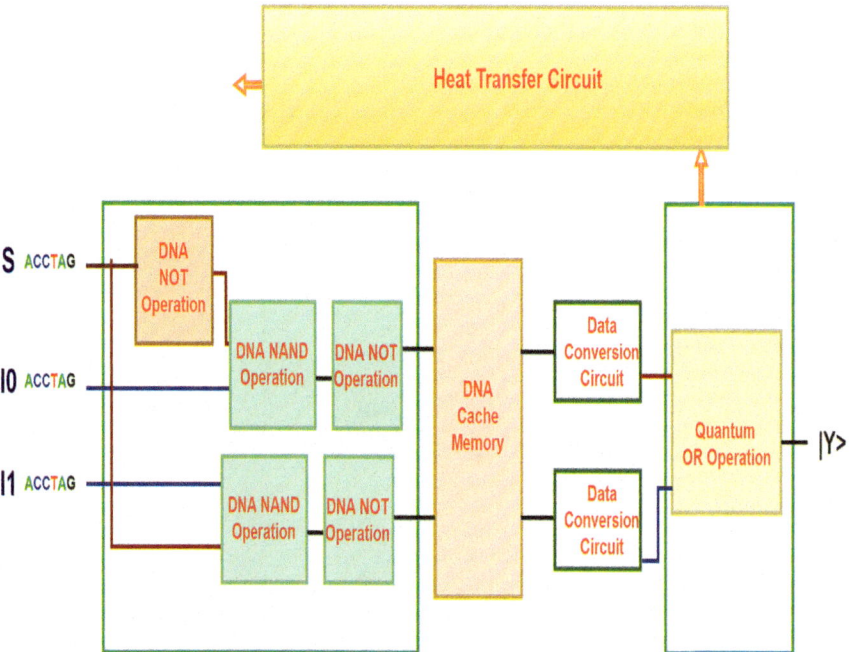

Fig. 11.1 Block diagram of a DNA-quantum 2-to-1 multiplexer

given in Table 11.1; and Table 11.2 shows the truth table of a 4-to-1 DNA-quantum multiplexer.

Figure 11.1 presents the block diagram of a 2-to-1 multiplexer. For this circuit, one DNA NOT operation, two DNA AND operations, and one quantum OR operation are needed.

Figure 11.2 presents the block diagram of a 4-to-1 multiplexer. In this circuit, two DNA NOT operations, eight DNA AND operations, and three quantum OR operations are needed.

Table 11.2 Truth table of a 4-to-1 DNA-quantum multiplexer

Inputs		Outputs
S1	S0	|Y>
TGGATC	TGGATC	$\|I_0>$
TGGATC	ACCTAG	$\|I_1>$
ACCTAG	TGGATC	$\|I_2>$
ACCTAG	ACCTAG	$\|I_3>$

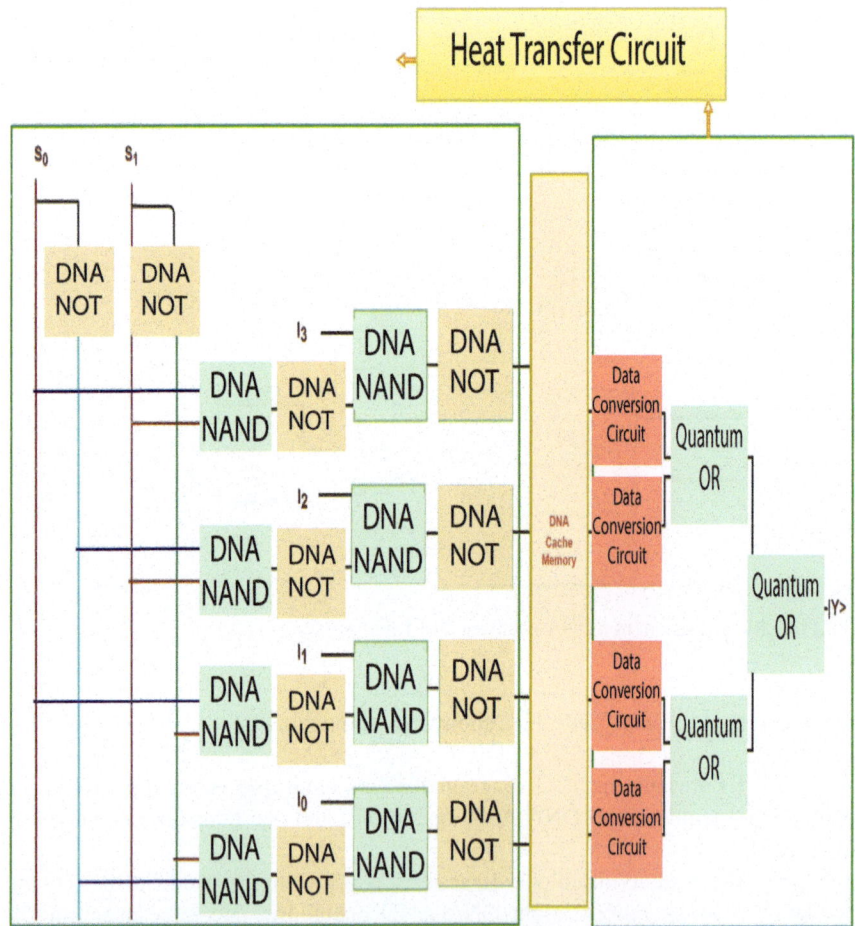

Fig. 11.2 Block diagram of a DNA-quantum 4-to-1 multiplexer

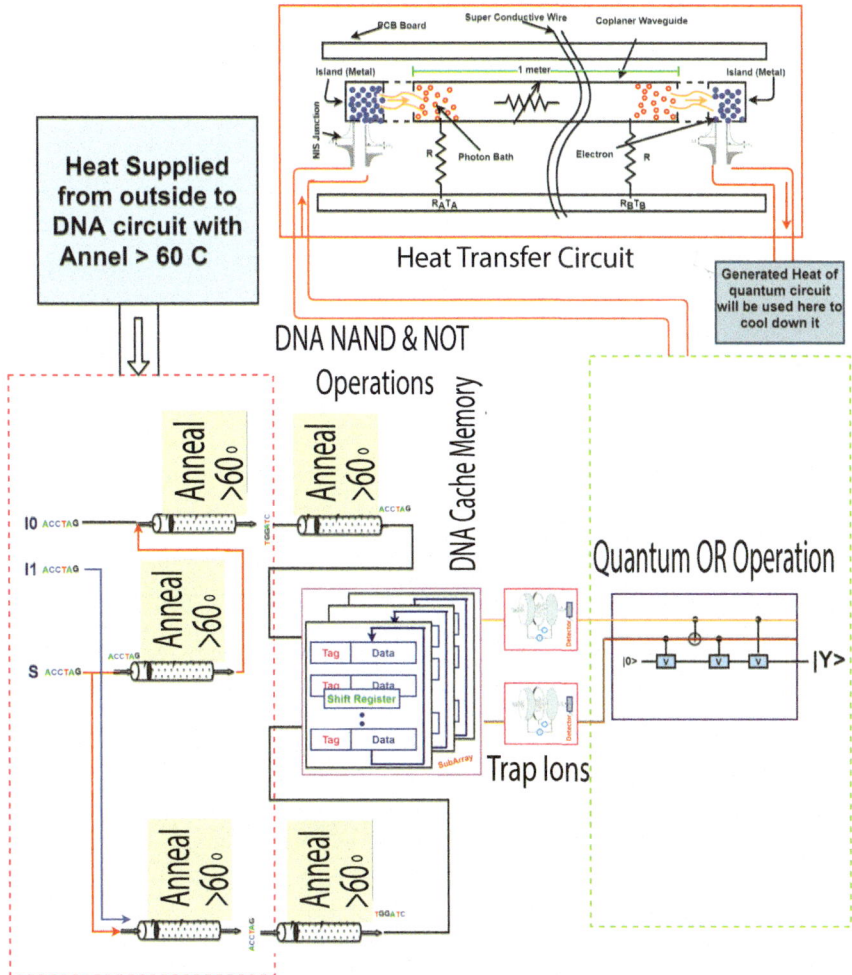

Fig. 11.3 DNA-quantum 2-to-1 multiplexer circuit

11.2.1.2 Circuit Architecture

Figure 11.3 depicts the DNA-quantum circuit of a 2-to-1 multiplexer. Firstly, each of inputs I_0 and I_1 separately is connected with two DNA AND operations. The input of the selection line, S, is connected directly to DNA AND-1 operation, whereas another input is connected to DNA AND-2 operation after completing the DNA NOT operation. Finally, both the outputs of DNA AND operation are passed through trap ion and perform quantum OR operation, and generate the value of $|Y>$.

Figure 11.4 gives the view of the 4-to-1 multiplexer DNA circuit. Firstly, each of inputs I0, I1, I2, and I3 separately is connected with four DNA AND operations.

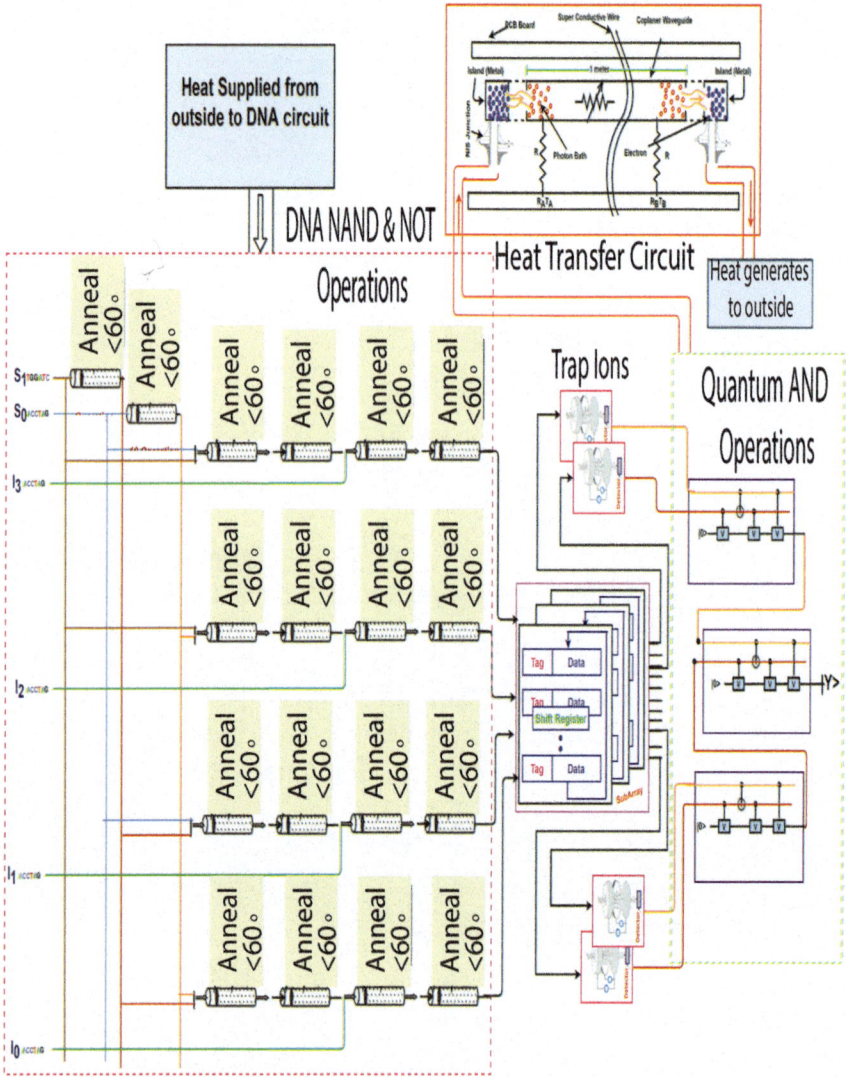

Fig. 11.4 DNA-quantum 4-to-1 multiplexer circuit

Then, the other inputs of these AND operations come from the output of another four DNA AND operations, which are executed through two selection lines, S0 and S1. Each DNA AND operation needs a DNA NAND and a DNA NOT operation to produce its desired output. Finally, the outputs of DNA NOT-2, NOT-4, NOT-6, and NOT-8 go through trap ion and after that, three quantum OR operations are required to generate the value of |Y>.

11.2.1.3 Working Principle

DNA-Quantum 2-to-1 Multiplexer: The DNA-quantum 2-to-1 multiplexer needs one selection line S0 and two inputs, I0 and I1. When inputs I0-I1 are "true", **ACC-TAG**, consider the following cases:

[i] The output |Y> = |I0>, when the input sequence S is "false" **TGGATC**.
[ii] The output |Y> = |I1>, when the input sequence S is "true" **ACCTAG**.

DNA-Quantum 4-to-1 Multiplexer: The DNA-quantum 4-to-1 multiplexer needs two selection lines S0 and S1, and four inputs, I0-I3. When all inputs I0-I3 are "true", **ACCTAG**, consider the following cases:

[i] The output |Y> = |I0>, when the input sequences S1 and S0 are "false" **TGGATC**.
[ii] The output |Y> = |I1>, when the input S1 is "false" **TGGATC** and input S0 is "true" **ACCTAG**.
[iii] The output |Y> = |I2>, when the input S0 is "false" **TGGATC** and input S1 is "true" **ACCTAG**.
[iv] The output |Y> = |I3>, when the inputs S0 and S1 are "true" **ACCTAG**.

11.2.2 DNA-Quantum Demultiplexer

A DNA-quantum demultiplexer is a combinational circuit [5], where the first part is a DNA circuit and the rest of the part is the quantum circuit. In the context of DNA and quantum computing, a demultiplexer (DEMUX) is a logic circuit that distributes data from a single input to multiple output lines, similar to a switch. Specifically, in DNA computing, demultiplexers are constructed using DNA sequences that can be selectively hybridized, allowing for the routing of information based on control signals.

11.2.2.1 Block Diagram

In 1-to-2 DNA-quantum demultiplexer, there are only two inputs, I0 and I1, one selection line, S, and one output, Y. The truth table of a 1-to-2 DNA-quantum demultiplexer is given in Table 11.3, and Table 11.4 shows the truth table of a 1-to-4 DNA-quantum demultiplexer.

Figure 11.5 presents the block diagram of a 1-to-2 demultiplexer. For this circuit, two DNA NOT operations, one DNA NAND operation, and one quantum AND operation are needed.

Figure 11.6 shows the block diagram of a 1-to-4 demultiplexer. In this circuit, six DNA NOT, four DNA NAND, and four quantum AND operations are needed.

Table 11.3 Truth table of a 1-to-2 DNA-quantum demultiplexer

Inputs	Outputs	
S	$|I_0>$	$|I_1>$
TGGATC	$|0>$	$|D>$
ACCTAG	$|D>$	$|0>$

Table 11.4 Truth table of a 1-to-4 DNA-quantum demultiplexer

Inputs		Outputs			
S1	S0	$|Y0>$	$|Y1>$	$|Y2>$	$|Y3>$
TGGATC	TGGATC	$|I>$	$|0>$	$|0>$	$|0>$
TGGATC	ACCTAG	$|0>$	$|I>$	$|0>$	$|0>$
ACCTAG	TGGATC	$|0>$	$|0>$	$|I>$	$|0>$
ACCTAG	ACCTAG	$|0>$	$|0>$	$|0>$	$|I>$

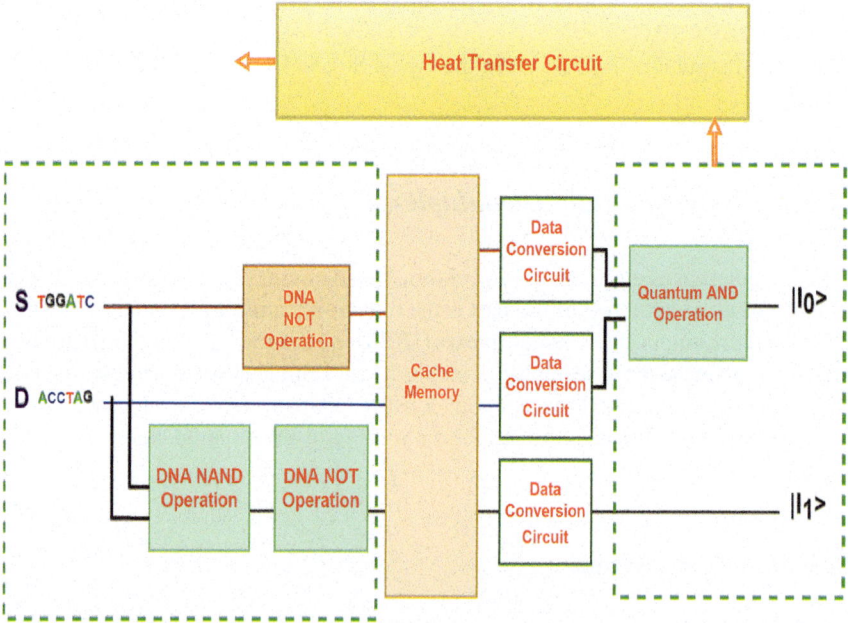

Fig. 11.5 Block diagram of a DNA-quantum 1-to-2 demultiplexer

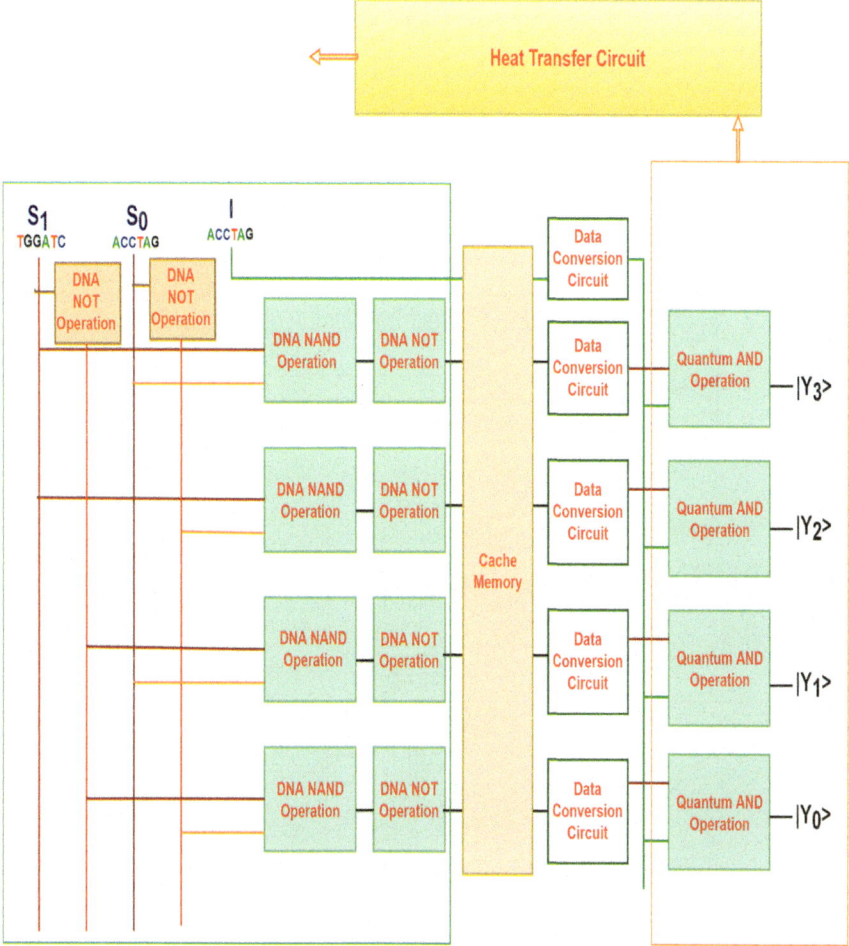

Fig. 11.6 Block diagram of a DNA-quantum 1-to-4 demultiplexer

11.2.2.2 Circuit Architecture

Figure 11.7 depicts the DNA-quantum circuit of a 1-to-2 demultiplexer. Firstly, inputs D and S are connected to DNA AND operations. The output of this operation then converts into qubit through trap ion and produces the result of |I1>. On the other hand, to get the result of |I0>, at first a DNA NOT operation is performed on S. Then, S′ and D perform data conversion using trap ion to generate qubit and perform a quantum AND operation to get |I0>. Based on the values of S and D, the output will be |I1>.

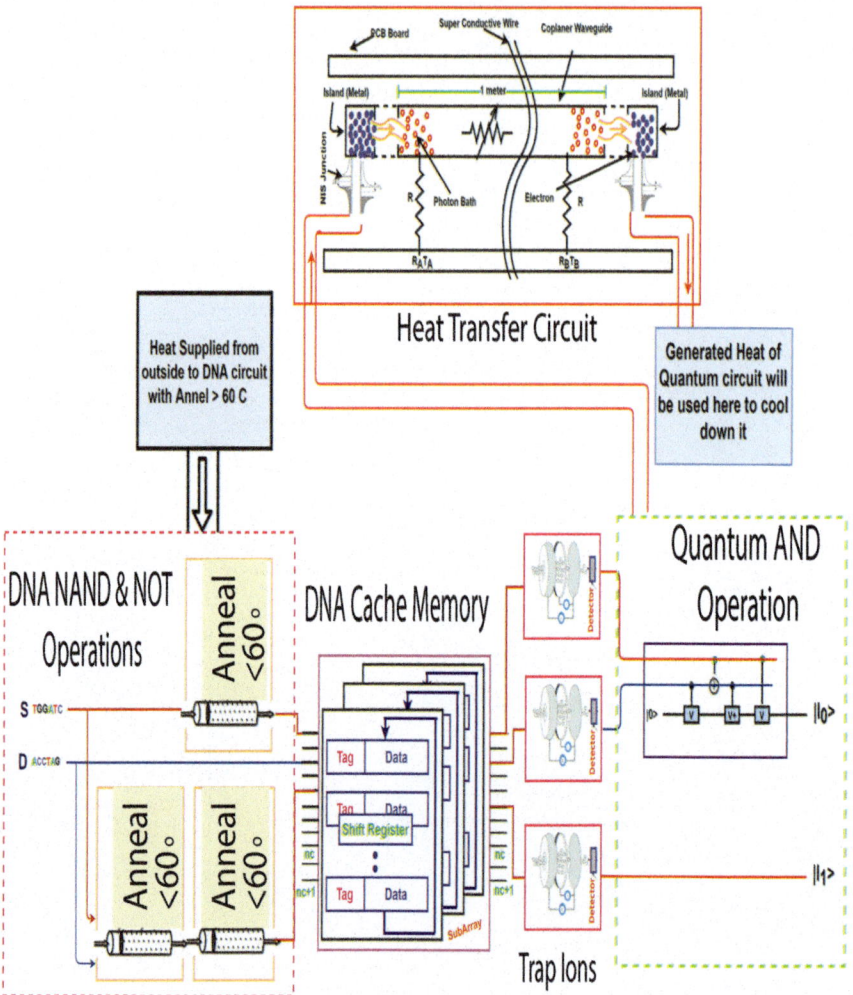

Fig. 11.7 DNA-quantum 1-to-2 demultiplexer circuit

For the 1-to-4 DNA-quantum demultiplexer circuit (Fig. 11.8), the input sequence is directly connected with four quantum AND operations after converting into qubit, |I>, by trap ion. The other inputs of these quantum AND operations come from the outputs of four DNA AND operations, which are executed through two selection lines, S0 and S1. To obtain the desired output, each DNA AND operation requires a DNA NAND and a DNA NOT operations. After this, the output of this DNA AND operation first performs a trap ion operation to convert into qubits. Then, these qubits perform quantum AND operation with input, |I>, separately to produce the results of |Y3>, |Y2>, |Y1>, and |Y0>.

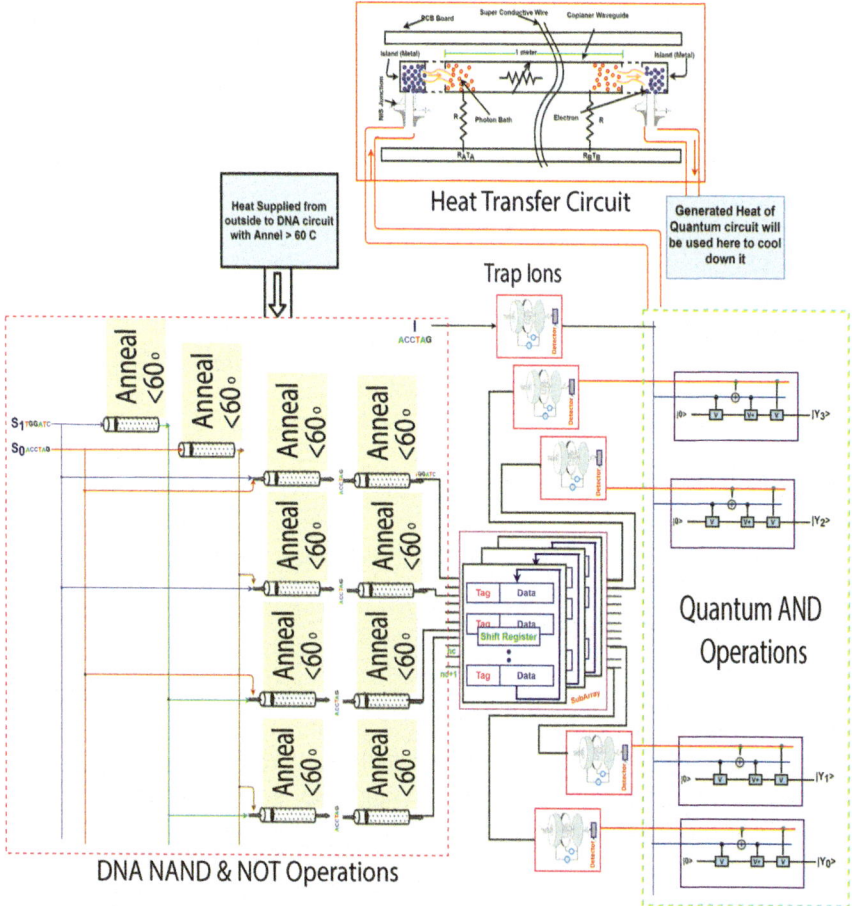

Fig. 11.8 DNA-quantum 1-to-4 demultiplexer circuit

11.2.2.3 Working Principle

DNA-Quantum 1-to-2 Demultiplexer: The DNA-quantum 1-to-2 demultiplexer needs one selection line S0 and one input, D. For DNA-quantum 1-to-2 demultiplexer, when input D is "true", **ACCTAG**, consider the following cases:

[i] The output $|I0> = S'.D$ is "true" $|1>$, when the input sequence S is "false" **TGGATC** and $|I1>$ will be "false" $|0>$.
[ii] The output $|I1> = S.D$ is "true" $|1>$, when the input sequence S is "true" **ACCTAG** and $|I0>$ will be "false" $|0>$.

DNA-Quantum 1-to-4 Demultiplexer: The DNA-quantum 1-to-4 demultiplexer needs two selection lines S0 and S1, and one input, I. For DNA-quantum 1-to-4 demultiplexer, when the input I is "true", **ACCTAG**, consider the following cases:

[i] The output |Y0> = S0'. S1'. I is "true" |1>, when the input sequences S1 and S0 are "false" **TGGATC**. The output line |Y0> will be "true" |1> and |Y1> to |Y3> will be "false" |0>.

[ii] The output |Y1> = S0.S1'.I is "true" |1>, when the input S1 is "false" **TGGATC** and input S0 is "true" **ACCTAG**. The output line |Y1> will be "true" |1>; and |Y0>, |Y2>, and |Y3> will be "false" |0>.

[iii] The output |Y2> = S0'.S1.I is "true" |1>, when the input S0 is "false" **TGGATC** and the input S1 is "true" **ACCTAG**. The output line |Y2> will be "true" |1>; and |Y0>, |Y1>, and |Y3> will be "false" |0>.

[iv] The output |Y3> = S0.S1.I is "true" |1>, when the inputs S0 and S1 are "true" **ACCTAG**. The output line |Y3> will be "true" |1> and |Y0> to |Y2> will be "false" |0>.

11.2.3 DNA-Quantum Encoder

The DNA-quantum encoder is a combinational circuit that takes 2^n inputs and generates n outputs. In the context of DNA-based quantum computing, an "encoder" refers to a mechanism that translates classical information (like binary data or genetic sequences) into a quantum state, typically using DNA as a physical substrate. This encoding process allows quantum algorithms to be applied to problems involving biological sequences or other data represented by DNA. This type of encoder focuses on representing biological sequences (like DNA or proteins) in a quantum state.

11.2.3.1 Block Diagram

In 2-to-1 DNA-quantum encoder, there are only two inputs Y0 and Y1, and one output A. Truth table of a 2-to-1 DNA-quantum encoder is given in Table 11.5, and Table 11.6 represents the truth table of a 4-to-2 DNA-quantum encoder.

Table 11.5 Truth table of a 2-to-1 DNA-quantum encoder

Inputs		Outputs
Y1	**Y0**	**|A>**
TGGATC	**ACCTAG**	|0>
ACCTAG	**TGGATC**	|1>

Table 11.6 Truth table of a 4-to-2 DNA-quantum encoder

Inputs				Outputs	
Y0	Y1	Y2	Y3	$|A1>$	$|A0>$
I	TGGATC	TGGATC	TGGATC	$|0>$	$|0>$
TGGATC	I	TGGATC	TGGATC	$|0>$	$|1>$
TGGATC	TGGATC	I	TGGATC	$|1>$	$|0>$
TGGATC	TGGATC	TGGATC	I	$|1>$	$|1>$

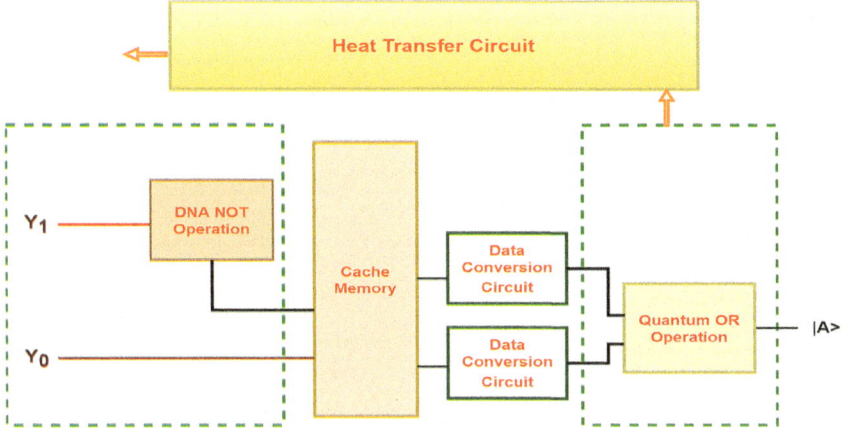

Fig. 11.9 Block diagram of a DNA-quantum 2-to-1 encoder

Figure 11.9 portrays the block diagram of a 2-to-1 encoder. For this circuit, one DNA NOT operation and one quantum OR operation are needed.

Figure 11.10 presents the block diagram of a 4-to-2 encoder. In this circuit, one DNA OR operation and one quantum OR operation are needed.

11.2.3.2 Circuit Architecture

In DNA-quantum 2-to-1 encoder circuit, two molecular sequences Y0 and Y1 act as the inputs (Fig. 11.11). Here, at first, Y0 performs a DNA NOT operation and produces Y0'. Then, both Y0' and Y1 sequences convert into quantum qubit by using trap ion. Finally, $|Y0'>$ and $|Y1>$ conduct quantum OR operation to produce the result of $|A>$.

In DNA-quantum 4-to-2 encoder circuit, three molecular sequences Y0, Y1, and Y2 act as the inputs (Fig. 11.12). Here, at first, Y2 and Y3 perform a DNA OR operation which then gives the result of $|A1>$ after the data conversion using trap ion. On the other side, Y0 and Y1 also perform trap ion to convert from molecular

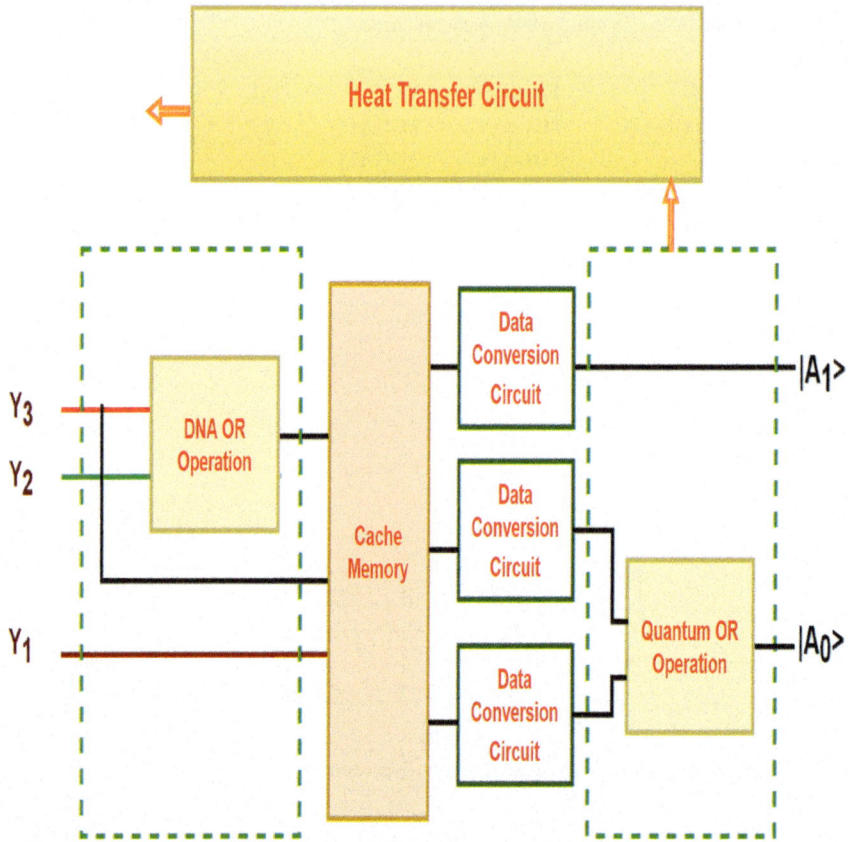

Fig. 11.10 Block diagram of a DNA-quantum 4-to-2 encoder

sequence to qubit, and these qubits, then act as the input of quantum AND operation to generate the outcome of |A0>.

11.2.3.3 Working Principle

DNA-Quantum 2-to-1 Encoder: The DNA-quantum 2-to-1 encoder needs two inputs Y0 and Y1, and one output A0. Consider the following cases:

[i] The output |A0> = Y0 is "true" |1>, when the input sequences Y1 is "false" **TGGATC** and Y0 is "true" **ACCTAG**.

[ii] The output |A0> = Y1 is "true" |1>, when the input sequences Y0 is "false" **TGGATC** and Y1 is "true" **ACCTAG**.

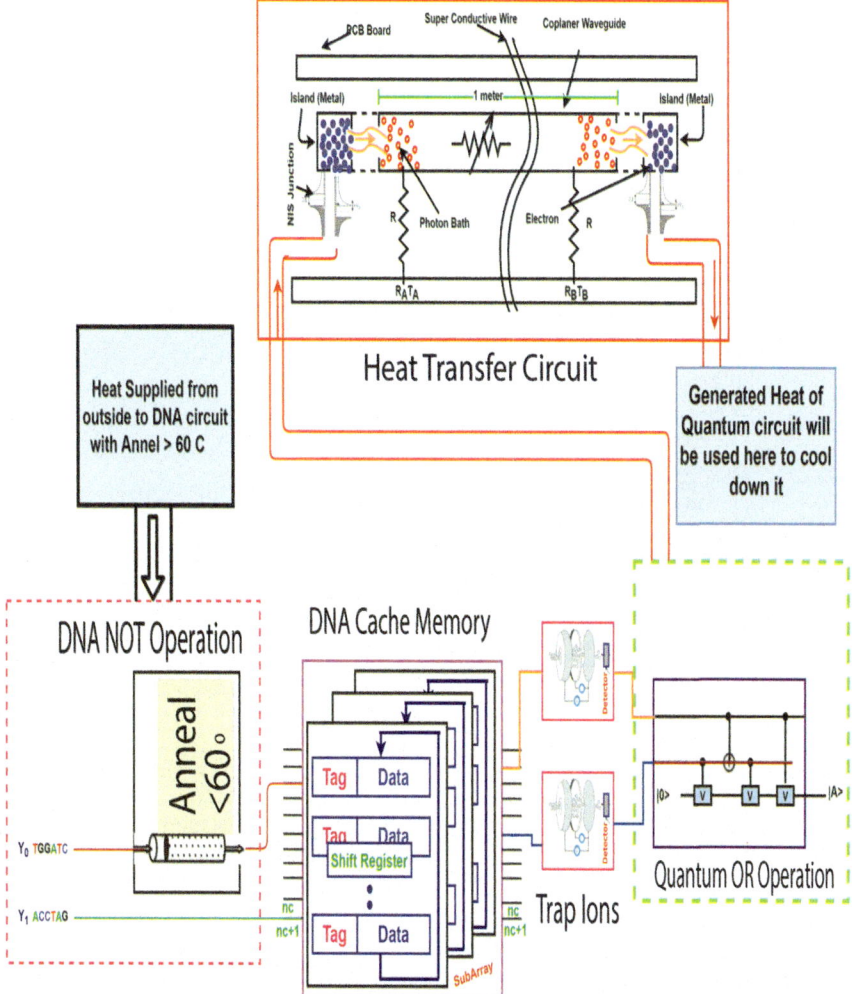

Fig. 11.11 DNA-quantum 2-to-1 encoder circuit

DNA-Quantum 4-to-2 Encoder: The DNA-quantum 4-to-2 encoder needs four inputs Y0, Y1, Y2, and Y3 and two outputs A0 and A1. Consider the following cases:

[i] The output $|A0> = Y1 + Y3$ is "true" $|1>$, when the input sequence Y1 or input sequence Y3 is "true" **ACCTAG**. The output $|A1> = Y2 + Y3$ is "true" $|1>$, when the input sequence Y2 or the input sequence Y3 is "true" **ACCTAG**.

[ii] The output $|A0> = Y1 + Y3$ is "false" $|0>$, when the both input sequences Y1 and Y3 are "false" **TGGATC**. The output $|A1> = Y2 + Y3$ is "true" $|1>$, when the input sequence Y2 or the input sequence Y3 is "true" **ACCTAG**.

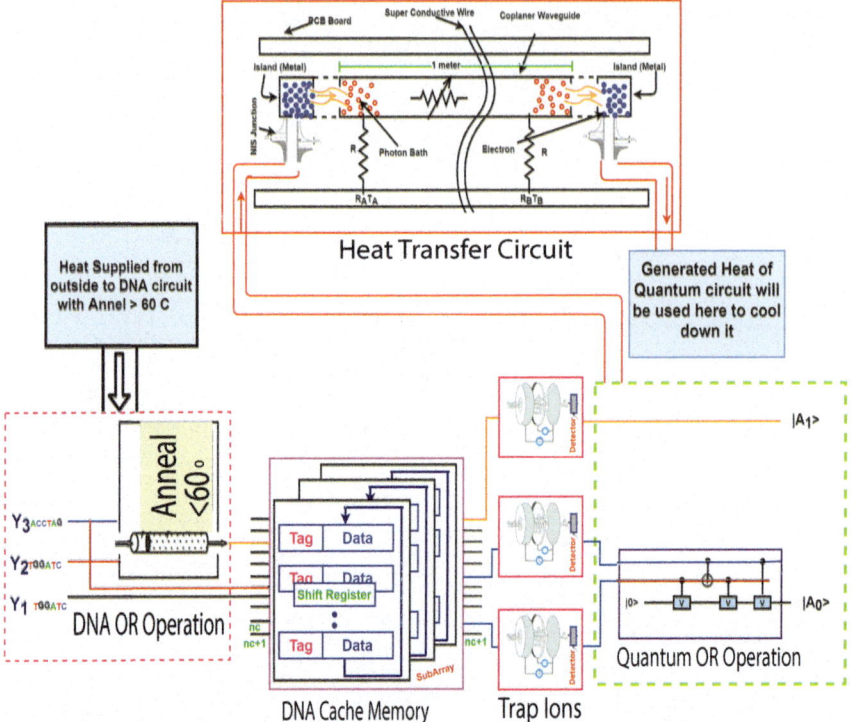

Fig. 11.12 DNA-quantum 4-to-2 encoder circuit

[iii] The output |A0> = Y1 + Y3 is "true" |1>, when the input sequence Y1 or the input sequence Y3 is "true" **ACCTAG**. The output |A1> = Y2 + Y3 is "false" |0>, when the both input sequences Y1 and Y3 are "false" **TGGATC**.

[iv] The output |A0> = Y1 + Y3 is "false" |0>, when the both input sequences Y1 and Y3 are "false" **TGGATC**. The output |A1> = Y2 + Y3 is "false" |0>, when the both input sequences Y1 and Y3 are "false" **TGGATC**.

11.2.4 DNA-Quantum Decoder

The DNA-quantum decoders are combinational circuits which are a combination of DNA decoder and quantum decoder. In the context of DNA-based quantum computing, a "decoder" refers to the process of converting the information encoded in DNA back into a quantum format that can be used for computation or further processing. This involves identifying and interpreting the specific DNA sequences or modifications that represent the data being stored or manipulated. A DNA-quantum decoder is a theoretical concept that explores using quantum phenomena to read and understand

Table 11.7 Truth table of a 2-to-4 DNA-quantum decoder

Enable	Inputs		Outputs			
E	S1	S0	$|Y0>$	$|Y1>$	$|Y2>$	$|Y3>$
TGGATC	X	X	$|0>$	$|0>$	$|0>$	$|0>$
ACCTAG	TGGATC	TGGATC	$|1>$	$|0>$	$|0>$	$|0>$
ACCTAG	TGGATC	ACCTAG	$|0>$	$|1>$	$|0>$	$|0>$
ACCTAG	ACCTAG	TGGATC	$|0>$	$|0>$	$|1>$	$|0>$
ACCTAG	ACCTAG	ACCTAG	$|0>$	$|0>$	$|0>$	$|1>$

the genetic code within DNA. It leverages quantum mechanics, specifically quantum tunneling, to differentiate between the four base pairs (A, T, C, G) that make up DNA. This approach could potentially revolutionize DNA sequencing, making it faster and more efficient. In a DNA-quantum decoder, quantum tunneling is used to probe the electrical conductivity of single DNA base pairs as they interact with nanoelectrodes. In essence, a DNA-quantum decoder is a futuristic idea that explores the possibility of using quantum physics to unlock the secrets hidden within our genetic code.

11.2.4.1 Block Diagram

Using the truth table in Table 11.7, Fig. 11.13 presents the block diagram of a 2-to-4 decoder. In this circuit, two DNA NOT operations, four DNA AND operations, and four quantum OR operations are needed.

11.2.4.2 Circuit Architecture

For 2-to-4 DNA-quantum decoder circuits (Fig. 11.14), firstly, the input sequence directly connects with four quantum AND operations after converting into qubits, $|E>$, by trap ion. The other input of these quantum AND operations comes from the output of four DNA AND operations which are executed between two selection lines, S0 and S1. To obtain the desired output, each DNA AND operation requires a DNA NAND and a DNA NOT operation. After this, the output of these DNA AND operations first performs trap ion operation in order to convert into qubits. Then, these qubits do quantum AND operation with input, $|E>$, separately to produce the result of $|Y3>$, $|Y2>$, $|Y1>$, and $|Y0>$.

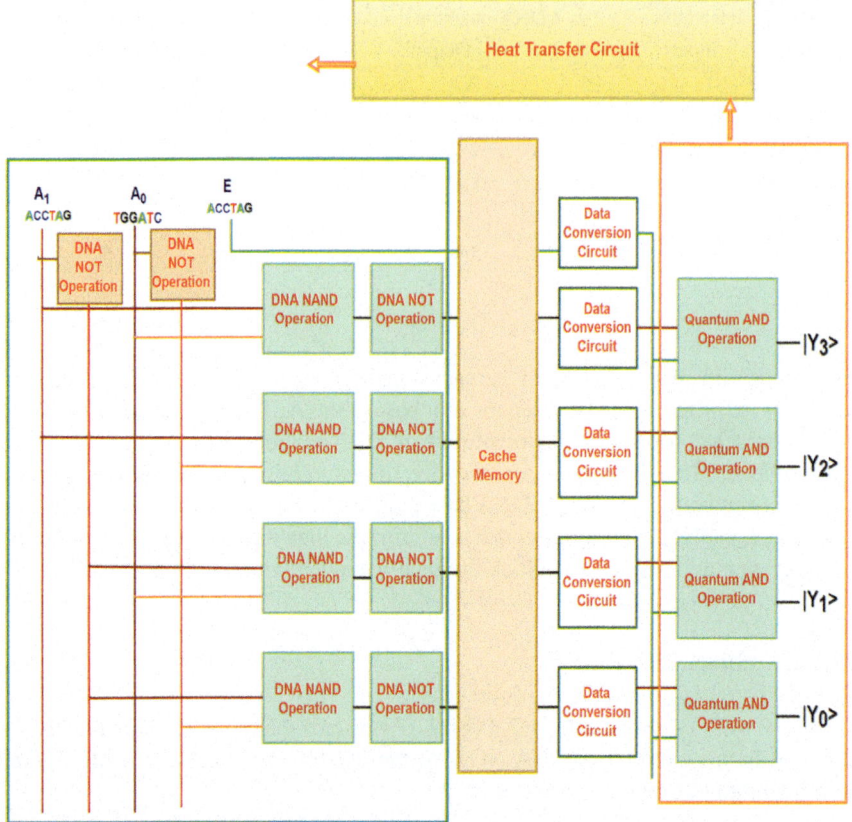

Fig. 11.13 Block diagram of a DNA-quantum 2-to-4 decoder

11.2.4.3 Working Principle

The DNA-quantum 2-to-4 decoder needs two inputs A1 and A0, and one enables input E, which activates the circuit. For the DNA 2-to-4 decoder, suppose E is "true", **ACCTAG**, consider the following cases:

[i] The output |D0> = A0'.A1' is "true" |1>, when the input sequences A1 and A0 are "false" **TGGATC**. The output line |D0> will be "true" |1> and |D1> to |D3> will be "false" |0>.

[ii] The output |D1> = A0.A1' is "true" |1>, when the input A1 is "false" **TGGATC** and input A0 is "true" **ACCTAG**. The output line |D1> will be "true" |1> and |D0>, |D2>, and |D3> will be "false" |0>.

[iii] The output |D2> = A0'.A1 is "true" |1>, when the input A0 is "false" **TGGATC** and input A1 is "true" **ACCTAG**. The output line |D2> will be "true" |1> and |D0>, |D1>, and |D3> will be "false" |0>.

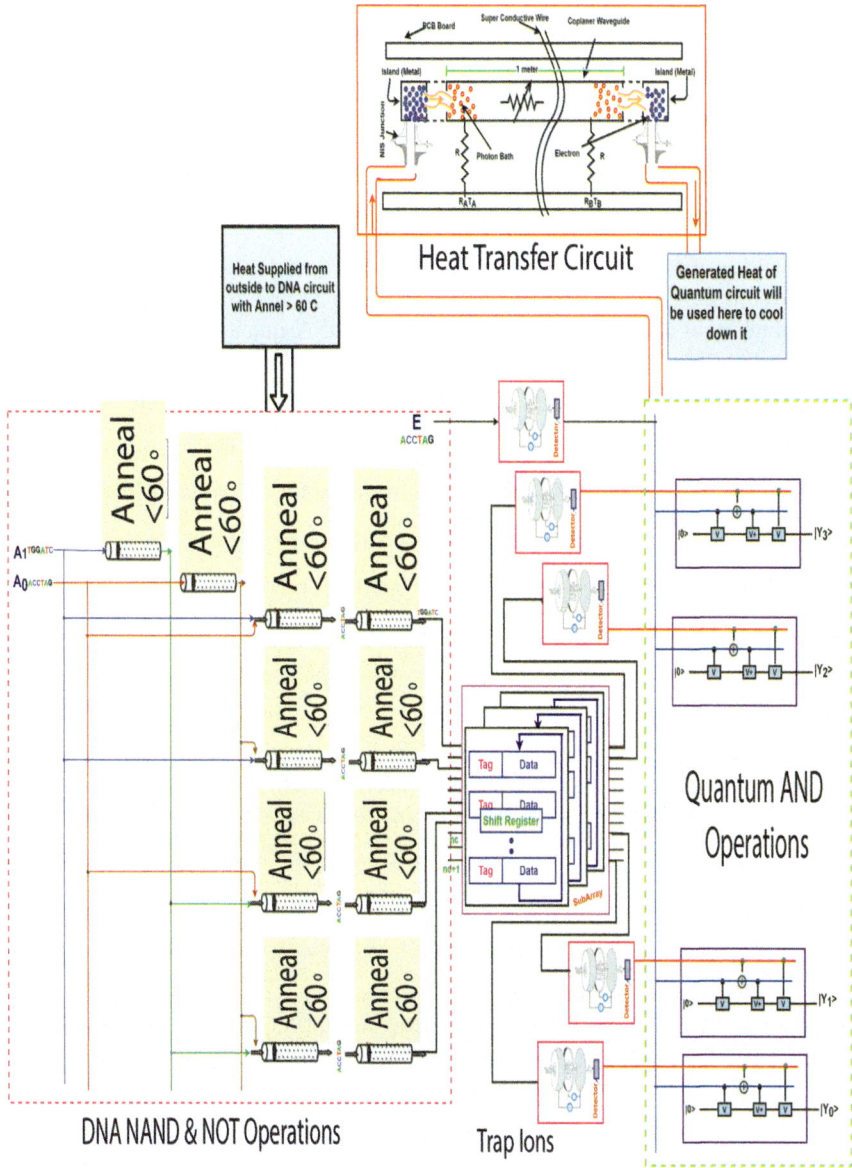

Fig. 11.14 DNA-quantum 2-to-4 decoder circuit

[iv] The output |D3> = A0.A1 is "true" |1>, when the inputs A0 and A1 are "true" **ACCTAG**. The output line |D3> will be "true" |1> and |D0> to |D2> will be "false" |0>.

11.3 Summary

The new idea of DNA-quantum computing is introduced in this chapter along with possible architectural concepts for creating multiple 2-valued combinational circuits. The architectural concepts for high-performance DNA-quantum combinational circuits are described in this chapter. Combinational circuits for 2-qubits and N-qubit, as well as their design approaches and operating principles, are also illustrated. Heat is required for the functioning of DNA circuits, whereas quantum circuits emit heat. The quantum part's excessive heat might ruin the entire link. To keep the circuit balanced, heat is delivered to the DNA section from an outside source, while excessive heat from the quantum part is sent to the cooler.

Bibliography

1. L.M. Adleman, Molecular computation of solutions to combinatorial problems. Science **266**(5187), 1021–1024 (1994)
2. K.J. Breslauer, R. Frank, H. Blöcker, L.A. Marky, Predicting DNA duplex stability from the base sequence. Proc. Nat. Acad. Sci. **83**(11), 3746–3750 (1986)
3. N. Isailovic, Y. Patel, M. Whitney, J. Kubiatowicz, Interconnection networks for scalable quantum computers, in *33rd International Symposium on Computer Architecture (ISCA'06)* (IEEE, 2006), pp. 366–377
4. L.B. Levitin, T. Toffoli, Z. Walton, Operation time of quantum gates. arXiv preprint quant-ph/0210076 (2002)
5. M. Morris Mano, *Digital Logic and Computer Design* (Pearson Education India, 2017)
6. J. Watada, DNA computing and its application, in *Computational Intelligence: A Compendium* (Springer, 2008), pp. 1065–1089
7. X. Zheng, J. Yang, C. Zhou, C. Zhang, Q. Zhang, X. Wei, Allosteric DNAzyme-based DNA logic circuit: operations and dynamic analysis. Nucl. Acids Res. **47**(3), 1097–1109 (2019)

Part IV
Sequential Circuits in Quantum Biocomputing

Overview

A sequential circuit is a special type of circuit that has a number of inputs and outputs. The outputs of the sequential circuits depend on the combination of current inputs and previous outputs. The previous output is considered the current state. In these circuits, their outputs depend not only on the combination of logic states of their inputs but rather on the logic states that previously existed. In other words, their outputs depend on the sequence of events occurring at the circuit's inputs. Examples of such circuits include clocks, flip-flops, stable circuits, counters, memory, and registers. Combinational circuits do not use memory. So, the previous input state does not affect the current state of the circuit. However, sequential circuits have memory, so the output can depend on the input. Quantum biocomputing sequential circuits or quantum biological sequential circuits combine quantum computing principles with the study of biological information processing. They utilize quantum circuits to represent and manipulate biological sequences (like DNA or protein sequences) and can be used for tasks like sequence comparison or understanding the evolution of biological systems. Quantum circuits can perform sequential operations on qubits, allowing for the manipulation and processing of biological information in a structured way. Quantum biological sequential circuits can be used for various biological tasks, such as sequence comparison, understanding evolution, and drug discovery, etc. While quantum computers offer potential advantages for certain biological calculations, they are not yet competitive with classical technologies in terms of scale. This type of circuit uses older inputs, outputs, clocks, and memory elements. This part has three chapters where sequential circuits in quantum computing, quantum-DNA computing, and DNA-quantum computing (quantum biocomputing) will be discussed. All sequential circuits will be described in these three computing modes. The block diagram, circuit architecture, and the working principle of all sequential circuits will be discussed here. Their examples and applications of sequential circuits are also discussed. D flip-flop, SR latch, SR flip-flop, JK flip-flop, T flip-flop, shift register, ripple counter, and synchronous counter will be discussed in different computing modes.

Chapter 12
Sequential Circuits in Quantum Computing

12.1 Introduction

Quantum computing is one of the most exciting and new scientific disciplines to emerge in the world today. Sequential quantum circuits are specialized quantum circuits designed to map between different gapped phases of quantum matter. They achieve this by applying unitary transformations to local regions of a system in a sequential manner. This structure preserves key properties like entanglement area law and the gapped nature of quantum states, while still enabling changes in long-range correlations and entanglement, and ultimately, the phase of the system. Quantum computers are still a pipe dream, and they are not yet scalable. In this research, an attempt is made to suggest a quantum sequential circuit which is titled as sequential circuits in quantum computing. Quantum D flip-flop, quantum SR latch, quantum SR flip-flop, quantum JK flip-flop, quantum T flip-flop, quantum shift register, quantum ripple counter, and quantum synchronous counters are all examples of sequential circuits. This quantum circuit makes the use of three basis operations, which are V+, V, and CNOT quantum basic operations, for its operation. This section contains in-depth descriptions of each circuit's overall structure and construction methods. When it comes to large-scale quantum computing, there are numerous architectural obstacles to overcome. On a tiny scale, these circuits are an attempt to fully resolve some forms of difficulty in quantum computing.

12.2 Quantum D Flip-Flop

A quantum D flip-flop is essentially a two-state timed flip-flop. In one clock cycle, the qubit inputs of a quantum D-type flip-flop are actuated with a delay. A delay flip-flop is another term for the quantum D flip-flop. In the context of quantum computing, a

D flip-flop is a type of flip-flop circuit that is used to build more complex quantum circuits. It's essentially a way to store and manipulate quantum information. D flip-flops, along with other flip-flops like JK, can be used to create sequential quantum circuits, which are important for building quantum computers and other quantum devices. In quantum computing, D flip-flops can be implemented using quantum gates. Two or more quantum gates can be used to design a D flip-flop circuit. Quantum D flip-flops allow for the creation of sequential logic, where the output of a circuit depends on the order in which the inputs are applied. This is crucial for implementing many quantum algorithms. These quantum flip-flops provide a reliable way to store quantum information (qubits), which is necessary for building quantum memories and registers.

The indeterminate input conditions of SET = "$|0>$" and RESET = "$|0>$" are banned in the basic quantum SR NAND Gate Bistable circuit, which is one of its fundamental drawbacks. This condition forces both qubit outputs to logic "$|1>$," overriding the feedback latching action, and whichever input goes to logic "$|1>$" first loses control, while the other input, which is still at logic "$|0>$," controls the latch's final state. However, an inverter may be connected between the "SET" and "RESET" qubit inputs to create a quantum Data Latch, quantum Delay flip-flop, quantum D-type Bistable, quantum D-type flip-flop, or simply a quantum D flip-flop as it is most often known.

By far the most essential of all, the quantum timed flip-flop is the quantum D flip-flop. The $|S>$ and $|R>$ inputs become complements of each other when a quantum inverter (quantum NOT operation) is added between the Set and Reset inputs, ensuring that the two inputs $|S>$ and $|R>$ are never equal ($|0>$ or $|1>$) to each other at the same time, allowing us to control the toggle action of the flip-flop with just one $|D>$ (Data) input.

The Data input, labeled "$|D>$," is then utilized in place of the "Set" signal, and the inverter is used to create the complementary "Reset" input, resulting in a level-sensitive quantum D-type flip-flop from a level-sensitive SR-latch, with $|S> = |D>$ and $|R> = |D>$.

The quantum D flip-flop circuit has just one qubit input, and the qubit input must be in a coherence state in order to conduct the quantum computational function. As a result, the circuit must exist in an environment that does not exist. If any particle emerges, the coherence state will be disrupted. The quantum D flip-flop will generate heat, which must be removed quickly in order to cool down the circuit and stabilize the coherence state.

Quantum D flip-flop operational circuit is constructed using the basic component called the quantum NAND operation circuit. Quantum D flip-flop's working principle shows if the clock qubit is $|0>$ it will not work but if this clock qubit is $|1>$ then it will SET. Quantum D flip-flop produces much heat unlike other quantum computing circuits and operational time is much faster compared to other systems.

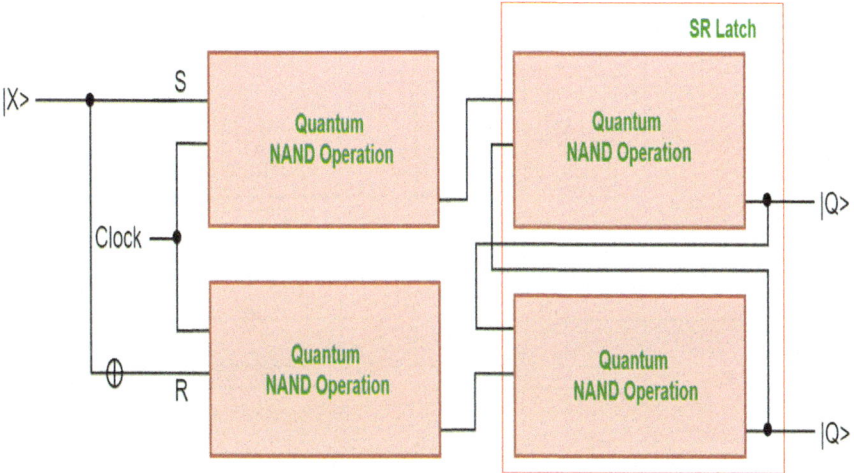

Fig. 12.1 Block diagram of a quantum D flip-flop

12.2.1 Block Diagram

The major operations of a D flip-flop are quantum NAND operations and quantum NOT operations. The quantum D flip-flop fundamentally has one qubit input, and in the core of the D flip-flop, there is an SR latch. When compared to other timed type flip-flops, the D flip-flop is one of the most significant memory devices. D flip-flop verifies that the two inputs of the SR flip-flop are never the same. Block diagram of a quantum D flip-flop is given in Fig. 12.1.

D flip-flops have two inputs: one for data and one for clock. D flip-flops have two outputs that are logically opposite to one another. The clock input aids in the circuit's synchronization with an external signal. The output of a D flip-flop can have two potential values. This block diagram shows that data input is sent to a quantum NAND operation circuit, while the reversal of data input is routed to another quantum NAND operation circuit. The clock pulse input is used by both NAND processes. The result of two quantum NAND operations is fed into the SR latch. The SR latch is used to build the D flip-flop. This attribute is utilized to create a delay in the data flow in the circuit. Two quantum NAND Operations create the SR latch. The remaining two outputs of a quantum SR latch function are discovered. The output of a D flip-flop can be of two sorts, one of which is logically inverse to the other. If the clock is enabled, the D flip-flop will continue to function; otherwise, the D flip-flop will be ceased to function.

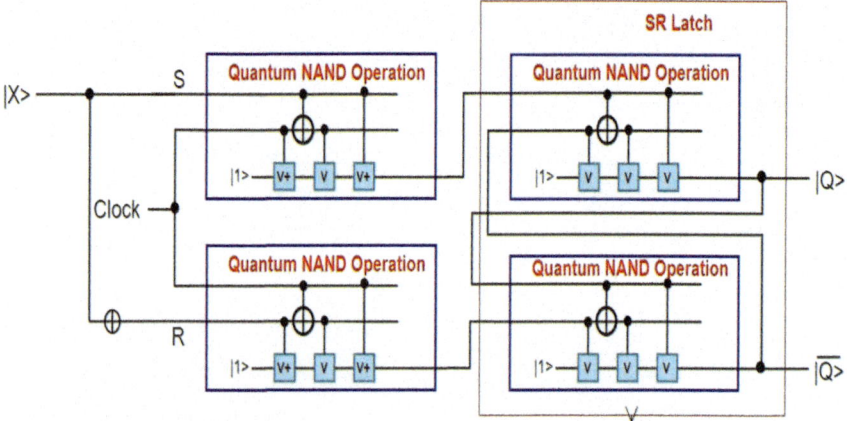

Fig. 12.2 Quantum D flip-flop

12.2.2 Circuit Architecture

The D flip-flop has a single qubit input and is developed using quantum NAND operations and a quantum SR latch. Circuit architecture of a quantum D flip-flop is shown in Fig. 12.2.

The clock qubit input affects the quantum D flip-flop. Seeing the diagram that the circuit has one qubit input. One line of this qubit input will be directed into the quantum NAND operation known as |S> input in a circuit. In this case, the |S> qubit input and the Clock qubit input are used in a quantum NAND operation.

When a |X> qubit traverses another line, it first undertakes a quantum NOT operation. This quantum NOT operation was dubbed R when it was entered into the quantum NAND operation. The R and Clock qubit inputs are used in this quantum NAND operation. The output of the quantum NAND operations is sent to the quantum SR latch as an input. With these two inputs, an SR latch will be done. The quantum SR latch will be used, and the output of the quantum SR latch will be the final output of |Q> and |O>. The conclusion of |O> will always be the inverse of |Q>.

12.2.3 Working Principle

There are two inputs in the quantum SR flip-flop: SET and RESET. Alternatively, in a D flip-flop, one input and the input's one line are referred to as a SET, and by coupling a quantum NOT operation toward the other line input, it may designate the D flip-flop as a RESET. This complement resolves the contradiction inherent in the SR latch when both inputs are LOW because that circumstance is no longer feasible. Quantum D flip-flops have a single input, which is sometimes alluded to as a data

Table 12.1 Truth table of a quantum D flip-flop

|Clk>	|x>	|Q>	|O>	Description
↓ >> |0>	X	Q	O	Memory no change
↑ >> |1>	|0>	|0>	|1>	Reset Q >> 0
↑ >> |1>	|1>	|1>	|0>	Set Q >>

input. If this data input is high, the flip-flop becomes SET; if the data input is low, such as |0>, the flip-flop changes state and becomes RESET.

However, this would be pretty futile because the output of the flip-flop's will always vary with each pulse delivered to this data input. To circumvent this, an extra input known as the "CLOCK" or "ENABLE" input is used to separate the data input from the latching circuitry of the flip-flop after the appropriate data has been stored. The result is that the |X> input condition is only replicated to the output |Q> while the clock input is active. This then serves as the foundation for yet another sequential gadget known as a D flip-flop.

As long as the clock input is HIGH, the "D flip-flop" will store and output any logic level that is applied to its data terminal. Once the clock input is changed to LOW, the flip-flop's "set" and "reset" inputs are both kept at logic level "|1>," preventing the flip-flop from altering the underlying and preserving whatever statistics are available on its output prior to the clock transition. In other words, either logic "|0>" or logic "|1>" latches the output. The truth table of quantum D flip-flop is given in Table 12.1.

12.2.4 Example

In quantum D flip-flop operational circuit one input is qubit input |x>. Assume this input is |0>. This input will perform quantum NAND operation with the clock input.

Then qubit input will perform quantum NOT operation, and after that, again it will perform quantum NAND operation.

Two quantum NAND operations produce the output which will be entered into an input in the quantum SR latch operational circuit. This quantum SR latch operational circuit output will count as a final output in quantum D flip-flop operational circuit.

This is the final output of quantum D flip-flop operational circuit and these outputs clarify that the correct output is produced by the quantum D flip-flop operational circuit.

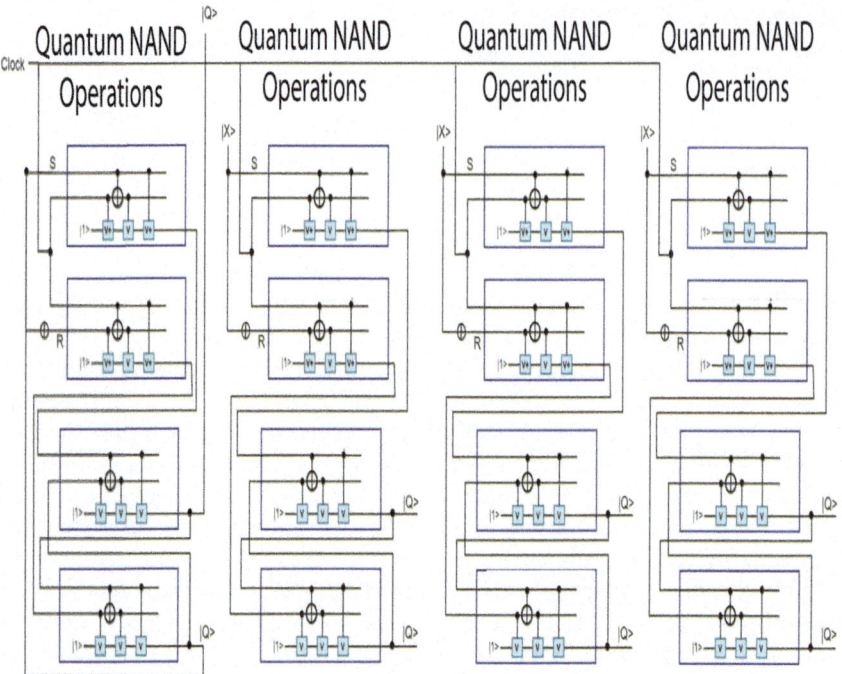

Fig. 12.3 Quantum asynchronous counter using D flip-flop

12.2.5 *Applications*

When the clock is triggered, the quantum D flip-flop is utilized to store a qubit. Quantum D flip-flops are used in a variety of applications, the most common of which is in processors. The primary application of a quantum D flip-flop is as a Frequency Divider. If the |Q> output of a quantum D flip-flop is linked directly to the |X> input, resulting in closed-loop "feedback," the successive clock pulses will cause the bistable to "toggle" once every two clock cycles. D flip-flops may be used to construct delay lines, which are commonly used in digital signal processing systems. The synchronous D flip-flop is an input that has been delayed by one clock cycle. Quantum asynchronous counter using D flip-flop is given in Fig. 12.3.

Quantum D flip-flops are also utilized in quantum shift registers and quantum ring counters. Two inputs are used in quantum D flip-flops, one for data and another for the clock. The outputs of quantum D flip-flops are logically opposite to one another. The circuit's synchronization with an external signal is aided by the clock input. A quantum D flip-flop's output can have two possible values. Data input is directed to a quantum NAND operation circuit in this block diagram, while data input reverse is routed to another quantum NAND operation circuit. Both NAND procedures use the clock pulse input. Quantum SR latch receives the result of two quantum NAND operations. The D flip-flop is constructed using the SR latch. This property is used to induce a delay in the circuit's data flow. Quantum SR Latch is made up of two quantum NAND Operations. The final two outputs of the quantum SR latch function are uncovered. A quantum D flip-flop may provide two types of output, one of which is logically inverse to the other. The quantum D flip-flop will continue to function if the clock is enabled; otherwise, the quantum D flip-flop will stop working.

Quantum D flip-flop operational circuits are also coupled through serial connection in the quantum shift register block diagram. The quantum D flip-flop operational circuit is the fundamental component of the quantum shift register.

Quantum sequence generators, quantum multiplexers, quantum D flip-flop-based counters, and other processor components are constructed using the quantum D flip-flop operational circuit. Quantum D flip-flop has a significant role in quantum processors since it can solve the quantum SR latch problem. Quantum D flip-flop can also be used in register, multiplexer, frequency divider, etc., quantum circuits. This quantum D flip-flop is very essential to prepare the quantum computing processor.

12.3 Quantum SR Latch

Based on the triggering that is suited to operate it, there are two types of memory elements. One of them is a latch, and the other one is a flip-flop. Latches operate with enable signal and level-sensitive, whereas flip-flops are edge sensitive. In the context of quantum computing, an SR latch is a fundamental building block for storing information. It's a quantum circuit designed to emulate the behavior of a traditional SR latch using quantum gates. Quantum SR latches allow for the preservation of information during operations, and can be implemented using quantum gates like Fredkin or Toffoli gates. In essence, the quantum SR latch provides a way to store and manipulate quantum information reliably within quantum circuits, enabling the development of more sophisticated quantum computing systems.

A quantum SR latch is an asynchronous tool. It operates without the use of control signals, relying solely on the state of the $|S>$ and $|R>$ inputs. Two quantum NAND operations can make a quantum SR latch. Nevertheless, two quantum NOR operations can also make a quantum SR Latch. In quantum SR latch, two-qubit inputs are swapped and negated. Quantum SR latch can be said as SET RESET latch. In quantum SR latch from two-qubit input, two outputs are generated. This output is reversed to one another. This circuit consists of a quantum SR Latch, and two quantum NAND operation circuits designed for this SR latch. The quantum SR latch has the input line swapped between two quantum NAND operations, but it is not negated. Quantum SR latch works as memory stuff in quantum computers, and it has several applications in a quantum processor. If some embedded systems have been designed using the quantum device, then quantum SR latch will be used on this device as a memory unit. Quantum SR latch is level sensitive and has few disadvantages, and this will be recovered by the quantum flip-flop.

Fig. 12.4 Block diagram of
a quantum SR latch

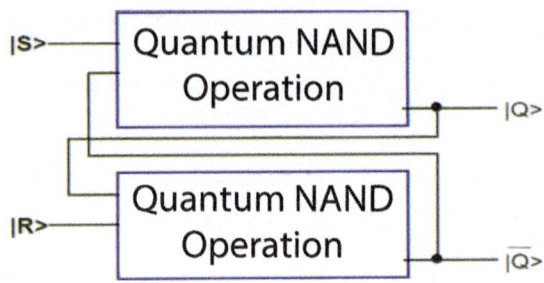

12.3.1 Block Diagram

The quantum SR latch is one of the most common memory devices, and it has an effect on the output as long as it is active. The essential properties of a quantum SR latch are that one qubit input behaves like a SET and another qubit input behaves like a RESET. Block diagram of a quantum SR latch is given in Fig. 12.4.

The quantum SR latch is made up of two fundamental processes, which are depicted in this block diagram of a quantum SR latch. There are two input lines in the quantum SR latch, one for |S> and the other for |R>. Two-qubit input generates two outputs |Q> and |O>. The output of the first quantum NAND operation is used as an input in the second quantum NAND operation, and the output of the second quantum NAND operation is used as an input in the first quantum NAND operation. If the input of |S> is |1>, the SR latch is activated; however, if the input of |R> is |1>, the SR latch has no influence on the output. In a quantum SR latch, a value of |1> cannot be used to activate two inputs.

12.3.2 Circuit Architecture

Quantum SR latches are level sensitive and are built using only one fundamental operation which is the quantum NAND function. The circuit architecture of quantum SR latch is depicted in Fig. 12.5.

As seen in the diagram above, the quantum SR latch has two-qubit inputs. In the first quantum NAND operation, the input |S> and the output of the second quantum NAND operation |O> are both entered as inputs. There are two V+ operations and one V operation in quantum NAND operation, as well as a CNOT gate. The output of this quantum NAND operation is mostly |Q>.

Secondly, the quantum NAND operation is performed on the inputs |R> and |Q>, yielding |O> as the output. The auxiliary qubit |1> is used in every quantum NAND operation. In quantum systems, this is a logical bit that is used to repair errors.

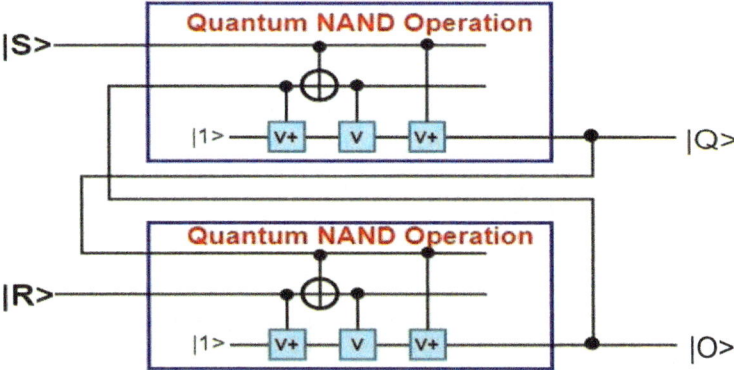

Fig. 12.5 Quantum SR latch

12.3.3 Working Principle

The state of the |S> and |R> inputs are important that matters in a quantum SR latch. It acts independently of control signals. |S> input behaves as if it is a SET instruction, and |R> input behaves as if it is a RESET instruction. If the SET input of a quantum SR latch is high, the output |O> will be |1> and the opposing output |0> will be the value of |Q>. When the RESET input is high, the value of |Q> is |1>, and when the RESET input is low, the value of |Q> is |0>. The latch's "memory" is basically reset. When both inputs are low, the latch "latches" stay in their previously set or reset state.

The actual problem comes when both the inputs SET and RESET go high.

The outputs |Q> and |O> will have the opposite value as shown in the circuit. When the SET and RESET inputs |1> are used together, the circuit creates a "race situation." In order for the device to be "metastable," which implies it will remain in an indeterminate state indefinitely, both gates must be identical. In reality, if the suggested circuit is to be manufactured, one gate will win; however, determining which gate won is difficult. Because of this, having both the SET and RESET inputs high is forbidden in a quantum SR latch.

The same thing happens when the device is switched on, since both outputs, |Q> and |O>, are low. The device will quickly leave the metastable state due to the differences between the two gates, but it's difficult to forecast which of |Q> and |O> will end up high. To avoid erroneous actions, SR flip-flops must always be set to a known starting state before being used; users should not assume that they would initialize to a low state. The truth table of quantum SR latch is given in Table 12.2.

The flip-flop discussed in the flip-flop chapter solves the difficulty of the quantum SR latch. The quantum SR latch, on the other hand, is still a vital component of a CPU or embedded device.

The quantum circuit generates a lot of heat, making it difficult to isolate the qubit into a superposition state. As a result, it is required to cool the circuit to isolate the

Table 12.2   Truth table of a quantum SR latch

|S>	|R>	|Q>	|O>
|0>	|0>	Latched	
|0>	|1>	|1>	|0>
|1>	|0>	|0>	|1>
|1>	|1>	Metastable	

qubit into a superposition for an able quantum circuit. Any type of external particle can disrupt the qubit's coherence and cause it to become decoherent. If all of these are preserved, the quantum SR latch can truly function.

12.3.4   Example

Presume that the qubits |0> and |1> are both present in the quantum SR Latch. One qubit input will be used for SET instructions, while the other will be used for RESET instructions. Here, |0> will be used as a SET instruction, and it will conduct the quantum NAND operation according to the suggested circuit idea.

As a result, the final output is |0>. The output principle of a quantum NAND operation is that if one of the inputs is |0>, the output will also be |0>. As a corollary, whether |Q> is |0> or |1>, the output will be |1>.

Qubit input |1> now functions as a reset instruction and is inserted into the circuit, as well as performing the quantum NAND operation.

Quantum SR Latch operates in the same mechanism with each qubit input. However, because all of the computations take place in the quantum superposition state, a lot of heat is generated throughout the process. All computations take place in the coherence state, and the result will be decoherent.

12.3.5   Applications

The circuits can be utilized as storage devices in power gating circuits and clocks since the quantum SR latch is a single-qubit storage element. Quantum SR latch operational circuits are utilized as a memory device in computers and, in certain cases, in IoT devices. Latches are used in the construction of memory devices like flip-flops. Quantum SR flip-flops are made utilizing the quantum SR latch operational circuit in quantum computing.

The quantum SR latch is an asynchronous circuit that has been employed as an input-output port in several quantum asynchronous systems. The quantum SR latch has employed at different times in the quantum Asynchronous Counter and Shift Register.

Four quantum SR flip-flops construct a quantum Asynchronous Counter. In a quantum computer, the quantum SR flip-flop primarily functions as a SET RESET flip-flop. Quantum SR latches are used to build quantum SR flip-flops. To store one qubit till the asynchronous counter employed a quantum SR latch as a memory device.

Quantum SR latches are operational circuits that may be employed in a variety of IoT and computational devices. They can be utilized as pulse latches, which accomplish the same function as flip-flops by rapidly pulsing the clock. Quantum SR latches can be employed for data storage as well as computation. Making is nearly difficult without a quantum SR latch, quantum computer and other devices.

12.4 Quantum SR Flip-Flop

Quantum Sequential Logic circuits, unlike quantum Combinational Logic circuits, include some types of built-in "Memory" that change the state depending on the real signals supplied to its inputs at the moment. Quantum SR flip-flops, for example, have a 1-qubit memory bistable. The SET and RESET inputs of the SR flip-flop are the same. The output of the SET input is a $|1>$, whereas the output of the RESET input is a $|0>$.

The Quantum SR flip-flop is often referred to as the SET RESET flip-flop. The reset input is used to restore the flip-flop to its starting state from the current state with an output. The NAND gate SR flip-flop is a basic flip-flop with both outputs providing feedback to its opposite input. This circuit is used to store a single data qubit in a memory circuit. The three inputs are SET, RESET, and a found output. A two-qubit model will be used since quantum SR flip-flops have two inputs that are mostly from the outside. Because of using two-qubit, it generates more heat at first than quantum D flip-flops. The computation time of this quantum SR flip-flop is determined by the fundamental gate in its middle. Quantum SR flip-flops may be found in a wide range of processors and embedded systems. Although the suggested flip-flop can generate some trash, an error correcting auxiliary qubit provides the desired output. The real-world implementation of the suggested quantum circuit will address a wide range of issues more quickly and effectively. An SR flip-flop, is a fundamental memory element in quantum computing. While there have been some research efforts exploring quantum-based flip-flops using concepts like quantum technology, quantum dot cellular automata (QCA). In quantum computing, the basic unit of information is the qubit, which can exist in a superposition of 0 and 1 states. Quantum computers use various physical systems (like superconducting circuits, trapped ions, or semiconductor devices) to implement qubits. Research has explored designs that incorporate flip-flop-like behavior at the quantum level, such as flip-flop qubits. These flip-flop qubits leverage the interaction between electron spins and nuclear spins in silicon to create a single qubit. Quantum SR flip-flops are commonly used in memory storage, data synchronization, and as building blocks for more complex memory devices.

Quantum SR flip-flop operational circuit is built with the basic component known as quantum NAND operation circuit. A quantum SR flip-flop is also known as a SET RESET flip-flop. Quantum SR flip-flop can be found in quantum register circuit, quantum multiplexer, and quantum frequency divisor. Quantum SR flip-flop though have less use in real applications compared to other flip-flops because these flip-flop circuits have illegal state which is very much tough to conduct.

12.4.1 Block Diagram

A SET-RESET flip-flop is a common name for a quantum SR flip-flop. As a result, it is evident that the quantum SR latch has a two-qubit input. Block diagram of a quantum SR flip-flop is given in Fig. 12.6.

The major inputs of a quantum SR flip-flop are |S> and |R>, as well as one clock input called |clock>. Firstly, a quantum NAND operation with clock input is performed on two qubits |S> and |R>. Then, the quantum NAND operation is performed on input |S> and input |clock>. A quantum NAND operation is also performed in parallel by the |R> and |clock> inputs. Because the two quantum NAND operations are performed in parallel, where they take the same amount of time. Then, using quantum NAND operations output from the |S> input lines, another quantum NAND operation generates the output |O>, where both produce the final output |Q>, similarly, quantum NAND operations generate output from the |R> input lines, and the final output |Q> conducts the quantum NAND operation and generates the |O>.

Actually, the created output enters the quantum SR latch and finds the |Q> and |O> after the first quantum NAND operations are completed. The values of output |Q> and output |O> are diametrically opposite. The block diagram depicts all of

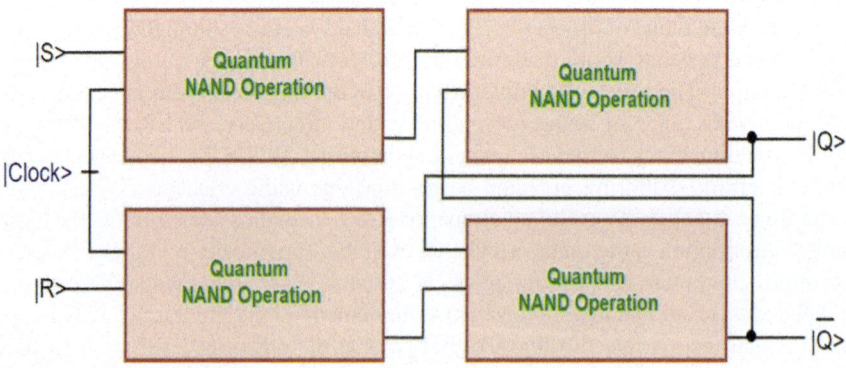

Fig. 12.6 Block diagram of a quantum SR flip-flop

the procedures involved in the quantum SR latch. During the operation, all qubits must be in the superposition state, which implies that they must be coherent. If any particle from the environment approaches close enough, the superposition state will evaporate.

12.4.2 Circuit Architecture

In quantum, SR flip-flop mainly works as a SET RESET flip-flop in a quantum computer. This quantum SR flip-flop circuit shown in Fig. 12.7 presents two inputs IS> and IR>, as well as a clock input. Four quantum NAND operation circuits make up the quantum SR. To begin, one quantum NAND operation is performed on the IS> and clock inputs. V, V+, and CNOT are the three main operations that make up the quantum NAND operating circuit. The quantum NAND operation is carried out by another input IR> and Iclk>. Two of the quantum NAND operations' outputs were used as inputs into the quantum SR latch. The quantum SR latch is controlled by these two inputs. One output from earlier quantum NAND operations comes from the IS> input line in a quantum SR latch, and one of the output SR flip-flops IQ> conducts the quantum NAND operation and provides the output IQ>. The output of the quantum NAND operation, which is based on IR> input and IQ> output, enters as an input in the quantum NAND operation and creates the output of the quantum SR flip-flop as IQ>, just as it did previously.

12.4.3 Working Principle

There are two-qubit inputs IS> and IR> in a quantum SR flip-flop. The quantum operation is performed once the qubit input becomes coherent and enters into a

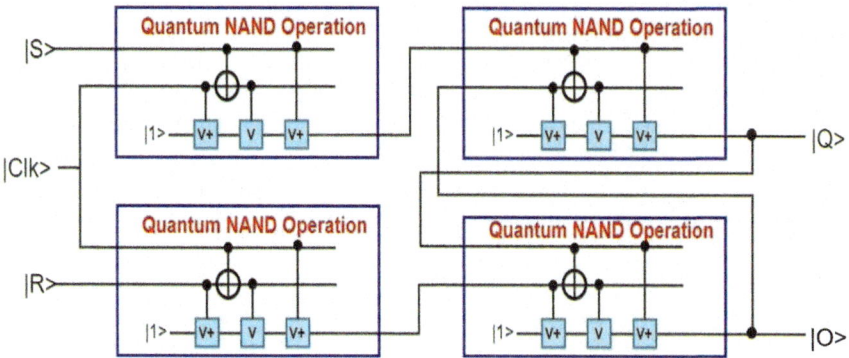

Fig. 12.7 Quantum SR flip-flop

Table 12.3 Truth table of a quantum SR flip-flop

IS>	IR>	IQ>	IO>
I0>	I0>	No change	
I0>	I1>	I0>	I1>
I1>	I0>	I1>	I0>
I1>	I1>	Invalid	

superposition state. It is required to transfer the heat from the quantum circuit since it creates a lot of it.

At first, the IS> input in a quantum SR flip-flop changes its state to superposition and becomes coherent. The quantum NAND operation is then performed using the clock qubit input Iclk> using the IS> input. It generates an output after completing the quantum NAND operation. The quantum NAND operation is also performed by the IR> input with the Iclk> input. The final outputs of both quantum NAND operations are used as inputs in the quantum SR latch. One input acts as if it is high, while the other acts as if it is low. The state of the IS> and IR> inputs is important in a quantum SR latch, which is independent of control signals. IS> input behaves as if it were a SET command, whereas IR> behaves as if it were a RESET command. The output IQ> will be I1> if the SET input of the quantum SR latches becomes high, and the opposing output I0> will be the value of IO>. When the RESET input is high, the value of IQ> is I1> and when the RESET input is low, the value of IQ> is I0>. The "memory" of the latch is basically reset. The latch "latches" stay in their previously set or reset state when both inputs are low.

The real issue arises when both the inputs SET and RESET go high. The outputs IQ> and IO> will have the opposite values as shown in the circuit. When the SET and RESET inputs I1> are used, the circuit creates a "race situation." Both gates should be identical for the device to be "metastable," which means it will be in an indeterminate state for an endless amount of time. In reality, if the suggested circuit is manufactured, one gate will win; however, determining which gate's winning is difficult. Because of this, having both the SET and RESET inputs high states in a quantum SR latch are unlawful.

When the device is turned on, both outputs, IQ> and IO>, are low, resulting in a similar situation. Because of the disparities between the two gates, the device will swiftly depart the metastable state, but it's hard to anticipate which of IQ> and IO> will end up high. It must always put SR flip-flops to a known starting state before using them to avoid spurious actions and must not presume that they will initialize to a low state. The truth table of quantum SR flip-flop is given in Table 12.3.

12.4.4 Example

Assume that this study effort receives the inputs |0> and |1> in order to ensure that the quantum SR flip-flop operational circuit produces the right output. The |0> input end is connected to the |S> input end, while the |1> input end is connected to the |R> input end. The fact that the |R> input represents |1> which indicates that it is for a reset operation. Assume that the clock is activated and the clock's input is |1>.

At first, do the quantum NAND operation using the |clk> input. If just one of the inputs is |0> in a quantum NAND operation, the result is |1>; otherwise, the output is |0>.

The quantum NAND operation is then performed using another |clk> and |R> inputs. In this case, the |clk> input is |1>, and the |R> input is also |1>. Now, consider the following:

Then, as the input to the quantum SR latch, these two input qubits, |1> and |0>, will be inputted. The quantum SR latch operation will be performed by them. A collection of quantum NAND operation circuits constructs the SR latch operation circuit.

For the inputs |0> and |1>, this quantum SR flip-flop circuit now has the outputs |0> and |1>. In a quantum SR flip-flop, this is the needed input corresponding to the provided input. Quantum SR flip-flops generate a lot of heat, yet this has no effect on the output.

12.4.5 Applications

A simple quantum NAND gate SR flip-flop circuit gives feedback from both of its outputs to its opposing inputs and is widely used to store a single data qubit in memory circuits. Many memories and IoT devices employ the quantum SR flip-flop operating circuit. The quantum SR flip-flop was primarily used to store one-qubit data. Quantum SR flip-flops are employed in quantum shift registers, quantum counters, asynchronous counter, and other memory devices.

The quantum shift register is also built using a quantum SR flip-flop. In a quantum computer, a quantum shift register is a particularly valuable memory device.

Although different quantum flip-flops can be utilized to create quantum shift registers, quantum SR flip-flops are sometimes employed. In some cases, quantum SR flip-flops are employed to build the quantum delay circuit. The hardware debouncing mechanism used by the quantum SR flip-flop also employs an S-R latch to eliminate bounces in the circuit, as well as the pull-up resistors. The quantum S-R flip-flop circuit is the most effective within all debouncing techniques. It can also function as a frequency divisor.

12.5   Quantum JK Flip-Flop

In flip-flop designs, the quantum JK flip-flop will be the most extensively utilized flip-flop. J and K have no abbreviated letters of other words, such as "S" for Set and "R" for Reset, but are independent letters chosen by the inventor Jack Kilby to identify the flip-flop design from others. Despite the fact that the digital electronics JK flip-flop was created by Jack Kilby. The functioning concept of the quantum JK flip-flop differs from that of the digital JK flip-flop. A JK flip-flop can be conceptually mapped to a quantum flip-flop (QFF) using three qubits. While a classical JK flip-flop is built with NAND gates, a quantum version leverages the principles of quantum bits (qubits). This allows for the development of quantum finite state machines (QFSMs) without the need for explicit external memory. A quantum JK flip-flop is utilizing qubits to store and manipulate quantum information. It can be implemented with three qubits, leveraging quantum gates (like CNOT) to control the qubits' superposition and entanglement. Three qubits are used to represent the flip-flop's state, potentially using the electron and nucleus of a phosphorus atom as qubits. By using qubits, QFFs can potentially reduce the resource requirements for certain sequential computations compared to classical flip-flops. QFFs can be used as building blocks for quantum sequential circuits, allowing for the development of quantum-based sequential logic. QFFs can also be used to implement QFSMs, which are computational models that can perform complex tasks using quantum information.

The quantum JK flip-flop's sequential operation is identical to that of the prior quantum SR flip-flop, with the same "Set" and "Reset" inputs. The distinction this time is that even though S and R are both at logic "1," the "quantum JK flip-flop" has no incorrect or prohibited quantum SR latch input states. It is evident that the quantum JK flip-flop does not solve the disadvantages of the quantum SR flip-flop.

The quantum JK flip-flop is essentially a gated quantum JK flip-flop with the addition of clock qubit input circuitry to avoid the unlawful or invalid output state that can arise when both inputs |J> and |K> are equal to logic level "1." A quantum JK flip-flop has four potential input combinations due to the extra timed input: "logic |1>", "logic |0>," "no change," and "toggle." A quantum JK flip-flop has the same symbol as a quantum SR Bistable Latch, as seen in the preceding chapter. The quantum JK flip-flop, like other flip-flops, generates a lot of heat, which must be dissipated in order to run properly. As compared to other quantum circuits, the quantum JK flip-flop will not require as much power. The qubit may simply conduct the operation when all of the qubits are in superposition state and coherence mode.

The quantum NAND operation circuit serves as the foundation for the quantum JK flip-flop operational circuit. The problem of quantum SR flip-flop is solved by a quantum JK flip-flop. The operation of a quantum JK flip-flop is demonstrated. Unlike other quantum computing circuits, the quantum JK flip-flop creates a lot of heat and has a much faster operational time than other systems. The quantum JK flip-flop has numerous applications in various devices. It can be utilized in quantum ripple counters, as well as many other counters, registers, and frequency division circuits that are built with quantum JK flip-flops. Quantum JK flip-flops are also employed in the construction of event detectors and many other devices.

12.5.1 Block Diagram

In the construction of quantum computers, the quantum JK flip-flop is the most often utilized flip-flop. |J> and |K> are two-qubit inputs in a quantum JK flip-flop. The block diagram of a quantum JK flip-flop is given in Fig. 12.8.

The qubit inputs |J> and |k> are used in the quantum JK flip-flop. Many quantum NAND operations construct this quantum JK flip-flop. After performing a pair of fundamental quantum NAND operations, the result of this operation is inserted into the JK flip-flop, yielding the outputs |Q> and | $\overline{Q}$ >. Firstly, the quantum NAND operation is performed using the |J> and |clk> inputs. Then, the output of the quantum NAND operation and the output of the quantum JK flip-flop |$\overline{Q}$> perform another quantum NAND operation and generate the designated output |J>. Because the |K> and |clk> inputs are shared, where the quantum NAND operation is performed on both of them. The quantum JK flip-flop's |Q> output executes another quantum NAND operation and generates the |R> output.

The quantum JK flip-flop accepts these |J> and |K> and produces two outputs, |Q> and |$\overline{Q}$>. There are four quantum NAND operations in the quantum JK flip-flop. The quantum NAND operation is performed by the |J> input and |clk> input. The quantum NAND operation is also performed by the |K> input and shared |clk> input. The result of these two NAND operations is used as input in the quantum JK latch. The quantum NAND operation is performed on quantum JK latches using two |Q> and one input, as well as |$\overline{Q}$> and another input. Finally, the quantum JK flip-flop's outputs |Q> and |$\overline{Q}$> are obtained after all of these quantum operations.

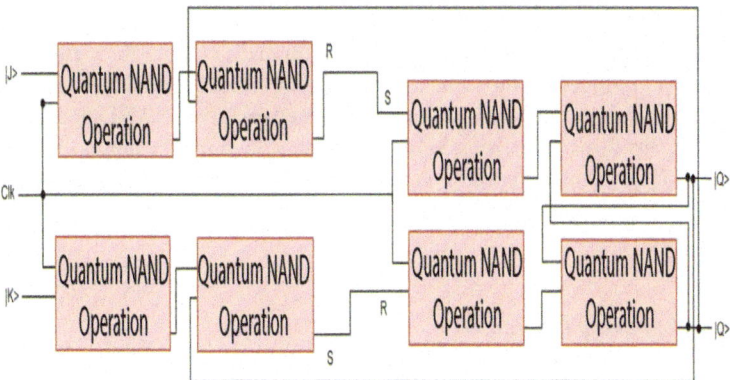

Fig. 12.8 Block diagram of a quantum JK flip-flop

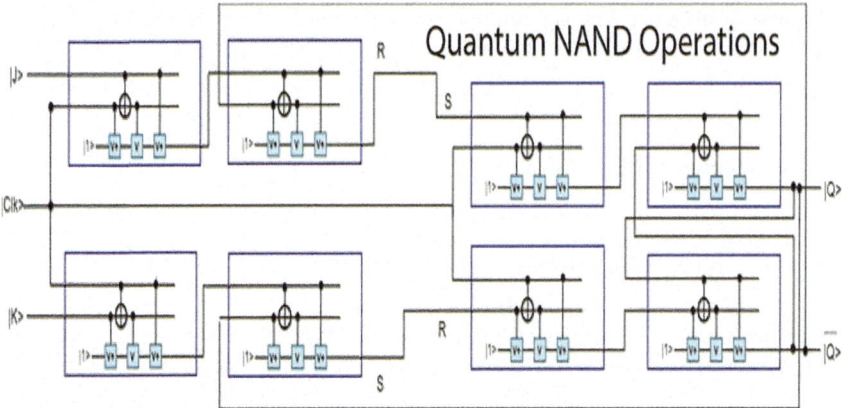

Fig. 12.9 Quantum JK flip-flop operation

12.5.2 Circuit Architecture

The two-qubit input and one clock input of the quantum JK flip-flop are shared. The clock affects the quantum JK flip-flop as well. The circuit will turn on if the clock is turned on, else it will not. The circuit architecture of a quantum JK flip-flop operation is depicted in Fig. 12.9.

Both qubit inputs |J> and input |K> will conduct the quantum NAND operation differently using the shared |clk> input. The quantum NAND operation is performed using the |J> and |clk> inputs. Quantum fundamental operations are also used to perform the NAND operation. The V, V+, and CNOT operations are the most common operations used in quantum computing. An auxiliary qubit is employed to rectify the problem in this case. The value of an auxiliary qubit in a quantum NAND operation is |1>. Acquiring an output qubit after the quantum NAND operation is performed by |J> and |clk>, and this output qubit together with the output of the flip-flop |Q>, is used to execute the quantum NAND operation and generate the output |J>. The |K> output is produced by the same technology that creates the |K>, |clk>, and |Q> inputs. The majority of these circuits are performed quantum NAND operations. The quantum SR flip-flop operation is conducted on the |S> and |R> inputs. The essential component of quantum computing is quantum NAND operation, which is also used in quantum SR flip-flop operation circuit design.

12.5.3 Working Principle

The two-qubit circuit is the quantum JK flip-flop. In quantum computing, the quantum JK flip-flop is the most often utilized flip-flop. The inputs |J> and |K> of a quantum JK flip-flop conduct two quantum processes in parallel. The quantum NAND

Table 12.4 Truth table of a quantum JK flip-flop

|J>	|K>	|Q>	$\overline{|Q>}$
|0>	|0>	No change	
|0>	|1>	|0>	|1>
|1>	|0>	|1>	|0>
|1>	|1>	|0>	|1>
|1>	|1>	|1>	|0>

operation is performed using |J> and shared input |clk>. The quantum NAND operations are then completed, and one of the quantum JK flip-flop's outputs executes the quantum NAND operation, producing |J>. The |K> and |clk > inputs are performed first in the quantum NAND operation. The output of the quantum JK flip-flop |Q>, as well as the result of the quantum first NAND operation, are then used to perform another quantum NAND operation, yielding |K>. The steps for creating |J> and |K> are carried out in simultaneously. One of the distinctive properties of quantum operations is that they may do several operations at the same time, and this is exactly what is happening. The quantum JK flip-flop operation is then conducted operations on these |J> and |K> inputs. The quantum NAND operation, which is employed here, is also used to make quantum SR flip-flops. Two outputs are discovered after conducting the quantum JK flip-flop operation. The one output is the opposite of the other. The quantum JK flip-flop truly solves the quantum SR flip-flop problem.

Initially, when a qubit is in a superposition state, it is being coherent. The flip-flop process will be maintained until the qubits are coherent. The process will be disturbed if a particle from outside breaks the coherence and causes the qubit to become decoherent. If the heat in the circuit is not reduced, the operation in this circuit will likewise come to a halt. The truth table of a quantum JK flip-flop is given in Table 12.4.

Assume that the clock input always enables the truth table of the quantum JK flip-flop. When any input is |0>, it acts as a quantum JK latch circuit, but when both inputs are |1>, it toggles to create the output according to the truth table.

The JK flip-flop is a timed SR flip-flop with better performance. However, the "race" issue still exists. When the state of the output |Q> is altered before the timing pulse of the clock input has time to go "Off," this issue occurs.

12.5.4 Example

For the purpose of testing the quantum JK flip-flop circuit, assume that the qubit inputs are |0> and |1>. If and only if the clock input is high or |1>, the quantum JK flip-flop will function. If the clock input is high, the quantum NAND operation will be performed by qubits |0> and |clk>.

In addition to performing the quantum NAND operation and producing the appropriate output, the |1> and |clk> inputs work in parallel.

These two intermediate qubit outputs are now used to conduct two independent quantum NAND operations. The quantum NAND operation is performed by the qubit |0> and the final output of the quantum JK flip-flop is $|Q'> (Q'$ is the complement of Q). Assume that the most recent state $|Q'>$ is |1>, where Q' is the complement Q.

So, now the produced output is |0>, which is referred to as |J>. The quantum NAND operation is then performed operation on the intermediate output |0> and the final output of the quantum JK flip-flop is |Q>.

According to the suggested circuit of a quantum JK flip-flop, these qubits labeled |J> and |K> will now conduct the quantum JK flip-flop operation.

Finally, a quantum memory element has been made. The JK flip-flop provides the necessary qubit inputs |0> and |1>, demonstrating that the circuit theoretically produces the ideal output.

12.5.5 Applications

A 1-qubit word may be stored in a single quantum flip-flop. Thus, by joining a group of quantum flip-flops, the storage capacity in terms of qubits may be increased. Quantum JK flip-flops are widely used in computers and Internet-of-Things devices. Quantum JK flip-flops may be utilized in a variety of applications, including registers, counters, frequency dividers, and event detectors. The big application is ripple counter. The output of a quantum JK flip-flop will be recorded as an output of the quantum ripple counter once the inputs have been processed in the flip-flop. The next quantum JK flip-flop of the quantum ripple counter will utilize the same output as its clock input. Using four quantum JK flip-flops, quantum ripple counter generates four qubit outputs.

Consider a quantum JK flip-flop with positive edge-triggered qubit inputs that are connected together and pushed high. The output of the quantum JK flip-flop will toggle for each positive edge of the clock signal in this condition.

Once developed, the quantum JK flip-flop can be used as a frequency divider. Quantum JK flip-flops can also help event detectors and registers. Quantum JK flip-flops can also help to design memory devices.

12.6 Quantum T Flip-Flop

The "quantum Toggle Flip–Flop" is another name for the quantum T flip–flop. To avoid the occurrence of the intermediate state in a quantum SR flip–flop, just one input, called the Trigger qubit input or Toggle input, should be sent to the flip–flop.

'Changing the next state output to complement the current state output' is referred to as toggling. By making modest changes to the quantum JK flip–flop, a quantum T flip–flop is created. Because the quantum T flip–flop is a single qubit input device, where a JK flip–flop is transformed into a quantum T flip–flop by linking the $|J>$ and $|K>$ inputs together and giving them a single input named $|T>$. In the context of quantum computing, "T flip-flop" refers to a type of qubit implementation, specifically the "T flip-flop qubit". It's a novel approach to building quantum computers, utilizing a pair of coupled qubits, typically in a silicon-based system. It aims to simplify the process of building and controlling qubits, potentially leading to more scalable quantum architectures. It allows for easier control and manipulation of the qubits, making it potentially more robust for building large-scale quantum circuits. The quantum T flip-flop design relies on the interaction and coupling between the two physical qubits, which is crucial for performing quantum computations. In essence, the flip-flop qubit is a physical realization of a quantum qubit, offering a novel approach to building and controlling qubits for quantum computation.

T flip-flops, like any quantum operation circuits, face the difficulty of producing additional heat. If and only if the heat is lowered, and the temperature is near zero, this quantum T flip-flop will learn. The most fundamental components of a quantum T flip-flop are the quantum AND and NOR operations. This fundamental components are made up of basic quantum operations. When compared to the functioning of classical computers, this quantum process is extremely quick. Only this flip-flop operation will be performed if the qubits are in a state of coherence; otherwise, this operation will not be performed. Any particle in the surroundings can disrupt the state of coherence, causing it to become decoherent.

The quantum AND and NOR operational circuits serve as the foundation for the quantum T flip-flop operational circuit. The quantum flip-flop is known as a toggle flip-flop because it toggles and changes its state. Here, a quantum T flip-flop is demonstrated using its operation. The quantum T flip-flop, unlike other quantum computing circuits, generates a lot of heat and operates at a significantly faster rate than other systems. The quantum T flip-flop has a wide range of applications in a variety of devices. It can be used in quantum ripple counters, as well as a variety of other counters, registers, and frequency division circuits made with quantum T flip-flops. Quantum T flip-flops are also employed in the construction of event multiple counters and many other devices. This chapter attempts to explain quantum T flip-flop architecture, operating time, and application, among other things.

12.6.1 Block Diagram

The quantum T flip-flop solves the problems of the quantum JK and SR flip-flops. One input $|T>$ is used in quantum T flip-flops. The block diagram of a quantum T flip-flop is shown in Fig. 12.10.

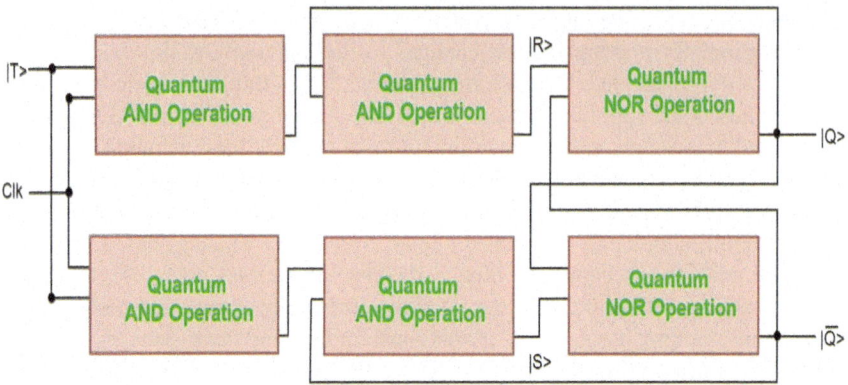

Fig. 12.10 Block diagram of a quantum T flip-flop

Two quantum AND computations are shared with one input |T>. The combination of the |clk> input and the |T> output yields two-qubit outputs by performing two quantum AND operations in tandem. As inputs, their outputs are fed into quantum SR flip-flops. Two quantum AND operations and two quantum NOR operations are used to create these quantum SR flip-flops. There are two inputs to this quantum SR flip-flop. Firstly, the quantum AND operation is performed using the output of the previous quantum AND operation and the output of the quantum SR flip-flop $|\overline{Q}>$. |S> is the name given to the output. The quantum AND operation is then performed on another output of the prior operation by using |Q>. |R> is the moniker given to the result of these quantum AND operations. The quantum NOR operation uses input |S> and input |R> to build a quantum SR latch. The quantum NOR operation is performed on the output of quantum SR latch |Q> and |S>. The quantum NOR operation is carried out in parallel by both |R> and |Q>. Finally, the quantum T flip-flop gives the outputs |Q> and |Q'> (Q' is the complement of Q). The outputs of these quantum T flip-flops are diametrically opposite.

12.6.2 Circuit Architecture

Quantum T flip-flop is an operation with one input named "T" that alleviates the JK flip-flop issue. Toggling is the primary use of this quantum flip-flop action. Quantum T flip-flop operational circuit is shown in Fig. 12.11.

Basically, it is some quantum AND operations and some quantum NOR operations in the quantum T flip-flop operation circuit. However, a detailed examination reveals that the quantum T flip-flop design is similar to the quantum JK flip-flop. Quantum JK flip-flop operation has various issues that are addressed by the quantum T flip-flop operation. Quantum JK flip-flops have a two-qubit input, whereas quantum T flip-flops have a shared input. A quantum SR flip-flop is sandwiched between

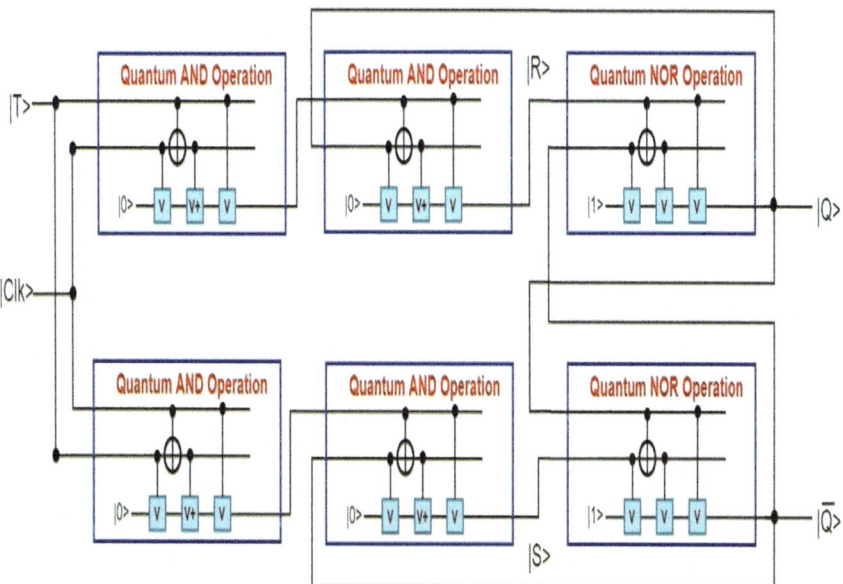

Fig. 12.11 Quantum T flip-flop operational circuit

two quantum T flip-flops. Two concurrent quantum procedures are used to create this flip-flop. When compare to the SR flip-flop discussed in the preceding chapter, the quantum SR flip-flop is slightly different. Two quantum AND operations and a quantum NOR SR latch were used to create the quantum SR latch in the quantum SR flip-flop's design. Firstly, two parallel quantum operations are built in the quantum T flip-flop. Then two quantum AND operations are connected in parallel in the SR flip-flop section, and two quantum NOR operations are set up in parallel in the SR latches. The quantum T flip-flop design keeps quantum parallelism's properties, although it's simply a theoretical circuit. Depending on the qubit origin, heat, and temperature, this circuit design can be modified.

12.6.3   Working Principle

The quantum T flip-flop differs from the quantum JK flip-flop in a few ways. When it is required to toggle, quantum T flip-flops can be created. This operating circuit only has one input, $|T>$, and one clock, $|clk>$. The circuit will be activated if the clock input is $|1>$; else, the circuit will be disabled. As a result, the quantum T flip-flop operation is enabled, and the fundamental operation begins to function. This circuit will operate if the qubit is in a coherent mood, just like any other circuit. Because any environment particle entering the circuit would disrupt coherence and make it decoherent. So, this circuit must be kept at a temperature close to 0 °F.

Table 12.5 Truth table of a
quantum T flip-flop

lT>	lQ>	lQ'> [Q' is the complement of Q]
l0>	l0>	l1>
l1>	l0>	l1>
l0>	l1>	l0>
l1>	l1>	l0>

The basic operation of quantum computing, known as the quantum AND operation, is performed initially in the quantum T flip-flop operating circuit. Because
the lT> and lclk> inputs are shared, two quantum AND operations run in parallel,
requiring the same amount of time. After the process, this output acts as an input for
a quantum SR flip-flop. First and foremost, the SR flip-flop conducts two quantum
AND operations in parallel in this quantum circuit. Two quantum AND operations
run in parallel, the first of which is the quantum SR flip-flop, and the output of which
is sent to the quantum SR latch as input. A quantum NOR SR latch is used here. Two
quantum NOR operations are used to build the quantum NOR SR latch. The two
NOR operations of the quantum SR latch run in parallel and take the same amount
of time.

The output of the quantum T flip-flop operation circuit is obtained after finishing
all of the operations. This procedure generates two outputs, one of which is the polar
opposite of the other. Truth table of a quantum T flip-flop is given in Table 12.5.

An ancillary bit will be used for error correction in a quantum T flip-flop, and this
flip-flop may create trash, although this is not the focus of our study.

12.6.4   Example

Presume that the clock input is l1> and the value of input lT> is entered as l0>. To
begin, two quantum AND operations will be performed simultaneously.

After conducting two quantum AND operations, two outputs are produced, both
of which are l0>.

As inputs, these two l0> are fed into a quantum SR flip-flop. Here, two quantum
AND processes run in parallel. Each quantum AND operation produces an output
l0> after conducting these procedures. Because it is known from the earlier section
that if the input to any quantum AND operation is l0>, the output will also be l0>.
Two outputs are then fed as inputs into the quantum NOR SR latch. The quantum
NOR operation is then carried out in parallel by these two inputs. Each of these
processes produces the result l0>. The functioning of quantum NOR is discussed in
in earlier chapters of this book.

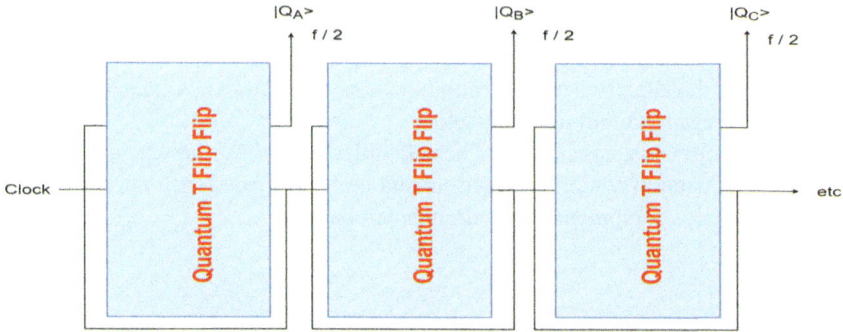

Fig. 12.12 Quantum frequency division circuit

12.6.5 Applications

A quantum T flip–flop is a quantum Toggle switch. Toggling means 'changing the next state output to complement the current state output.' Simple modifications to the JK flip–flop can result in quantum T flip–flops. Because the quantum T flip-flop can store data, it can be used in a variety of memory devices. Quantum T flip-flop operational circuits can address some of the shortcomings of quantum JK flip-flops. In quantum T flip-flops, a couple of applications are as follows:

1. Frequency Division Circuit; and
2. 2-qubit Parallel Load Registers

By feeding back the complementary output $|Q'>$ to the $|T>$ input, a quantum T flip–flop may be employed as a 'quantum Frequency Divider Circuit.' A frequency divider employing a quantum T flip–flop is depicted in the logic symbol as shown in Fig. 12.12.

If the clock frequency of the quantum T flip-flop is 'f' Hz, the frequency of the pulse at output $|Q>$ is 'f/2' Hz. This can be used to cascade a series of frequency divider circuits, further dividing the frequency. This circuit can be used to store data in the form of registers and shift registers. Size, on the other hand, is always an important factor for memory components such as registers. As a result, instead of quantum 4-bit registers, quantum 2-bit parallel load registers can be used.

When building a parallel quantum load register, there are two procedures which are as follows:

1. Hold the Data; and
2. Parallel Load the Data.

To hold the output of the quantum T flip–flop, just set the input $|T>$ to $|0>$. The Parallel Load, on the other hand, is the most difficult component. Getting the value $|X>$ from the flip-flop output is referred to as "parallel load." Flip-flop is to accomplish this, XOR the $|X>$ input and current state output before sending it to the

2-to-1 MUX. The other MUX input is a fixed value $|0>$ (logic low). The output of the MUX is connected to the quantum T flip-flop's input. Due to the fact that it is a quantum 2-bit register, two of these combinations are required. A quantum 2-qubit parallel load register circuit is given below.

These two circuits generate a lot of heat. In order to put these circuits into practice, the requisite quantum computing environment must be created. In comparison with digital electronics equipment, these circuits will be faster.

12.7 Quantum Shift Register

A single qubit of two-valued qubit data ($|1>$ or $|0>$) can be stored in a quantum flip-flop. However, many quantum flip-flops are required to store multiple qubits of data. To store n qubits of data, N quantum flip-flops must be coupled in a certain order. A quantum register is a gadget that stores this type of data. It consists of a sequence of quantum flip-flops used to store multiple qubits of data. In quantum computing, a shift register is a fundamental data structure used to store and manipulate qubits. It's essentially a sequence of qubits where the values can be shifted (moved) to the left or right. This is crucial for various quantum algorithms and operations, such as arithmetic calculations and bitwise manipulations. Quantum shift registers can be used to implement quantum addition, subtraction, or other arithmetic operations by shifting and manipulating the data. Shift registers can also be used to build quantum counters, which are useful for tracking iterations in quantum algorithms. In addition, shift registers are a fundamental data structure in quantum computing, enabling the manipulation and processing of quantum information.

Quantum shift registers enable the information stored in these quantum registers to be transmitted. A quantum shift register is a collection of flip-flops that stores several qubits of information. By applying clock pulses to the qubits contained in such quantum registers, they may be made to move inside them and in and out of them. By linking n quantum flip-flops, each of which stores a single qubit of data, an n-qubit quantum shift register may be built. "Quantum Shift left registers" are quantum registers that will shift the qubits to the left. "Quantum Shift right registers" are quantum registers that will shift the qubits to the right.

Quantum shift registers are basically of 4 types. These are as follows:

1. Quantum Serial-In Serial-Out shift register;
2. Quantum Serial-In parallel Out shift register;
3. Quantum Parallel In Serial-Out shift register; and
4. Quantum Parallel In parallel Out shift register.

In this chapter, a shift register is built utilizing a quantum D flip-flop operational circuit to convert serial data into quantum data. The quantum Serial-In Serial-Out shift register is a type of quantum shift register that permits serial input one qubit at a time over a single data line and outputs a serial output. The data exits the quantum shift register one qubit at a time in a serial pattern since there is only one qubit output. Thus, the term is quantum Serial-In Serial-Out (QSISO) Shift Register. Four quantum D flip-flops are linked in a serial fashion in this circuit. Because the same clock signal is supplied to each quantum flip-flop, they are all synchronized with one another. The circuit above is an example of a quantum shift right register, which accepts serial data from the quantum flip-flop's left side. A QSISO's main function is to operate as a delay element.

12.7.1 Block Diagram

Four quantum D flip-flop operational circuits are used to make a quantum shift register. As a fundamental component, a quantum shift register is utilized for data shift, and a quantum D flip-flop is used to make it happen. The block diagram of a quantum shift register is shown in Fig. 12.13.

Two inputs are used in quantum D flip-flops, one for data and another for the clock. The outputs of quantum D flip-flips are logically opposite to one another. The circuit's synchronization with an external signal is aided by the clock input. A quantum D flip-flop's output can have two possible values. Data input is directed to a quantum NAND operation circuit in this block diagram, while data input reverse is routed to another quantum NAND operation circuit. Both NAND procedures use the clock pulse input. Quantum SR latch receives the result of two quantum NAND operations. The D flip-flop is constructed using the SR latch. This property is used

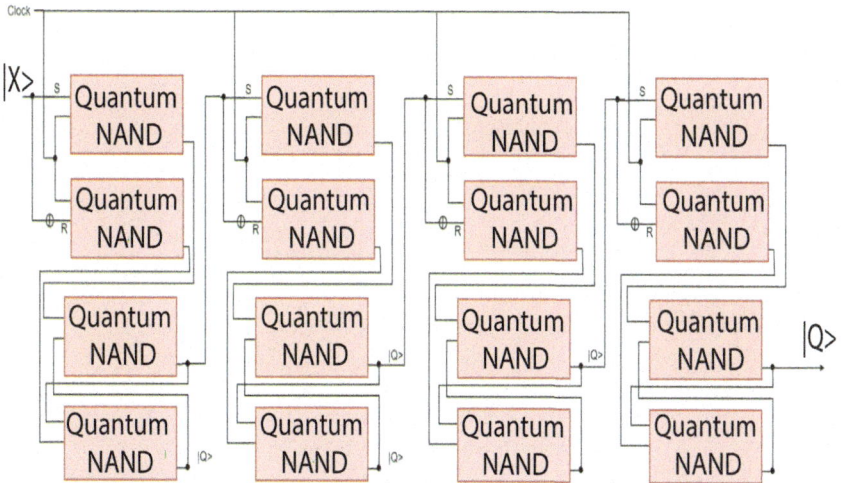

Fig. 12.13 Block diagram of a quantum shift register

to introduce a delay in the circuit's data flow. Quantum SR latch is made up of two quantum NAND operations. The final two outputs of the quantum SR latch function are uncovered. A quantum D flip-flop may provide two types of outputs, one of which is logically inverse to the other. The quantum D flip-flop will continue to function if the clock is enabled; otherwise, the quantum D flip-flop will stop working.

Quantum D flip-flop operational circuits are also coupled through serial connection in the quantum shift register block diagram. The quantum D flip-flop operational circuit is the fundamental component of the quantum shift register.

12.7.2   Circuit Architecture

Four D flip-flops and a quantum AND gate are used in the quantum shift register. A shift register generates four qubit outputs using these components. The D flip-flop has a single qubit input and is developed using quantum NAND and quantum SR latch operations. The simplified block diagram of a quantum shift register is given in Fig. 12.14.

The clock qubit input is required for the quantum D flip-flop and the circuit has one qubit input. One line of this qubit input will be directed to a quantum NAND operation termed S input in the circuit. Quantum NAND is performed here using S qubit input and clock qubit input.

$|X>$ qubit input's another line, which performs a quantum NOT operation first. This quantum NOT operation was included in the quantum NAND operation with the designation R. The two-qubit inputs R and clock are used in this quantum NAND operation. The output of the quantum NAND operations is sent to the quantum SR latch as an input. These two inputs will be used to produce an SR latch. The output of the quantum SR latch will be the final outputs of $|Q>$ and $|Q' >$ (Q' is the complement of Q). $|Q>$ will always produce the opposite of $|Q' >$. The quantum SR latch is discussed in the chapter titled quantum SR latch earlier in the book.

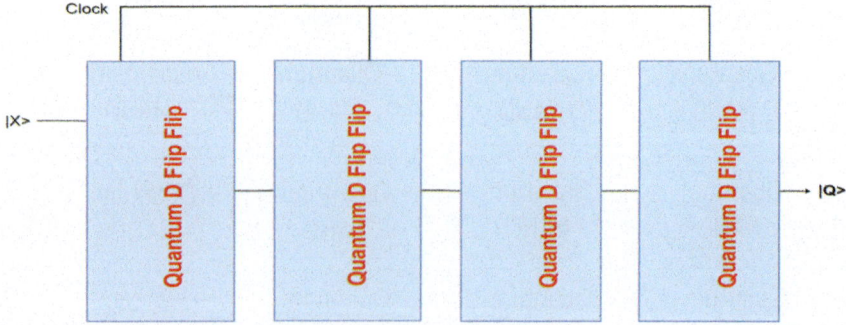

Fig. 12.14 Simplified block diagram of a quantum shift register

It produces one output after processing the inputs in a quantum D flip-flop. The output of the quantum D flip-flop is utilized as a clock input for the following quantum D flip-flop. As a result, a quantum shift register generates a single qubit output.

12.7.3   Working Principle

Quantum shift registers are registers in which both qubit data loading and data retrieval to and from the quantum shift register perform operation in serial mode at times. The positive edge of the clock pulse is sensitive in this synchronous quantum SISO shift register. At the qubit input of the first quantum flip-flop, the data word to be stored is fed qubit by qubit. Additionally, the outputs of the preceding ones influence the qubit inputs of all subsequent flip-flops, for example, the input of quantum D flip-flop number-2 is driven by the output of quantum D flip-flop number-1. Finally, the data from the quantum register are serially retrieved from the output pin of the nth quantum D flip-flop.

To begin, all the quantum flip-flops in the quantum register are cleared by setting their clear pins to high. Following that, the input data word is serially fed to quantum D flip-flop number-1.

As soon as the first leading edge of the clock occurs, the qubit arriving at the first pin is stored in quantum D flip-flop number-1. At the second clock tick, B1 is stored in quantum D flip-flop number-2, and a new bit is inserted into quantum flip-flop number-2.

Every rising edge of the clock pulse results in a similar shift in data qubits. This means that the data in the quantum register shift a single qubit to the right with every clock pulse. Following the qubit data transmission, the first qubit of an input word appears at the output of the nth flip-flop during the nth clock tick, as discussed previously. The serial output of the next succeeding qubits of the qubit input data word can be obtained by applying additional clock cycles. Table 12.1 presents the truth table of a quantum D flip-flop.

It creates one output after processing the inputs in a quantum D flip-flop. The output is fed into the following quantum D flip-flop as a clock input. As a result, a quantum shift register produces a single qubit output.

12.7.4   Applications

The quantum shift register is a very much useful circuit in quantum computing. It can be used in counter, data format convertor, data processor, etc.

Data Format Converters of Quantum Shift Registers

Because of the economic worth of the cables utilized, serial data transmission is chosen for long-distance communications. This demands parallel-to-serial conversion at

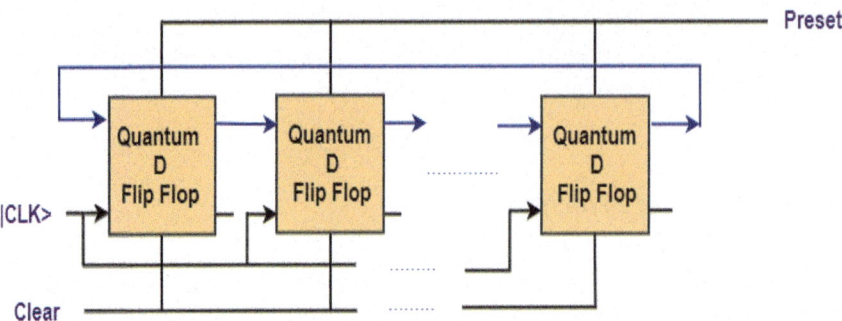

Fig. 12.15 Quantum n-qubit shift register-based quantum ring counter

the sender end, which can be accomplished with Quantum Parallel In Serial Out shift registers (QPISO). However, many quantum microprocessor-based systems prefer a parallel form of data-in, in which the transmitted qubit data is translated into parallel mode using a serial-to-parallel converter such as the Quantum Serial In Parallel Out shift register (QSIPO).

Counters of Quantum Shift Registers

The two shift-register-based counters that are widely utilized in digital applications are the quantum ring counter and the quantum Johnson counter. In quantum ring counters, the output of the last step is fed back into the first stage as a qubit input. This causes the data recorded in the quantum shift register to continuously circulate within it. For example, a 4-qubit ring counter storing a data word |0001> has a repeated sequence with four definite states viz., |0001>, |1000>, |0100>, and |0010>. Quantum Johnson Counter is similar to quantum ring counter except for the fact that the complement of the output at the last stage of the quantum shift register is sent as an input to the first stage. Quantum n-qubit shift register-based quantum ring counter is shown in Fig. 12.15; and Fig. 12.16 presents the quantum n-qubit shift register-based quantum Johnson counter.

Pseudo-Random Pattern Generator of Quantum Shift Registers

Quantum shift registers can be used to generate pseudo-random patterns which are used for testing. In order to achieve this, the outputs of a few stages in the quantum shift registers are XORed and connected as an input to the first stage of it.

The number of patterns generated depends upon the number of points that are tapped to be provided as quantum XOR operation inputs. If tapped appropriately, the maximum number of patterns that can be generated using an n-stage shift register is $(2n-1)$. Quantum pseudo-random pattern generator is shown in Fig. 12.17.

Quantum shift register has also some other applications. Quantum shift register can be used in various quantum devices. Quantum shift register is faster than digital

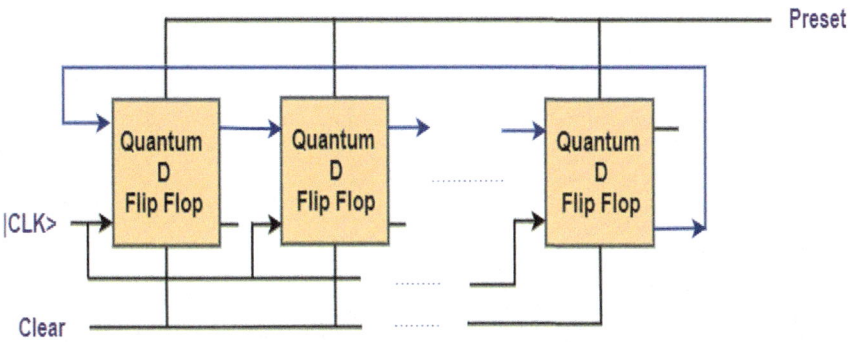

Fig. 12.16 Quantum n-qubit shift register-based quantum Johnson counter

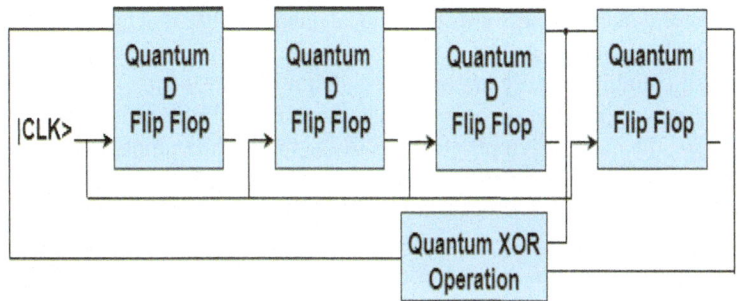

Fig. 12.17 Quantum pseudo-random pattern generator

shift register. So, it will be more useful as well as it is easier to handle than other devices.

12.8 Quantum Ripple Counter

A quantum counter is basically used to count the number of clock pulses applied to a quantum flip-flop. In the context of quantum computing, a ripple counter, also known as an asynchronous counter, uses a cascaded arrangement of quantum latches where the output of one latch triggers the clock input of the next. Unlike synchronous counters, which use a common clock, ripple counters rely on the sequential propagation of the clock pulse through the latches. Ripple counters in quantum computing typically employ quantum T-latches or other suitable quantum latches that can store and manipulate quantum information. Quantum latches in the counter circuit are connected in a chain, where the output of one latch acts as the clock input for the subsequent latch. Researchers have explored various design techniques to optimize the quantum cost and minimize the propagation delay of quantum ripple counters. It can also be used for quantum frequency divider, quantum time measurement, quantum frequency measurement, quantum distance measurement, and also for generating square waveforms. In this section, the quantum flip-flops are discussed which are useful to construct quantum asynchronous counters and are supplied with different clock signals, there may be a delay in producing output. Also, a few numbers of quantum logic gates are needed to design asynchronous counters. So, they are basic components in the design and also are less expensive.

An n-qubit ripple counter can count up to $2n$ states. It is also called as MOD n counter. It is known as a ripple counter because of the way the clock pulse ripples its way through the flip-flops. It is an asynchronous counter. Different flip-flops work with a different clock pulse. All the flip-flops are used in toggle mode. Only one flip-flop is applied with an external clock pulse and one flip-flop clock is obtained from the output of the previous flip-flop. The flip-flop applied with an external clock pulse acts as LSB (Least Significant Bit) in the counting sequence. A counter may be an up counter that counts upwards or can be a down counter that counts downwards or can do both, i.e., count up and count downwards, depend on the input control. The sequence of counting usually gets repeated after a limit.

Quantum ripple counter is constructed by four JK flip-flops. Using these quantum JK flip-flops, the quantum ripple counter creates four qubit outputs. Here in JK flip-flop, J and K are not shortened abbreviated letters of other words, such as "S" for Set and "R" for Reset, but are autonomous letters chosen by its inventor Jack Kilby to distinguish the flip-flop design from other types. Quantum ripple counter is an asynchronous counter. It is created using quantum JK flip-flops; and these flip-flops are only controlled by clock pulse input.

Quantum ripple counter produces much heat to keep the qubits in superposition state and also produces some garbage values.

Quantum ripple counter uses four quantum JK flip-flops to create four qubit outputs, where the quantum JK flip-flop has the two-qubit inputs named as $|J>$ and $|K>$.

This quantum JK flip-flop consists of many quantum NAND operations. At first the basic quantum NAND gate performs a couple of operations and then the outputs of these operations are entered into the JK flip-flop and it produces the outputs $|Q>$ and $|\overline{Q}>$. First of all, $|J>$ and $|clk>$ inputs perform the quantum NAND operation. The output of the quantum NAND operation and output of the quantum JK flip-flop $|\overline{Q}>$ perform another quantum NAND operation and produce the output named as $|J>$. The $|clk>$ input is shared. So, $|K>$ and $|clk>$ inputs also perform the quantum NAND operation. This quantum NAND operation's output and $|Q>$ output of the quantum JK flip-flop perform another quantum NAND operation and it produces the output named as $|K>$.

These $|J>$ and $|K>$ enter into the quantum SR flip-flop and produce two outputs $|Q>$ and $|\overline{Q}>$. In quantum JK flip-flop, it has four quantum NAND operations. $|J>$ input and $|clk>$ input perform quantum NAND operation as well as $|K>$ input and the shared $|clk>$ input also perform the quantum NAND operation. These two NAND operations outputs entered into the quantum JK latch as input. In quantum JK latches two $|Q>$ and one input as well as $|\overline{Q}>$ and other inputs perform the quantum NAND operation. Finally, the quantum JK flip-flop's final outputs are $|Q>$ and $|\overline{Q}>$.

After processing the inputs in a quantum JK flip-flop, the output of the flip-flop is stored as an output of the quantum ripple counter. Quantum ripple counter can be performed as an up counter and also a down counter. Here every quantum JK flip-flop's output |Q> enters into another quantum JK flip-flop as a clock pulse. This |clk> will decide whether the quantum JK flip-flop operational circuit will perform or not. Every quantum JK flip-flop operational circuit will produce the final output.

12.8.1 Circuit Architecture

Quantum ripple counter uses four quantum JK flip-flops to create four qubit outputs. Quantum JK flip-flop has a two inputs and one shared clock input. Quantum JK flip-flops also depend on the clock. If the clock is enabled then the circuit will be enabled, otherwise not. Figure 12.18 shows simplified block diagram of a quantum ripple counter.

In quantum ripple counter there's a one clock input and also a logic input which are shared in both |J> and |K> input ports. Input |J> and input |K> will be performing the quantum NAND operation with the shared |clk> input differently. |J> and |clk> inputs perform the quantum NAND operation. NAND operation is made by using quantum basic gates. Basic operations in quantum computing are V, V+, and CNOT operations. For error correction an ancillary qubit is used. In quantum NAND operation the value of an ancillary qubit is |1>. After |J> and |clk> performing the quantum NAND operation it produces an output qubit. This output qubit and the output of the flip-flop $|\overline{Q}>$ generate the quantum NAND operation and finally produce the output |J>. Like the same procedure of |J>, |clk> and |Q> inputs generate the |K> output. The operational circuits are mainly quantum NAND operations. The |J> and |K> inputs are performed the quantum JK flip-flop operation. In quantum JK flip-flop, the operational circuit is made by the basic component of quantum computing which is the quantum NAND operation.

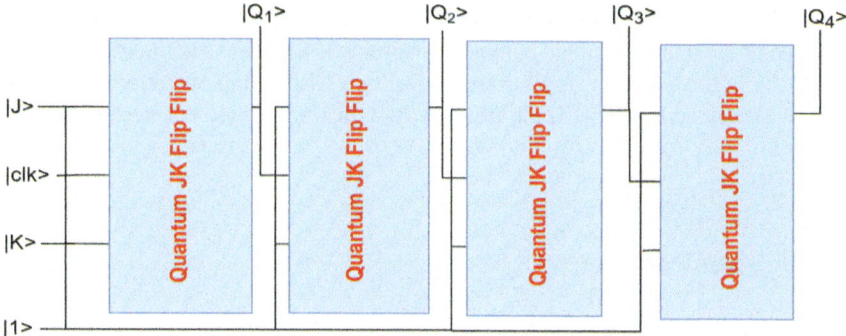

Fig. 12.18 Simplified block diagram of a quantum ripple counter

After processing the inputs in a quantum JK flip-flop, the output of the flip-flop is going to be stored as an output of the quantum ripple counter. In quantum ripple counter, the intermediate architectures of the four quantum JK flip-flops are connected in serial connection using the logical qubit input. Every clock input as |clk> enters like a qubit and from the 2nd quantum JK flip-flop |clk> input along with the immediate previous quantum JK flip-flop's first output |Q>. With the same architecture the quantum ripple counter can be performed as a up counter or as a down counter but the clock pulse as |clk> needs to have sometimes a positive edge triggered or a negative edge triggered.

12.8.2 Working Principle

In the quantum ripple counter, there are four quantum JK flip-flop operational circuits. These quantum JK flip-flop operational circuits are connected in serial connection. In this quantum ripple counter, there is one clock input and a logic input which are shared with the ports |J> and |K> of quantum JK flip-flop operational circuit. The inputs |J> and |K> of a quantum JK flip-flop conduct two quantum processes in parallel. The quantum NAND operation is performed using |J> and shared input |clk>. When the quantum NAND operations are completed, one of the quantum JK flip-flop's outputs executes the quantum NAND operation and it is producing one output. The |K> and |clk > inputs are performed first in the quantum NAND operation. The output of the quantum JK flip-flop |Q>, as well as the result of the quantum first NAND operation, are then used to perform another quantum NAND operation.

Here if the |clk> is |0> then the quantum JK flip-flop will not be triggered but if |clk> is |1> the quantum JK flip-flop will be triggered and it will toggle the output. First of all the |clk> value of quantum JK flip-flop operational circuit is |1> which toggles the output value from the previous state value. Then the output of the initial quantum JK flip-flop will be |clk> input of next quantum JK flip-flop. If the |clk> value is |1> then the output value will be toggled, otherwise the output will be the previous state output. Maintaining the same procedure every quantum JK flip-flop operated in the quantum ripple counter. Quantum JK flip-flop is toggled very much, that's why in the quantum ripple counter quantum JK flip-flop operational circuit is used as a basic component. The truth table of a quantum ripple counter is shown in Table 12.6.

12.8.3 Applications

The counter in which the external clock is only given to the first flip-flop and the succeeding flip-flops are clocked by the output of the preceding flip-flop is called

Table 12.6 Truth table of a quantum ripple counter

|clk>	|Q_0>	|Q_1>	|Q_2>	|Q_3>
|0>	|0>	|0>	|0>	|0>
|1>	|1>	|1>	|1>	|1>
|1>	|0>	|0>	|0>	|0>

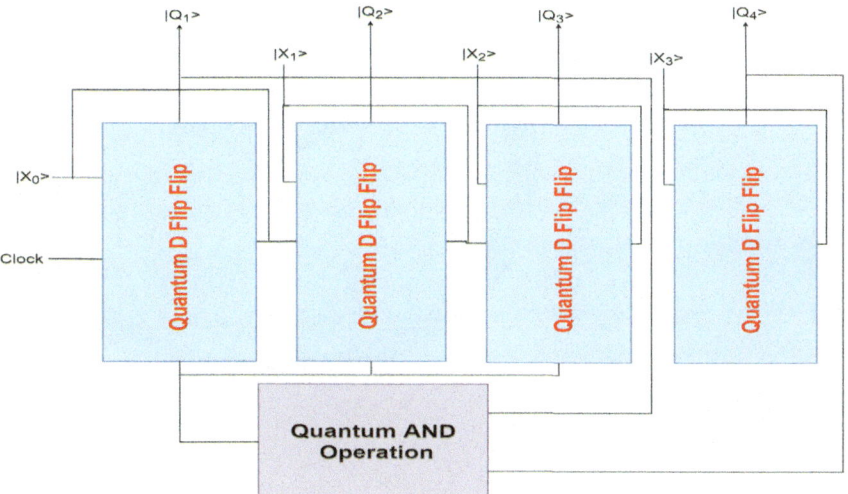

Fig. 12.19 Simplified block diagram of a quantum BCD counter

asynchronous counter or ripple counter. The name ripple counter is because the clock signal ripples its way from the first stage of flip-flops to the last stage.

Quantum BCD (Binary coded decimal) counter is a decade counter which has Mod = 10. Mod means the number of states the counter has. Quantum BCD counter counts decimal numbers from 0 to 9 and resets back to default 0. With each clock pulse, the counter counts up a decimal number. Quantum ripple BCD counter is the same as quantum ripple up-counter, the only difference is when the quantum BCD counter reaches count 10, it resets its flip-flops. Figure 12.19 shows the simplified block diagram of a quantum BCD counter.

Different types of flip-flops are used different clock pulses in the quantum ripple counter. It is an example of an asynchronous counter. The flip-flops are used in toggle mode in a quantum ripple counter. The external clock pulse (such as alarm clock shown in Fig. 12.20) is applied to only one flip-flop. The output of this flip-flop is treated as a clock pulse for the next flip-flop in the quantum ripple counter. In counting sequence, the flip-flop in which the external clock pulse is passed acts as LSB in quantum ripple counter operational circuits.

Quantum ripple counter is an asynchronous counter. This kind of counter can also be used for constructing an alarm clock. Quantum ripple counter has wide use and this circuit is also used in quantum computing processor.

Fig. 12.20 Alarm clock

12.9 Quantum Synchronous Counter

A quantum counter is a quantum device which can count any particular event on the basis of how many times the particular event(s) has occurred. In a quantum logic system or computers, this quantum counter can count and store the number of times any particular event or process has occurred, depending on a quantum clock signal. Most common type of quantum counter is a sequential quantum logic circuit with a single clock input and multiple qubit outputs. The qubit outputs represent two-valued decimal numbers. Each clock pulse either increases the number or decreases the number.

Quantum synchronous circuit generally refers to something which is coordinated with others based on time. Quantum synchronous signals occur at the same clock rate and all the clocks follow the same reference clock. Quantum asynchronous counter has shown that the qubit output of that quantum counter is directly connected to the input of next subsequent counter and making a chain system, and due to this chain system, the propagation delay occurs during counting stage and creates counting delays. In a quantum synchronous counter, the clock qubit input across all the quantum flip-flops uses the same source and creates the same clock signal at the same time. So, a quantum counter which is using the same clock signal from the same source at the same time is called a quantum synchronous counter. In other words, a quantum synchronous counter is a counter circuit where all flip-flops are triggered simultaneously by a single clock signal. In contrast to asynchronous counters, where flip-flops are triggered by the output of the previous flip-flop, all flip-flops in a synchronous counter operate in unison, ensuring precise timing and counting accuracy. While the basic principle of a synchronous counter remains the same, quantum technology can be used to implement such circuits, potentially leading to ultra-low power consumption and smaller devices. In essence, a quantum synchronous counter leverages the principles of synchronous counting but incorporates quantum-based components like quantum logic devices. These quantum implementations can offer benefits in terms of energy efficiency and physical size compared to traditional CMOS-based implementations.

Quantum synchronous counter consists of four quantum JK flip-flops and two quantum AND operations. Using these quantum JK flip-flops and quantum AND operations, the quantum synchronous counter creates four qubit outputs. A quantum synchronous counter produces much heat and this circuit operation needs to be occurred in the required environment of quantum computing.

Quantum synchronous counter uses four quantum JK flip-flops to create four qubit outputs. This JK flip-flop has the two-qubit input named as $|J>$ and $|K>$. The description and diagram of a JK flip-flop have been discussed earlier.

12.9.1 Circuit Architecture

Quantum synchronous counter uses four quantum JK flip-flops to create four-qubit outputs. Quantum JK flip-flop has a two-qubit input and one clock shared input which also depends on the clock. If the clock is enabled then the circuit will be enabled, otherwise not. Figure 12.21 shows simplified block diagram of a quantum synchronous counter.

Input $|J>$ and input $|K>$ both the values will perform the quantum NAND operation with the shared $|clk>$ input differently. $|J>$ and $|clk>$ inputs perform the quantum NAND operation. NAND operation is made by using basic quantum operations. Basic operations of quantum computing are V, V+, and CNOT gates. For error correction here an ancillary qubit is used. In quantum NAND operation the value of an ancillary qubit is $|1>$. After $|J>$ and $|clk>$ performing the quantum NAND operation the output qubit is obtained. Like the same procedure of $|J>$, $|clk>$ and $|K>$ inputs produce the another output. These operational circuits are mainly quantum NAND operations. These $|J>$ and $|K>$ inputs are performed the quantum JK flip-flop operation. In quantum JK flip-flop operation, the circuit architecture is also made by the basic component of quantum computing which is quantum NAND operation.

After processing the inputs in a quantum JK flip-flop, the output of the flip-flop is going to be stored as an output in the quantum synchronous counter. Thus, the quantum synchronous counter creates four qubit outputs using four quantum JK flip-flops. The clock inputs for all of the four quantum JK flip-flops come from the same source. For this, all of the flip-flops work synchronously, where one output of each of the second and the third flip-flops goes through quantum AND operations.

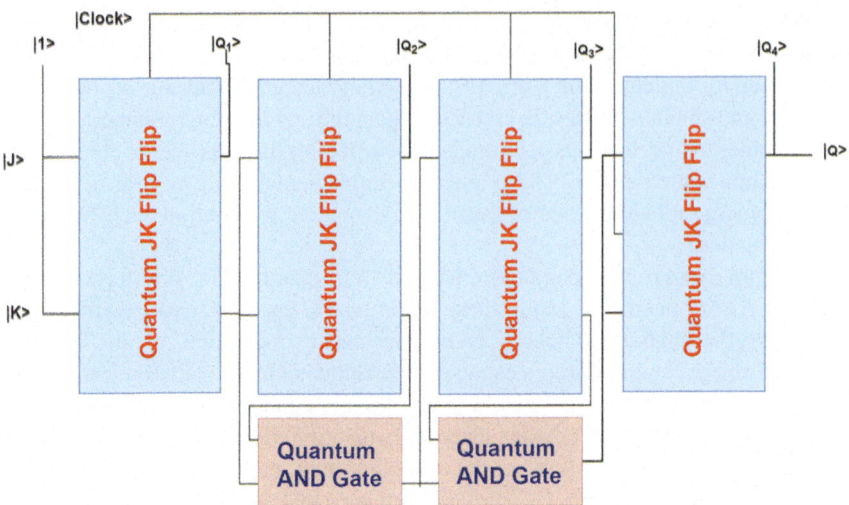

Fig. 12.21 Simplified block diagram of a quantum synchronous counter

12.9.2   Working Principle

Quantum synchronous counter has one logical qubit input and one clock input. Quantum synchronous counter is constructed by the basic component which is a quantum JK flip-flop operational circuit. Quantum JK flip-flop is the two-qubit circuit. Quantum JK flip-flop is mostly used flip-flop in quantum computing. Quantum JK flip-flop's inputs |J> and |K> perform two quantum operations through parallelism. |J> and shared input |clk> perform the quantum NAND operation. Then this quantum NAND operation's result and one output of the quantum JK flip-flop perform the quantum NAND operation and produce the output. Then the output of quantum JK flip-flop and the result of the quantum NAND operation execute again another quantum NAND operation and produce another output. After performing the quantum JK flip-flop operation, two outputs are found. One output is the opposite of another. This quantum JK flip-flop actually removes the problem of quantum SR flip-flop. Quantum SR flip-flop is described already.

Quantum JK flip-flop is triggered when the value of clock is |1>. So, in quantum synchronous counter clock needs to be always high. According to the principle, counters need to be toggled, that's why quantum JK flip-flop is perfect for quantum synchronous counters. Quantum JK flip-flop operational circuit is firstly triggered, and then the output of quantum JK flip-flop will be shared the input with the second quantum JK flip-flop operational circuit. Again, the first quantum JK flip-flop's output and the second quantum JK flip-flop's output enter into the quantum AND operation circuit and perform the quantum AND operation. The produced output from quantum AND operation performs the quantum JK flip-flop operation. Then again the produced output from the first quantum AND operation and the third quantum JK

Table 12.7 Truth table of a JK flip-flop

|clk>	|Q_3>	|Q_2>	|Q_1>	|Q_0>
|0>	|0>	|0>	|0>	|0>
|1>	|0>	|0>	|0>	|1>
|1>	|0>	|0>	|1>	|0>
|1>	|0>	|0>	|1>	|1>
|1>	|0>	|1>	|0>	|0>
|1>	|0>	|1>	|0>	|1>
|1>	|0>	|1>	|1>	|0>
|1>	|0>	|1>	|1>	|1>
|1>	|1>	|0>	|0>	|0>
|1>	|1>	|0>	|0>	|1>
|1>	|1>	|0>	|1>	|0>
|1>	|1>	|0>	|1>	|1>
|1>	|1>	|1>	|0>	|0>
|1>	|1>	|1>	|0>	|1>
|1>	|1>	|1>	|1>	|0>
|1>	|1>	|1>	|1>	|1>
|1>	|0>	|0>	|0>	|0>

flip-flop operation's output perform another quantum AND operation. Hence the previous quantum AND operation's output is shared with the quantum JK flip-flops as two inputs and performs the quantum JK flip-flop operational circuit. Quantum synchronous counter circuit mainly performs operations like a finite counter. Table 12.7 shows the truth table of the JK flip-flop.

12.9.3 Applications

As the name implies, the synchronous counter consists of flip-flops which are all in sync with each other, i.e., their clock inputs are connected together and are triggered by the same external clock signal. This implies that all the flip-flops update its value at the same time. Synchronous counters have many classifications such as up counter, down counter, etc. which are used in serval areas in the real life.

Quantum synchronous counter can be widely used in lots of other designs as well as processors, calculators, and real-time clock. The camera timer setting which can be made by quantum synchronous counter which is shown in Fig. 12.22.

Quantum synchronous counter can also be used in pulse counting machine. In quantum computing process this counter circuit has a vast amount of applications. Common uses are in home appliances like washing machine, microwave oven, time schedule led indicator, key board controller, etc.

Fig. 12.22 Camera timer setting which can be made by quantum synchronous counter

Fig. 12.23 Washing
machine

In washing machine (shown in Fig. 12.23), when it is necessary to count the time for washing the quantum synchronous counter can be used there to perform the required operation.

12.10 Summary

This chapter has presented all sequential circuits in quantum computing. Necessary figures and working principles are also shown with appropriate explanations. Some applications and examples are also presented at the end of each section. Quantum computing is a challenging and exciting field for the researchers. This chapter describes a little but completely a new discover of quantum computing. Sequential circuits in classical computer is known to all. But sequential circuits in quantum computing are a new research for the modern science and technology. It defines many new things and becomes a turning point to the new inventions for all. In other words, a sequential quantum circuit is a type of quantum circuit where unitary transformations are applied to parts of a system in a sequential manner. This approach is used to map between states of different gapped phases and preserves important properties like entanglement area law and the gappedness of the quantum states. Sequential circuits can also alter long-range correlations and entanglement. Quantum gates are the building blocks of the circuit, performing unitary transformations on the qubits. These gates manipulate the qubit's state, leading to superposition (existing in multiple states simultaneously) and entanglement (correlating the states of multiple qubits). Quantum gates are the building blocks of the circuit, performing unitary transformations on the qubits. These gates manipulate the qubit's state, leading to superposition (existing in multiple states simultaneously) and entanglement (correlating the states of multiple qubits). They can be used for various tasks, including generating non-trivial gapped states, performing renormalization group transformations, and creating defects in higher-dimensional topological states. In essence, sequential quantum circuits provide a way to manipulate quantum states in a controlled and structured manner, enabling transformations between different phases and altering their long-range properties while preserving key characteristics of the quantum system.

Bibliography

1. D. Greenberger, K. Hentschel, F. Weinert, *Compendium of Quantum Physics: Concepts, Experiments, History and Philosophy* (Springer Science & Business Media, 2009)
2. N. Isailovic, Y. Patel, M. Whitney, J. Kubiatowicz, Interconnection networks for scalable quantum computers, in *33rd International Symposium on Computer Architecture (ISCA'06)* (IEEE, 2006), pp. 366–377
3. R. Konik, Quantum coherence confined. Nat. Phys. **17**(6), 669–670 (2021)
4. T.S. Metodi, D.D. Thaker, A.W. Cross, F.T. Chong, I.L. Chuang, A quantum logic array microarchitecture: scalable quantum data movement and computation, in *38th Annual IEEE/ACM International Symposium on Microarchitecture (MICRO'05)* (IEEE, 2005), p. 12
5. M. Partanen, K.Y. Tan, J. Govenius, R.E. Lake, M.K. Mäkelä, T. Tanttu, M. Möttönen, Quantum-limited heat conduction over macroscopic distances. Nat. Phys. **12**(5), 460–464 (2016)

6. K. Saitoh, H. Yamamoto, K. Kawasaki, Y. Fukuda, H. Tanaka, M. Okada, H. Kitaguchi, Development of cryogenic probe system for high-sensitive NMR spectroscopy. J. Phys. Conf. Ser. **97**, 012141 (2008)
7. D.D. Thaker, T.S. Metodi, A.W. Cross, I.L. Chuang, F.T. Chong, Quantum memory hierarchies: efficient designs to match available parallelism in quantum computing, in *33rd International Symposium on Computer Architecture (ISCA'06)* (IEEE, 2006), pp. 378–390
8. X. Zheng, J. Yang, C. Zhou, C. Zhang, Q. Zhang, X. Wei, Allosteric DNAzyme-based DNA logic circuit: operations and dynamic analysis. Nucl. Acids Res. **47**(3), 1097–1109 (2019)

Chapter 13
Sequential Circuits in Quantum-DNA Computing

13.1 Introduction

In the context of quantum-DNA computing, sequential circuits are circuits where the output at any given time depends not only on the current input, but also on the sequence of past inputs and the circuit's internal state, which is essentially its memory. This contrasts with combinational circuits, whose output is solely determined by the current input. Sequential circuits, unlike combinational circuits, incorporate memory elements. These elements store the circuit's current state, which is a function of previous inputs and outputs. Sequential logic is used in various applications, such as finite-state machines, which are fundamental building blocks in digital circuits. A common example is the channel selection in a television, where the current channel is determined by past input (channel up/down buttons). In the realm of quantum-DNA computing, sequential circuits can be implemented using DNA-based logic gates and memory elements, where the "state" of the circuit is encoded in the structure or hybridization of DNA molecules. In simpler terms, think of a vending machine. The machine has a memory of how many items you've selected so far (that's its state). When you press a button to choose an item, the output (the item dispensed) depends not only on which button you pressed, but also on what items you've chosen before. That's a sequential circuit in action. In quantum-DNA computing, similar logic can be implemented using DNA molecules instead of traditional electronic circuits. This is one of the most intriguing new scientific topics to emerge in the time: quantum biocomputing or quantum-DNA computing. It is difficult to scale quantum-DNA computers at this point in time, even if they were possible. The purpose of this study and research is to build quantum-DNA processors, memory devices, and other devices, all of which are based on this technology. The quantum-DNA sequential circuit is one of ten chapters devoted to circuits in this section. Quantum-DNA fundamental operations AND, OR, and NOT are used in this circuit's functionality. This section goes into great details on the various aspects of the general design, construction, heat measurement, and operational time calculation. Large-scale quantum-DNA computing has a number of technical challenges to overcome. Computer architecture must comply with specific geometrical limitations in order to ensure smooth functioning. To put it another way, it's an attempt to entirely demolish a barrier to quantum-DNA computing.

13.2   Quantum-DNA D Flip-Flop

A quantum-DNA D flip-flop is essentially a two-state timed flip-flop. In one clock cycle, the qubit inputs of a quantum D-type flip-flop are actuated with a delay. A delay flip-flop is another term for the quantum-DNA D flip-flop. In other words, D flip-flop, a fundamental circuit, can be realized in various computing paradigms, including quantum-DNA computing. While a traditional D flip-flop uses electronic components, quantum biological computing explores the possibility of using biological components like DNA or quantum devices to build similar functionalities. The basic principle remains the same: capturing a data input (the "D" input) at a specific clock signal edge and holding that value as the output until the next clock edge. This field explores the use of DNA strands as qubits (quantum bits) and for implementing logic operations. D flip-flops, along with other sequential circuits, can be designed using quantum-DNA techniques. In summary, D flip-flops are fundamental building blocks in quantum-DNA logic that can be realized using various quantum and biological technologies. The exploration of D flip-flops in quantum-DNA computing focuses on using non-traditional materials and processes to achieve similar functionalities with potential benefits in miniaturization, energy efficiency, and novel computational architectures.

The indeterminate input's conditions of SET = "|0>" and RESET = "|0>" are banned in the basic quantum SR NAND gate bistable circuit, which is one of its fundamental drawbacks. This condition forces both DNA molecule sequence outputs to logic "ACCTAG," overriding the feedback latching action, and whichever the input goes to logic "ACCTAG" first loses control, while the other input, which is still at logic "TGGATC," controls the latch's final state. However, an inverter may be connected between the "SET" and "RESET" qubit inputs to create a quantum-DNA data latch, quantum-DNA delay flip-flop, quantum-DNA D-type bistable, quantum-DNA D-type flip-flop, or simply a quantum-DNA D flip-flop as it is most often known.

By far the most essential of all the quantum-DNA timed flip-flops is the quantum-DNA D flip-flop. The |S> and |R> inputs become complements of each other when a quantum inverter (quantum NOT operation) is added between the Set and Reset inputs, ensuring that the two inputs |S> and |R> are never equal (|0> or |1>) to each other at the same time, allowing us to control the toggle action of the flip-flop with just one |D> (Data) input.

The data input, labeled "|D>," is then utilized in place of the "Set" signal, and the inverter is used to create the complementary "Reset" input, resulting in a level-sensitive quantum D-type flip-flop from a level-sensitive SR-latch, with |S> = |D> and |R> = $|\overline{D}>$.

The quantum-DNA D flip-flop circuit has just one qubit input, and the qubit input must be in a coherence state in order to conduct the quantum computational function. As a result, the circuit must exist in an environment that does not exist. If any particle emerges, the coherence state will be disrupted. The quantum D flip-flop will generate heat, which must be removed quickly in order to cool down the circuit and stabilize the coherence state.

Hence, quantum-DNA flip-flops have two portions based on the principle of the quantum-DNA circuit. Quantum-DNA D flip-flop operational circuit's first portion is constructed using quantum principle and the second portion is constructed using DNA computing principle. Quantum circuit's portion produces qubit and stores into quantum cache memory. These qubits perform NMR relaxation process operation and make them normal molecules. Then these molecules perform DNA computing operations. DNA computing is very good for memory storage while quantum computing has a super-fast computation speed.

13.2.1 Block Diagram

The major components of a quantum-DNA D flip-flop are quantum NAND operations, DNA NAND Operations and quantum NOT operations. The quantum-DNA D flip-flop fundamentally has one qubit input, and in the core of the quantum-DNA D flip-flop there is a quantum-DNA SR latch. When compared to other timed type flip-flops, the quantum-DNA D flip-flop is one of the most significant devices. Quantum-DNA D flip-flop verifies that the two inputs of the quantum-DNA SR flip-flop are never the same. Block diagram of a quantum-DNA D flip-flop is shown in Fig. 13.1.

In quantum-DNA D flip-flop operational circuit there is one input which is |x>. There is another |clk> input which is also there. Quantum-DNA D flip-flops have two output molecule sequences that are logically opposite to one another. The clock qubit input aids in the circuit's synchronization with an external signal. The output of a quantum-DNA D flip-flop can have two potential values. This block diagram of a quantum-DNA D flip-flop operational circuit shows that the data input is sent to a quantum NAND operation circuit, while the reversal of data input is routed to another quantum NAND operation circuit. The clock pulse input is used by both quantum NAND operation processes. The result of two quantum NAND operations is fed into quantum cache memory. Quantum cache memory is made by using some quantum registers and it saves the quantum qubit data. When the accurate time appears this qubit goes to the NMR machine and it performs NMR relaxation process as well as it makes molecules from qubits. The DNA SR latch is used to build the quantum-DNA D flip-flop. This attribute is utilized to create a delay in the data flow in the circuit. Two DNA NAND Operations create the DNA SR latch. The circuit discovers the remaining two output molecules of the DNA SR latch function. The output of a quantum-DNA D flip-flop can be of two parts, one of which is logically inverse to the other. If the clock is enabled, the quantum-DNA D flip-flop will continue to function; otherwise, the quantum-DNA D flip-flop will cease to function.

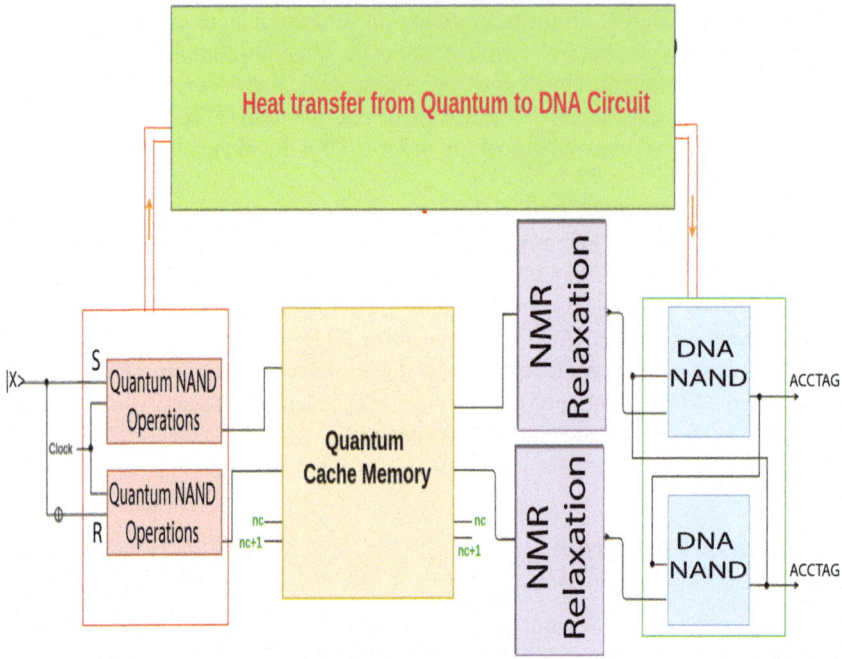

Fig. 13.1 Block diagram of a quantum-DNA D flip-flop

13.2.2 Circuit Architecture

The quantum-DNA D flip-flop has a single qubit input and is developed using quantum NAND operations, DNA NAND operations, and a DNA SR latch. Figure 13.2 shows the quantum-DNA D flip-flop circuit.

The clock qubit input affects the quantum-DNA D flip-flop. It is clear from the diagram that the circuit has one qubit input. One line of these qubit inputs will be directed into the quantum-DNA NAND operation known as |S> input in the circuit. In this case, the |S> qubit input and the clock qubit input are used in a quantum NAND operation.

When a |X> qubit traverses another line, it first undertakes a quantum NOT operation. This quantum NOT operation was dubbed R, when it was entered into the quantum NAND operation. The R and clock qubit inputs are used in this quantum NAND operation. The output of these quantum NAND operations is getting stored into quantum cache memory by using the line. Quantum cache memory is made by shift register, where the qubit data is stored in some sub-arrays. When the actual time appears, the quantum cache memory serves the qubit to the NMR relaxation process using a line. The qubit, first of all performs the NMR relaxation process,

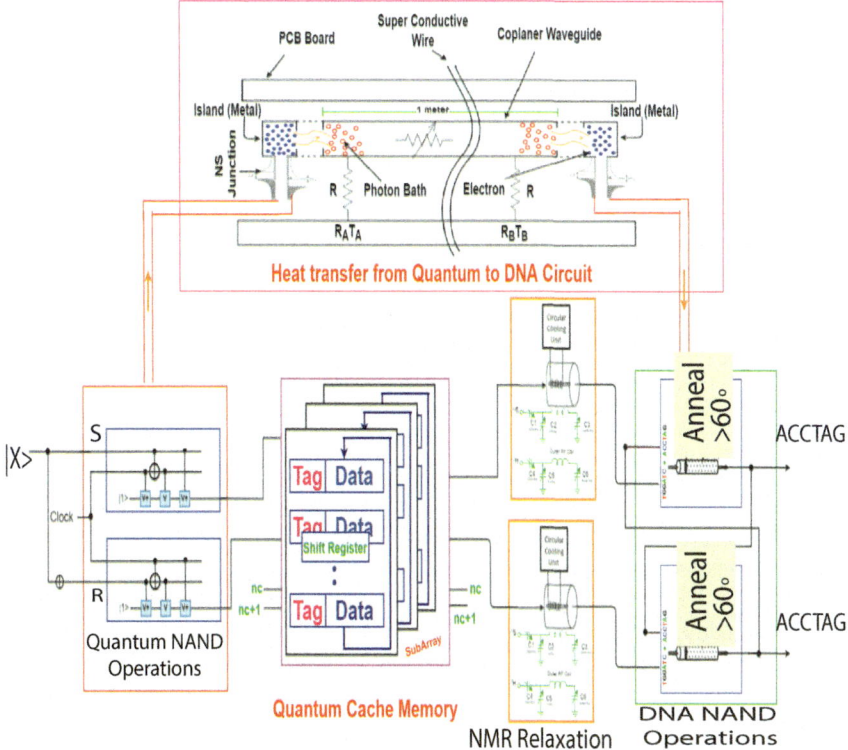

Fig. 13.2 Quantum-DNA D flip-flop circuit

where EMR emission is totally prohibited. Hence the qubit is first converted into a molecule sequence and then performs the DNA SR latch operation. In DNA SR latch operation two DNA NAND operations perform with parallelism and produce the sequence of the final molecules.

Quantum circuit produces more heat and the DNA circuit needs the heat to process the input. For that reason, it would be good to transfer the heat into the DNA circuits. The quantum portion transfers the heat using the heat transfer circuit into DNA circuit's portion. In the heat transfer circuit, two junctions are connected to quantum circuits and DNA circuits. A maximum of one can transfer heat to this circuit. This heat transfer circuit has superconductive ware, photon batch, and PCB board to make the architecture. This circuit mainly transfers the heat using the photon bath. With this circuit, the quantum-DNA T flip-flop circuit is constructed fully.

13.2.3 Working Principle

There are two inputs in the quantum-DNA SR flip-flop: SET and RESET. Alternatively, in a quantum-DNA D flip-flop, one input and the input's one line are referred to as a SET, and by coupling a quantum-DNA NOT gate toward the other line input, the circuit may designate the quantum-DNA D flip-flop as a RESET. This complement resolves the contradiction inherent in the quantum-DNA SR latch when both inputs are LOW because that circumstance is no longer feasible. Quantum-DNA D flip-flops have a single qubit input, which is sometimes alluded to as a data qubit input. If this qubit data input is high, the quantum-DNA flip-flop becomes SET; if the data input is low, such as $|0>$, the flip-flop changes to state and becomes RESET.

However, this would be pretty futile because the output of the flip-flop will always vary with each pulse delivered to this data input. To circumvent this, an extra input known as the "CLOCK" or "ENABLE" input is used to separate the data input from the latching circuitry of the flip-flop after the appropriate data has been stored. The result is that the $|X>$ input condition is only replicated to the output $|Q>$ while the clock input is active. This then serves as the foundation for yet another sequential gadget known as a quantum-DNA D flip-flop.

As long as the clock input is HIGH, the "quantum-DNA D flip-flop" will store an output with any logic level that is applied to its data terminal. Once the clock input is changed to LOW, the flip-flop's "set" and "reset" inputs are both kept at logic level "$|1>$," preventing the flip-flop from altering the underlying and preserving whatever statistics are available on its output prior to the clock transition. In other words, either logic "$|0>$" or logic "$|1>$" latches the output. Table 13.1 shows the truth table of a quantum-DNA D flip-flop.

In this quantum-DNA D flip-flop operational circuit at first half of the circuit is constructed using the quantum computing principle and the rest of the portion is constructed using the DNA computing principle. First of all two quantum NAND operations are performed and produced two outputs which are stored in quantum cache memory. The quantum cache stores the qubit into an array. Quantum cache memory stores the qubit data and when required it serves the data into the NMR relaxation process. In this quantum-DNA D flip-flop, two-qubit is stored in quantum cache memory. This qubit performs the NMR relaxation process because DNA circuits need molecular sequence. NMR relaxation process removes the superposition state and makes the qubit into a molecular sequence. This molecular sequence performs DNA SR latch operation. DNA SR latch operation has basic components:

Table 13.1 Truth table of a quantum-DNA D flip-flop

$|Clk>$	$|x>$	Q	$\overline{Q}$	Description
$\downarrow \gg |0>$	X	Q	$\overline{Q}$	Memory no change
$\uparrow \gg |1>$	$|0>$	TGGATC	ACCTAG	Reset Q $\gg$ 0
$\uparrow \gg |1>$	$|1>$	ACCTAG	TGGATC	Set Q $\gg$

DNA NAND operations. DNA NAND operation performs in a parallel way to a DNA SR latch operation and produces two required outputs. Quantum cache memory is mainly used because quantum operation performs so fast and DNA operation is performed very slowly. Quantum cache memory works as an intermediate process where a qubit is just stored and when needed cache memory serves the qubit.

Quantum cache memory works here as the intermediate process. Quantum circuit produces much heat and here two quantum NAND operations perform which produce much heat. But quantum circuit needs to be close to zero Kelvin temperature to perform the operation. So, from the quantum circuit, the circuit needs to reduce the temperature to maintain the superposition state of the qubit. Hence in DNA circuit operation, it needs much heat in several steps. Mainly in melting and annealing require much heat. This topic is briefly described in the earlier chapter named quantum-DNA circuit operation. For that reason, this circuit transfers the excessive heat from the quantum circuit portion to the DNA circuit portion using a heat transfer circuit. This heat transfer circuit using the junction captures the heat from the quantum circuit and using photon bath heat flows through the circuit and gives this into the DNA circuit. This circuit can transfer heat maximum in approximately 1 m distance and in quantum-DNA flip-flop distance is less than one meter between quantum and DNA portions. This heat transfer circuit cannot transfer full excessive heat which is produced from the quantum circuit and this heat is not enough to perform the DNA circuit operation. But this heat transfer can optimize the cost of heat based on the needs. After completion of all of this operation and architecture fully, the quantum-DNA D flip-flop can produce two output molecular sequences.

13.2.4 Applications

The quantum D flip-flop operational circuit is the fundamental component of the quantum shift register. After processing the qubit input in the quantum shift register, the output of the shift register is stored in quantum cache memory. Quantum-DNA shift registers enable the information which is stored in these quantum registers to be transmitted. A quantum-DNA shift register is a collection of flip-flops that stores several qubits of information. By applying clock pulses to the qubits contained in such quantum-DNA registers, they may be arranged to move inside them and in and out of them. By linking n quantum flip-flops and n numbers of DNA flip-flops, each of which stores a single qubit of data or molecule sequence of data, where an n quantum-DNA shift register may be built. "Quantum-DNA shift left registers" are quantum-DNA registers that will shift the qubits or molecules to the left. "Quantum-DNA shift right registers" are quantum-DNA registers that will shift the qubits or molecule sequences to the right.

Quantum-DNA sequence generators, quantum-DNA multiplexers, quantum-DNA D flip-flop-based counters, and other processor components are constructed using the quantum-DNA D flip-flop operational circuits. Quantum-DNA D flip-flop has a big role in quantum processors since it can solve the quantum-DNA SR latch problem with forbidden stage.

The circuits can be utilized as storage devices in power gating circuits and clocks since the quantum-DNA SR latch is a single-bit storage element. Quantum-DNA SR latch operational circuits are utilized as a memory device in computers and, in certain cases, in IoT devices. Latches are used in the construction of memory devices like quantum-DNA flip-flops. Quantum-DNA SR flip-flops are built utilizing the quantum-DNA SR latch operational circuit in quantum-DNA computing.

The quantum-DNA SR latch is an asynchronous circuit that has been employed as an input-output port in several quantum asynchronous systems. The quantum-DNA SR latch was employed at different times in the quantum asynchronous counter and shift register.

13.3   Quantum-DNA SR Latch

Based on the triggering that is suited to operate it, there are two types of memory elements. One of them is a quantum-DNA latch, and the other one is a quantum-DNA flip-flop. Quantum-DNA latches operate with the enabled signal and level-sensitive, whereas the quantum-DNA flip-flops are edge sensitive. A quantum-DNA SR latch is an asynchronous tool. It operates without the use of control signals, relying solely on the state of the $|S>$ and $|R>$ inputs. Two quantum NAND operations can make a quantum-DNA SR latch. Nevertheless, two quantum NOR operations can also make a quantum SR latch. In other words, an SR latch, a fundamental sequential logic circuit, could potentially be explored in the context of quantum-DNA computing. While not directly used in current quantum computing implementations, the concept of a latch, a circuit that can store information, could be adapted to biological systems using quantum principles. An SR latch is a basic building block in quantum-DNA computing, used for storing a single qubit of information. In summary, while an SR latch as a classical circuit isn't directly applicable in current quantum computers, the concept of a latch (a circuit for storing information) could be explored in the context of quantum-DNA computing, potentially leading to new ways of understanding and manipulating information within biological systems.

In quantum-DNA SR latch operational circuits, the half of the portion of the circuit is constructed using the quantum computing principle and the rest of the portion is constructed by DNA computing principle. This quantum-DNA SR latch operational circuit is built by one quantum NAND operational circuit and one DNA NAND operational circuit. Quantum computing computation speed is so high compared to DNA computing and for this reason, this circuit has intermediate cache memory to store the quantum qubit data. Quantum computing produces qubit which performs NMR relaxation process, where EMR emit is prohibited to make qubit molecules.

In quantum-DNA SR latch, two-qubit inputs are swapped and negated. Quantum-DNA SR latch can be called as SET-RESET latch. In quantum-DNA SR latch from two-qubit input, this circuit produces two outputs. These outputs are reversed to one another. For this quantum-DNA SR latch, two quantum NAND operation circuits are designed. This quantum-DNA SR latch has the input line swapped between two operational circuits where one is quantum NAND operational circuit and other is DNA NAND operational circuit, but it is not negated. Quantum-DNA SR latch works as memory stuff in quantum-DNA computers, and it has several applications in a quantum-DNA processor. If this circuit is used to design some embedded systems using the quantum-DNA operational device, then the quantum-DNA SR latch will be used on this device as a memory unit. Quantum-DNA SR latch is level-sensitive and has few disadvantages, but these will be recovered by the quantum-DNA flip-flop.

13.3.1 Block Diagram

The quantum-DNA SR latch is one of the most common memory devices, and it has an effect on the output as long as it is active. The essential properties of a quantum-DNA SR latch are that one qubit input behaves like a SET and another qubit input behaves like a RESET. The block diagram of a quantum-DNA SR latch is shown in Fig. 13.3.

The quantum SR latch is made up of two fundamental processes, which are depicted in this block diagram of the quantum SR latch. There are two input lines in the quantum SR latch, one for $|S>$ and the other for $|R>$. Two outputs are obtained from this two-qubit input such as $|Q>$ and $|Q'>$. The output of the first quantum NAND operation is used as an input in the second quantum NAND operation, and the output of the second quantum NAND operation is used as an input in the first quantum NAND operation. If the input of $|S>$ is $|1>$, the SR latch is activated; however, if the input of $|R>$ is $|1>$, the SR latch has no influence on the output. In a quantum SR latch, a value of $|1>$ cannot be used to activate two inputs.

In DNA computing, the molecular sequence needs to go through many processes during the operation. In DNA computing, melting and annealing are very important and these steps require a vast amount of heat. That's why in quantum-DNA SR latch, the quantum portion transfers a small amount of heat which is produced from the quantum circuits and it is transferred to the DNA circuits. Though this amount of heat is not enough, this heat or temperature will help to perform DNA computation. In DNA computing, DNA NOR SR latch performs its operation, where two DNA NOR gates perform in parallel way and the inputs come from quantum circuit portion. After that operation, this circuit finally finds the output molecular sequence from the quantum-DNA SR latch.

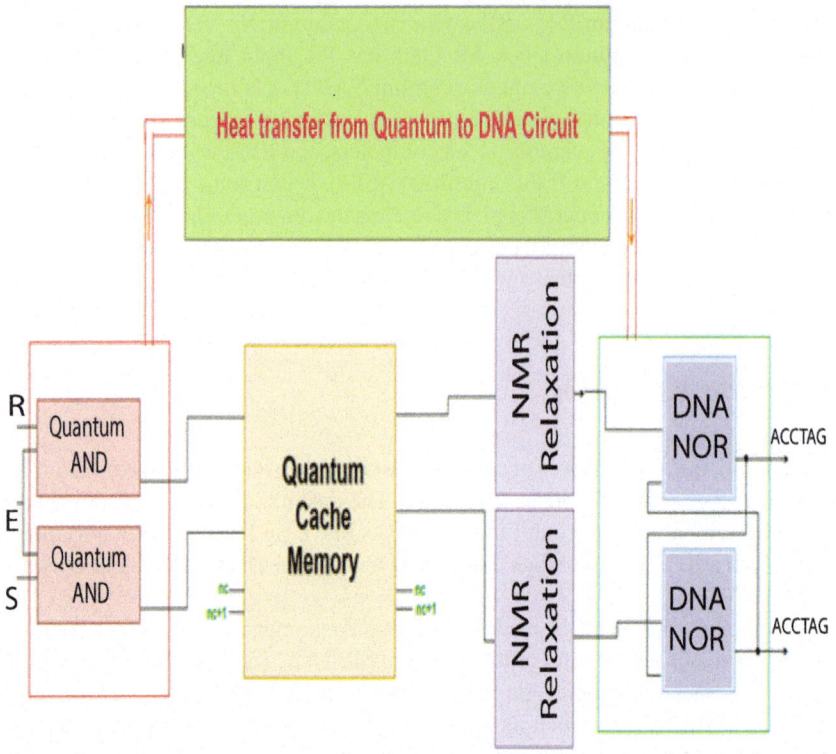

Fig. 13.3 Block diagram of a quantum-DNA SR latch

13.3.2 Circuit Architecture

Quantum SR latches are level sensitive and are built using only one fundamental operation which is, the quantum NAND operation. The architecture of a quantum SR latch is shown in Fig. 13.4.

The circuit of the quantum-DNA SR latch operation is constructed using one principle, that is, the first portion of this circuit will be constructed by using quantum circuit and the second portion will be constructed by using DNA circuit. This circuit also has quantum cache memory and a heat transfer circuit to achieve the best output at the end.

In this circuit, the input is qubit and the output is the molecular sequence. There are two inputs in the quantum NAND gate. In this circuit, those inputs are |S>, IR>; and |E> as clock qubit input, which are shared into the quantum NAND operational circuit and perform DNA NAND operations through an intermediate process. Two operations are performed in parallelism. The output lines of these quantum NAND operations circuit are connected with the quantum cache memory. This is the first portion of the circuit and is fully designed by using the principle of quantum

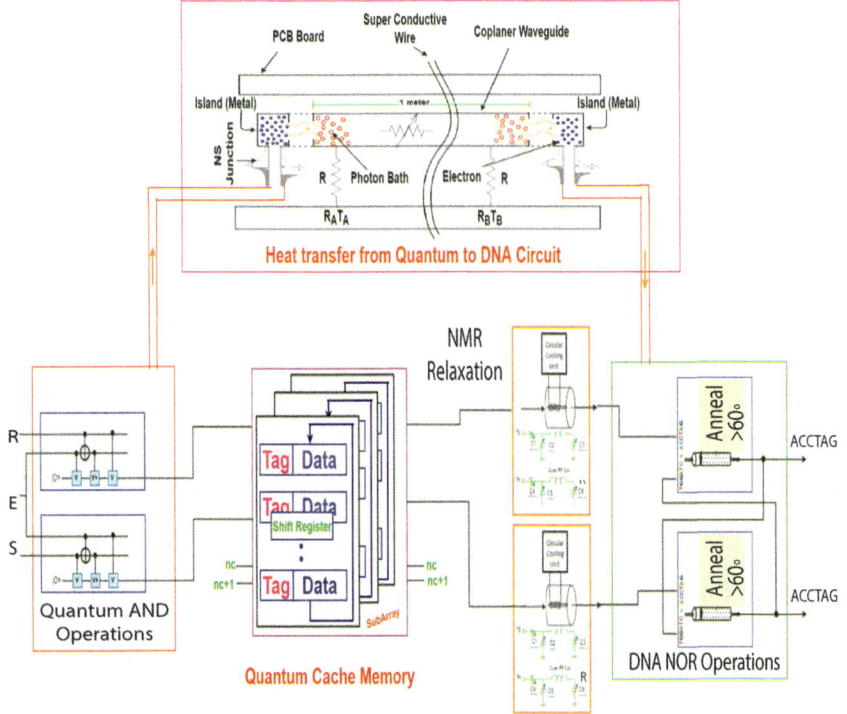

Fig. 13.4 Circuit architecture of a quantum SR latch

computing. Quantum cache memory is made by using some quantum shift registers and in this circuit, it saves data to the quantum array. When the time arrives, the quantum cache memory supplies the data to the "NMR Relaxation" process. The supply line of the quantum cache memory is fully connected to NMR relaxation process as well as here the EMR emit is fully prohibited. Now, the output line of the NMR relaxation process is connected to the DNA NAND gate. Two outputs of the process are connected to two different DNA NAND gates. In DNA NAND operations, one input comes from the produced output of quantum NAND operation using the NMR relaxation process and another input comes directly from the direct qubit |R> and by the intermediate process it becomes a molecule sequence and performs DNA NAND operation.

Quantum circuit produces more heat and the DNA circuit needs heat to process the input. For that reason, it would be good to transfer the heat into DNA circuits. Quantum portion transfers heat using the heat transfer circuit to the DNA circuit's portion. In the heat transfer circuit, two junctions are connected to quantum circuits and DNA circuits. Maximum of one meter can be transfered heat by this circuit. This heat transfer circuit has superconductive ware, photon batch, and PCB board to make the architecture. It mainly transfers the heat using the photon bath. With this circuit, the quantum-DNA SR latch circuit is constructed fully.

13.3.3   Working Principle

Quantum-DNA SR latch performs its operation by using two principles: quantum computing principle and the DNA computing principle. There are three inputs in the quantum NAND operation in this circuit and those inputs are $|S>$, $|R>$, and $|E>$, which are shared between the quantum NAND operational circuit and DNA NAND operation. Quantum circuit operation will be performed in close to zero temperature because the qubits need a coherence state. The superposition state will be stable if this circuit is full of prohibited particles from other environmental particles. The quantum NAND operations and the DNA NAND operation will be performed in parallel. These quantum NAND operations produce some garbage values but in this circuit these values are avoidable. The quantum NAND operation has one output. These qubit outputs go to the quantum cache memory which is basically made of shift registers. The quantum cache memory is built using the rules of quantum computing. This cache memory will store the qubit output which is received from the NAND operation and it will serve it when needed.

Quantum cache stores the qubit into an array. Quantum cache memory stores the qubit data and when required, it serves the data to the "NMR relaxation" process. In this quantum-DNA SR latch, two qubits are stored in quantum cache memory. This qubit performs the "NMR relaxation" process because DNA circuits need molecular sequence. This process removes the superposition state and makes the qubit into a molecular sequence. This molecular sequence performs the DNA NAND operation. Quantum cache memory is mainly used because quantum operation performs so fast while DNA operation is performed very slowly. The quantum cache memory works as an intermediate process, where qubit is just stored and when needed the cache memory serves the qubit.

Quantum cache memory works here as the intermediate process. Quantum circuit produces much heat and here two quantum NAND operations perform. So it produces much heat. But the quantum circuit needs to be close to zero Kelvin temperature to perform the operation. So, from the quantum circuit, it needs to reduce the temperature to maintain the superposition state of the qubit. Hence, in DNA circuit operation, it needs much heat in several steps. Mainly in melting and annealing require much heat. For that reason, the quantum circuit transfers the excessive heat to the DNA circuit portion using a heat transfer circuit. This heat transfer circuit using the junction captures the heat from the quantum circuit and using photon bath, the heat flows through the circuit and gives this to the DNA circuit. This circuit can transfer heat maximum in the 1 meter distance and in quantum-DNA flip-flop, the distance is less than one meter between the quantum and DNA circuits. This heat transfer circuit cannot transfer full excessive heat which is produced from the quantum circuit and it is not enough to perform the DNA circuit operation. But this heat transfer can

Table 13.2   Truth table of a quantum-DNA SR latch

|S>	|R>	|Q>	|Q'> [Q' is the complement of Q]
|0>	|0>	Latched	
|0>	|1>	**ACCTAG**	**TGGATC**
|1>	|0>	**TGGATC**	**ACCTAG**
|1>	|1>	**Metastable**	

optimize the cost of the heat which is based on the needs. After the completion of all these operations and architecture fully, the quantum-DNA SR latch can produce two output molecular sequences.

The quantum-DNA SR latch, on the other hand, is still a vital component of a quantum CPU or a quantum-DNA-based embedded device.

As the quantum circuit generates a lot of heat, it is difficult to isolate the qubit from a superposition state. As a result, it needs to cool the circuit to isolate the qubit from a superposition of a quantum circuit. Any type of external particle can disrupt the qubit's coherence and cause it to become decoherent. If all of these are preserved, the quantum-DNA SR latch can truly function. Table 13.2 shows the truth table of a quantum-DNA SR latch.

13.3.4   Applications

A simple quantum NAND operation using SR flip-flop gives feedback from both of its outputs to its opposite inputs and is widely used to store a single data qubit in memory circuits. Many memories and IoT devices employ the quantum-DNA SR flip-flop circuits. The quantum-DNA SR flip-flops are primarily used to store one-qubit data. Quantum-DNA SR flip-flops are also employed in quantum-DNA shift registers, quantum-DNA counters, and other memory devices.

13.4   Quantum-DNA SR Flip-Flop

Quantum-DNA sequential logic circuits, unlike quantum-DNA combinational logic circuits, include some built-in "Memories" that change the state depending on the real signals supplied to its inputs. Quantum-DNA SR flip-flops, for example, have a |1> qubit memory bistable. The SET and RESET inputs of the quantum-DNA SR flip-flop are the same. The output of the SET input is a |1>, whereas the output of the RESET input is a |0>.

The quantum-DNA SR flip-flop is often referred to as the SET-RESET flip-flop. The reset input is used to restore the flip-flop to its starting state from the current state with an output. In the quantum NAND operational circuit, the SR flip-flop is a basic flip-flop, where both outputs provide feedback to its opposite input. This circuit is used to store a single data qubit in a memory circuit. The three inputs are: SET, RESET, and a found output. A two-qubit model will be used since the quantum-DNA SR flip-flops have two inputs that are mostly from the outside. Because of two qubits, it generates more heat than quantum-DNA D flip-flops. The computation time of this quantum-DNA SR flip-flop is determined by the fundamental gates used in this circuit. Quantum SR flip-flops may be found in a wide range of processors and embedded systems. Although the suggested flip-flop can generate some trash, an error correcting auxiliary qubit provides the desired output. The real-world implementation of the suggested quantum circuit will address a wide range of issues more quickly and effectively. In other words, an SR flip-flop, a fundamental building block of sequential circuits, can be explored within the context of quantum-DNA computing, particularly when considering the design of memory devices and other quantum-DNA circuits. While traditionally implemented in classical electronics, the principles of SR flip-flops can be adapted to utilize quantum phenomena and biological molecules. This emerging field explores the use of quantum principles and biological molecules (like DNA) to perform computations. In essence, the SR flip-flop serves as a foundational concept for building memory and sequential logic in emerging fields like quantum-DNA computing.

13.4.1   Block Diagram

Quantum-DNA SR flip-flop toggles the input independently. It takes two qubit inputs |S> and |R>. In quantum SR flip-flop, the output of one operation is used by the other operation and its block diagram is shown in Fig. 13.5.

Quantum-DNA SR flip-flop needs two-qubit input and one shared input which is also a qubit. In one quantum NAND operation, |R> and in another quantum NAND operation, |S> are taken as inputs. In addition, a qubit input |E> is shared by the both NAND operations.

The quantum NAND operations perform in parallel according to the quantum computing principle. The outputs produced by these operations will be temporarily stored in a quantum cache memory. As the quantum NAND operations and the DNA NAND operations are held in different times, it is necessary to store the outputs of the quantum NAND operations in a quantum cache memory. As the block diagram depicts that the final output will be the DNA molecule sequence. So, the outputs from the quantum cache memory will go through the "NMR Relaxation" process and after that, one DNA molecule will go to DNA NAND operation as input. Like the same other output, the qubit from the quantum cache memory will go through NMR Relaxation Process and come as molecular sequence in DNA NAND operation. Then, finally two DNA NAND operations will be performed and it is the last operation of DNA computing according to the quantum-DNA SR flip-flop circuit's block diagram. In quantum computing portion, the conversion of the qubit happens after getting

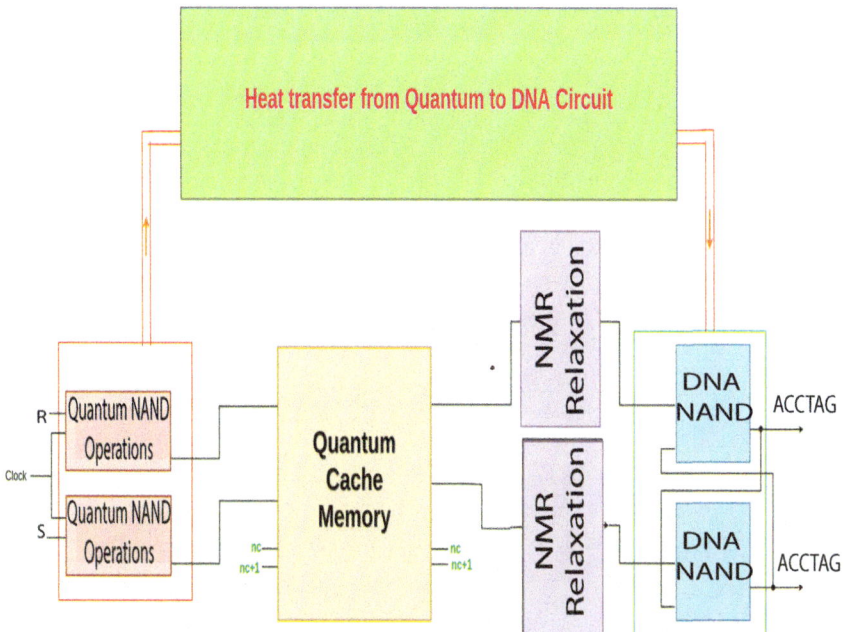

Fig. 13.5 Block diagram of a quantum-DNA SR flip-flop circuits

out from the quantum cache memory operation and the output becomes the DNA molecular sequence.

In DNA computing, the molecular sequence needs to go through many processes during the operation. In DNA computing, melting and annealing are very important and these steps require a vast amount of heat. That's why in quantum-DNA SR flip-flop, the quantum portion transfers a small amount of heat to the DNA circuits which is produced from the quantum circuits. Though this amount of heat is not enough, the heat or temperature will help to perform DNA computation. In DNA computing, DNA NAND SR flip-flop performs two DNA NAND operations parallelly, where the input comes from the quantum circuit portion. After that operation, the output molecular sequence is found from the quantum-DNA SR flip-flop.

13.4.2   Circuit Architecture

Quantum-DNA SR flip-flop circuit is constructed using one principle, that is, the first portion will be constructed by using quantum circuit and the second portion will be constructed by using DNA computing circuit. These circuits also have quantum cache memory and a heat transfer circuit to achieve the best output at the end. Quantum-DNA SR flip-flop is shown in Fig. 13.6.

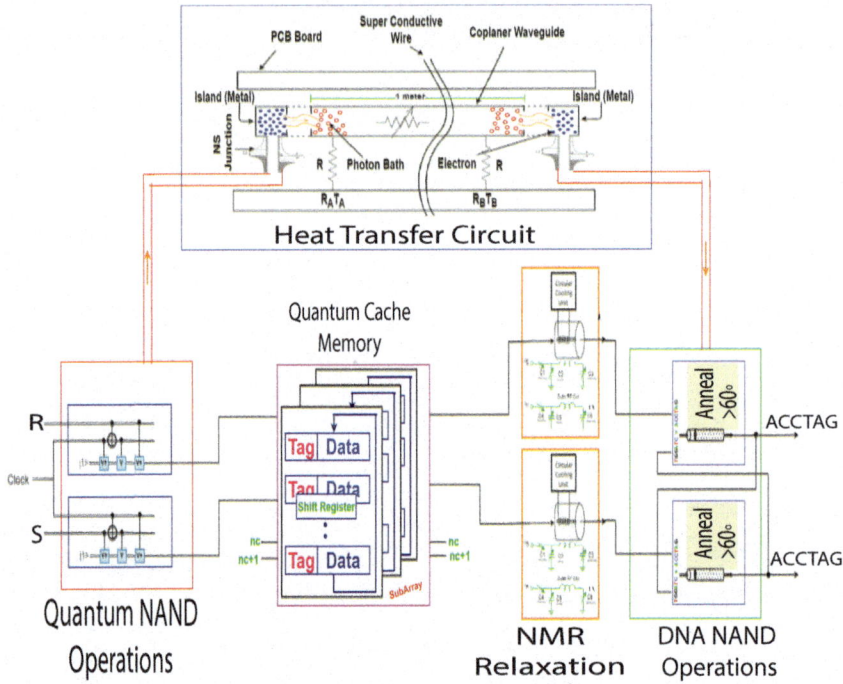

Fig. 13.6 Quantum-DNA SR flip-flop

In this circuit, first the inputs are qubits and the outputs are the molecular sequence. There are two inputs in the quantum NAND circuit, where the inputs are |S>, |R>, and |E>, which are shared between the two quantum NAND gates. Two quantum operations are performed parallelly. These two quantum NAND operations circuit's output lines are connected to the quantum cache memory. This is the first portion of the circuit and is fully designed by using the principle of quantum computing. Quantum cache memory is made by using some quantum shift registers and in this circuit it saves data to the quantum array. When the time arrives, the quantum cache memory supplies the data to the "NMR Relaxation" process. Quantum cache memory supply line is fully connected to "NMR Relaxation" process as well as here the EMR emission is fully prohibited. Then the output line of the "NMR Relaxation" process is connected to the DNA NAND operation. Two outputs of the process are connected to two different DNA NAND operations. The DNA NAND operations take its another inputs from the output which are ACCTAG and TGGATC molecular sequences. Hence, the final outputs of the DNA NAND operations correspond to the final output of quantum-DNA SR flip-flop.

Quantum circuit produces more heat and the DNA circuit needs heat to process the input. For that reason, it would be better to transfer the heat to DNA circuits. The quantum portion transfer the heat using the heat transfer circuit to the DNA portion. In the heat transfer circuit, two junctions are connected to quantum circuits and DNA circuits. Maximum of one meter can be transfered heat by this circuit. This heat transfer circuit has superconductive ware, photon batch, and PCB board to make the circuit architecture. This circuit mainly transfers the heat using the photon bath.

13.4.3 Working Principle

Quantum-DNA SR flip-flop works using two principles: the quantum computing principle and the DNA computing principle. There are two inputs in the quantum NAND operation in this circuit which are $|S>$, $|R>$, and $|E>$, where these inputs are shared with two NAND operations. The quantum circuit operation will be happening in close to zero temperature because qubits need to coherence state. The superposition state will be stable if this circuit is full of prohibited from other environment particles. The two quantum NAND operations will be performed parallelly. This quantum NAND operation produces some garbage value but in this circuit this topic is avoidable. Each of the quantum NAND Operations has one output. These qubit outputs go to the quantum cache memory which are basically built by the shift registers which follow the rules of quantum computing. This cache memory will store the qubit output from the NAND operation and it will serve when it is needed.

Quantum cache stores the qubit into an array. The cache memory stores the qubit data and when requires it serves the data to the "NMR Relaxation" process. In this quantum-DNA SR flip-flop, two qubits are stored in quantum cache memory. This qubit performs the "NMR Relaxation" process because DNA circuits need molecular sequence. This process removes the superposition state and makes the qubit into a molecular sequence. This molecular sequence performs the DNA NAND operation parallelly. The structure is basically a DNA NAND flip-flop operation. Quantum cache memory is mainly used because quantum operation performs so fast while the DNA operation is performed very slowly. Quantum cache memory works as an intermediate process where qubit is just stored and when needed the cache memory serves the qubit.

The truth table of a quantum SR flip-flop is given in Table 13.3.

Table 13.3 Truth table of a quantum SR flip-flop

$|S>$	$|R>$	$|Q>$	$|Q'>$ [Q' is the complement of Q]
$|0>$	$|0>$	No change	
$|0>$	$|1>$	$|0>$	$|1>$
$|1>$	$|0>$	$|1>$	$|0>$
$|1>$	$|1>$	Invalid	

13.4.4   Applications

A simple quantum NAND operation using SR flip-flop gives feedback from both of its outputs to its opposite inputs and is widely used to store a single data qubit in memory circuits. Many memory and IoT devices employ the quantum-DNA SR flip-flop circuits. The quantum-DNA SR flip-flop was primarily used to store one-qubit data. Quantum-DNA SR flip-flops are also employed in quantum-DNA shift registers, quantum-DNA counters, and other memory devices.

13.5   Quantum-DNA JK Flip-Flop

In flip-flop designs, the quantum JK flip-flop will be the most extensively utilized flip-flop. J and K are not abbreviated letters of other words, such as "S" for Set and "R" for Reset, but are independent letters chosen by the inventor Jack Kilby to identify the flip-flop design from others. Despite the fact that the digital electronics JK flip-flop was created by Jack Kilby. The functioning concept of the quantum JK flip-flop differs from that of the digital JK flip-flop. In other words, the concept of a quantum-DNA flip-flop explores implementing this logic in the context of quantum computing and DNA computing, potentially using qubits and DNA strands to represent the flip-flop's states. This concept proposes building a flip-flop, where

  i) Qubits (quantum bits) could represent the JK flip-flop's two states;
 ii) DNA strands could be used to encode the logic of the flip-flop (e.g., representing J and K inputs); and
iii) Quantum algorithms could be used to manipulate these qubit/DNA representations to perform flip-flop operations.

Research into quantum-DNA computing and the development of quantum flip-flops (like the Quantum JK Flip-Flop (QFF)) is ongoing. This area explores the potential of combining the strengths of quantum computing and DNA computing to create novel computational systems.

The quantum JK flip-flop's sequential operation is identical to that of the prior quantum SR flip-flop, with the same "Set" and "Reset" inputs. The distinction this time is that even though S and R are both at logic "1," the "quantum JK flip-flop" has no incorrect or prohibited quantum SR latch input states. It is evident that the quantum JK flip-flop does not solve the disadvantages of the quantum SR flip-flop.

The quantum JK flip-flop is essentially a gated quantum SR flip-flop with the addition of clock qubit input circuitry to avoid the unlawful or invalid output state that can arise when both inputs $|S>$ and $|R>$ are equal to logic level "$|1>$." A quantum JK flip-flop has four potential input combinations due to the extra timed inputs: "$|1>$," "$|0>$", "no change," and "toggle." The JK flip-flop has the same symbol as a quantum SR bistable latch, as seen in the preceding chapter. The quantum JK flip-flop, like other flip-flops, generates a lot of heat, which must be dissipated to run it properly.

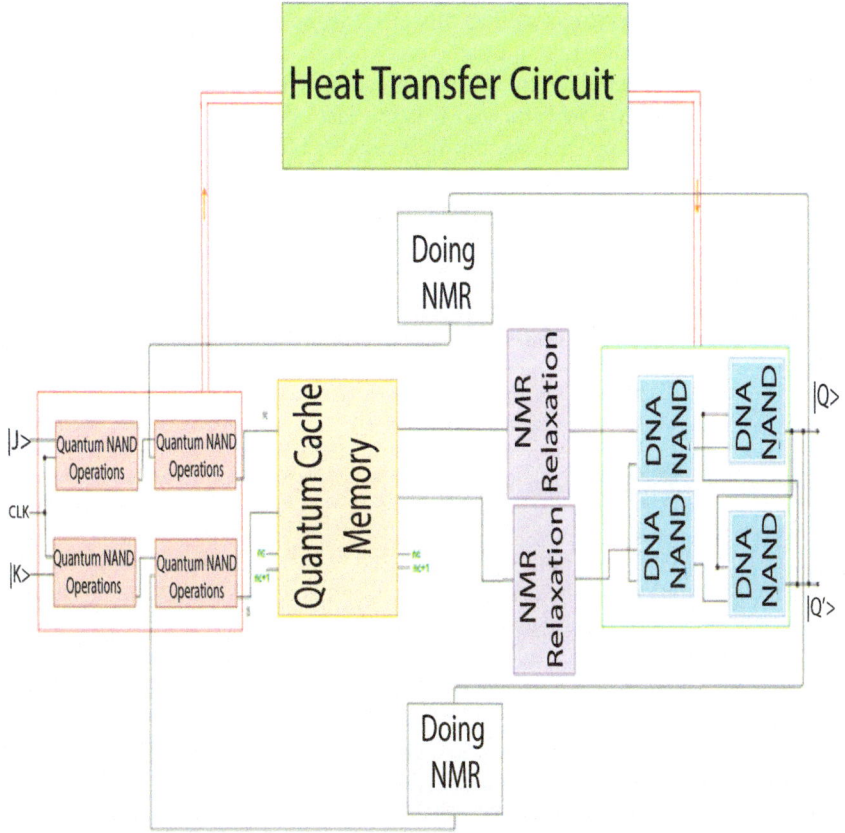

Fig. 13.7 Block diagram of a quantum-DNA JK flip-flop circuits

As compared to other quantum circuits, the quantum JK flip-flop will not require as much power. The qubit may simply conduct the operation once all of the molecules are in superposition state and coherence mode.

Quantum-DNA JK flip-flop toggles which has two-qubit inputs and clock qubit input. It relies on clock qubit input. The block diagram of a quantum-DNA JK flip-flop is shown in Fig. 13.7.

Quantum-DNA JK flip-flop needs one qubit input and one clock input which is also a qubit. The qubit input |J> and the qubit input |K>, which will enter into two different NAND operations. One |clk> input is shared by both of the quantum NAND operations. These quantum NAND operations perform parallelly according to the quantum computing principle. The output lines of the quantum NAND operations are entered as input lines into another couple of quantum NAND operations. The block diagram depicts that the final output will be the DNA molecule sequence. So, this qubit will be moved through the NMR process and the NMR will make this qubit into a DNA sequence as well as this DNA sequence will enter into the DNA NAND operation as input. Like the same other output, the molecular sequence will come as input in another quantum NAND operation. Then finally two DNA NAND operations will parallelly perform and it is the last operation of DNA computing according to the quantum-DNA JK flip-flop circuit's block diagram. It is important to note that two quantum NAND operations produce two-qubit which will be stored in a quantum cache memory. Then this quantum cache memory serves the qubit into the DNA computing circuit's portion. Thus, these qubits need to be converted into molecular sequences. So, qubits are performed the NMR relaxation process. In the NMR relaxation process, emitting EMR is strictly prohibited. After the NMR relaxation process, qubits are relaxing their states and are converted into the molecular sequences.

13.5.1   Circuit Architecture

Quantum-DNA JK flip-flop circuit is constructed using one principle, that is, in this circuit, the first portion will be constructed by using quantum circuit and the second portion will be constructed by using DNA computing circuit. This circuit also has quantum cache memory and a heat transfer circuit to achieve the best output at the end. Quantum-DNA JK flip-flop operation circuit is shown in Fig. 13.8.

In this circuit, the input is qubit and the output is the molecular sequence. The clock input |clk> is shared by two quantum NAND operations and two quantum operations are performed parallelly. The quantum NAND operations' outputs and the final output of the quantum-DNA JK flip-flop are performed another two quantum operations parallelly. The output lines of the two quantum NAND operations circuit are connected with the quantum cache memory. This is the first portion of the circuit and is fully designed by using the principle of quantum computing. Quantum cache memory is made by using some quantum shift registers and in this circuit it saves data in the quantum array. When the time arrives, the quantum cache memory supplies the data to the NMR relaxation process. Quantum cache memory supply line fully connected to NMR relaxation process, where the EMR emit is fully prohibited. The NMR relaxation process output line is connected to the DNA NAND operation. Two outputs of the NMR relaxation process are connected to two different DNA NAND operations. DNA NAND operations another input comes from the outputs which are molecular sequences. Hence this DNA NAND operations' final output lines are the final outputs of quantum-DNA JK flip-flop.

13.5.2   Working Principle

Quantum-DNA JK flip-flop is working using two principles: the quantum computing principle and the DNA computing principle. Quantum-DNA JK flip-flop has two

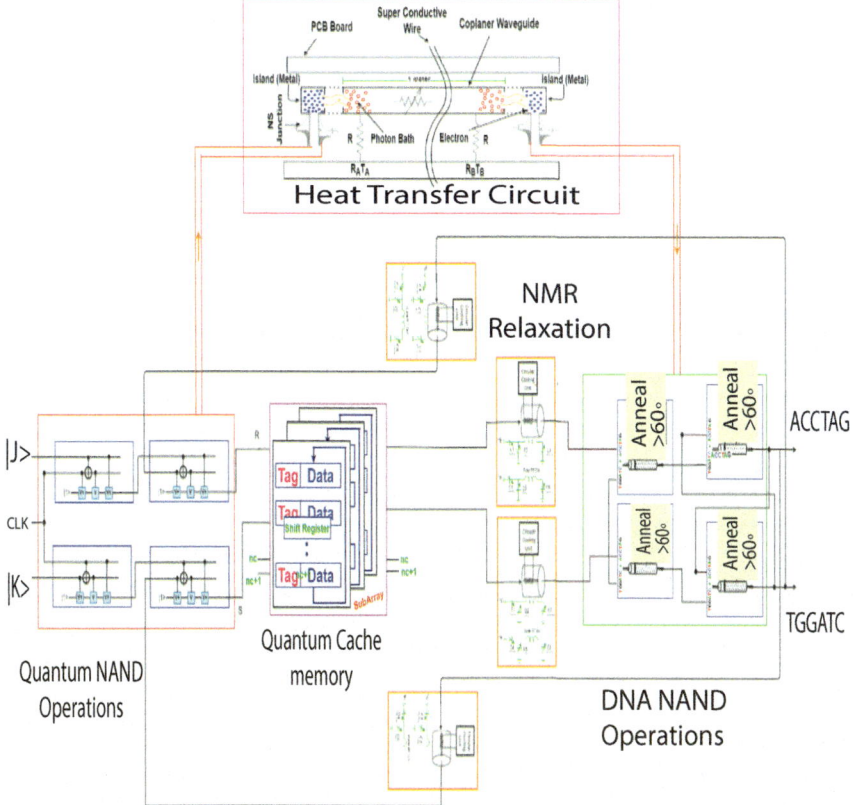

Fig. 13.8   Quantum-DNA JK flip-flop operation circuit

inputs which are |J> and |K> and the clock input which is |clk>. The JK flip-flop won't enable if the clock input is not |0>. Hence the clock input enables this circuit to start to work and first of all, it will perform the quantum circuit operation. This operation will be happening in close to zero temperature because qubits need a coherence state. The superposition state will be stable if this circuit is full of prohibited particles from other environmental particles. The |clk> input will be shared and performs two quantum NAND operations parallelly. This quantum NAND operation produces some garbage values but in this circuit this topic is avoidable. This NAND operations produce two output qubits at the same amount of time. These qubits, one of each will performs two quantum NAND operations parallelly. The first qubit and final output of quantum-DNA JK flip-flop will perform quantum NAND operation. Another qubit and the another output of the quantum-DNA JK flip-flop will perform the quantum NAND operation. These two operations will be performed parallelly. But the final output of quantum-DNA JK flip-flop is a molecular sequence.

Table 13.4 Truth table of a quantum-DNA JK flip-flop

|T>	Q (DNA sequence)	Q′ [Q′ is the complement of Q] (DNA sequence)
|0>	TGGATC	TGGATC
|1>	TGGATC	ACCTAG
|0>	ACCTAG	TGGATC
|1>	ACCTAG	TGGATC

Thus, the two outputs of two qubits of two quantum NAND operations enter into the quantum cache memory.

Quantum cache stores the qubit into an array, and it serves the data to the NMR relaxation process. In this quantum-DNA JK flip-flop, two-qubit is stored in quantum cache memory. This qubit performs the NMR relaxation process because DNA circuits need molecular sequence. NMR relaxation process removes the superposition state and makes the qubit into a molecular sequence. This molecular sequence performs the DNA NAND operation. The final output sequence and molecular sequence from the cache perform DNA NAND operation. Another final output sequence and previous output from the cache also perform another DNA NAND operation. These two NAND operations will be performed parallelly. This structure is basically a DNA NAND latch operation. Quantum cache memory is mainly used because quantum operation performs so fast while on the other hand DNA operation performs very slowly. Quantum cache memory works as an intermediate process where qubit is just stored and when needed cache memory serves the qubit. Table 13.4 shows the truth table of a quantum-DNA JK flip-flop.

13.5.3   Applications

A 1-qubit word may be stored in a single quantum flip-flop. Thus, by joining a group of quantum-DNA flip-flops, the storage capacity in terms of qubits may be increased. Quantum-DNA JK flip-flops are widely used in computers and Internet-of-Things devices. Quantum-DNA JK flip-flops may be utilized in a variety of applications, including registers, counters, frequency dividers, and event detectors.

It can also be utilized as a frequency divider. Event detectors and registers can also benefit from quantum-DNA JK flip-flops. Memory devices can also be benefited from quantum-DNA JK flip-flops.

13.6   Quantum-DNA T Flip-Flop

Quantum-DNA T flip-flop is also known as "Quantum-DNA Toggle Flip-Flop". A "T flip-flop" is a type of quantum-DNA memory circuit that toggles its output state on the rising edge of a clock signal when the toggle input is high. It is a single-input version of the JK flip-flop, where both inputs of the JK are tied together. In the context of quantum-DNA computing, exploring T flip-flops within the framework of quantum devices and DNA strands is an area of research, particularly for building memory elements and sequential circuits. The use of T flip-flops in quantum-DNA computing, particularly within the context of QCA, is an area of active research. It focuses on utilizing these memory elements to build fundamental building blocks for quantum-inspired computing and sequential circuits at the nanoscale. To avoid the occurrence of the intermediate state in quantum-DNA SR flip-flop, only one input is provided to the flip-flop called the Trigger qubit input or Toggle input. Toggling means 'Changing the next state output to complement the present state output'. The quantum-DNA T flip-flop can be designed by making simple modifications in the quantum-DNA JK flip-flop. The quantum-DNA T flip-flop is a single qubit input device and hence by connecting $|J>$ and $|K>$ inputs together and giving them with single input called $|T>$, a quantum-DNA JK flip-flop can be converted into a quantum-DNA T flip-flop.

Hence quantum-DNA flip-flops have two portions based on the principle of the quantum-DNA circuit. In quantum-DNA T flip-flop circuit one portion is made by quantum computing principle and the rest of the portion is made by DNA computing principle. Quantum computing and DNA computing are connected via the NMR relaxation process. For time consistency, the quantum cache memory in this circuit can be used. The quantum circuit portion produces so much heat according to the quantum computing principle that's why this circuit can transfer a small amount of heat to the DNA computing circuit portion. According to DNA computing, it needs a vast amount of heat to perform the calculation. In quantum-DNA T flip-flop, the circuit has one input which is a qubit input named $|T>$. This input performs first of all quantum operations and is then stored in quantum cache memory. When the perfect time arrived it will perform the NMR relaxation and convert into DNA molecular sequence as well as it performs the DNA computing operation. Finally, after all of these processes, the quantum-DNA T flip-flop produces the output as DNA molecular sequence. Quantum-DNA T flip-flop toggles and it requires much time compared to quantum T flip-flop but requires less compared to DNA T flip-flop. Quantum-DNA T flip-flop circuit's quantum circuit portion performs so fast but DNA circuit portion much slower. In quantum-DNA T flip-flop, a bunch of quantum AND operations perform according to quantum computing principle and a bunch of DNA NOR operations perform in DNA computing portion in the circuit. Quantum-DNA T flip-flop removes the problem of quantum-DNA SR flip-flop.

13.6.1 Block Diagram

Quantum-DNA T flip-flop toggles the input, where the quantum-DNA JK flip-flop's extended version is quantum-DNA T flip-flop architecturally. Quantum-DNA T flip-flops have one qubit input and clock qubit input. Quantum T flip-flop relies on clock qubit input. Quantum-DNA T flip-flop circuit's block diagram is shown in Fig. 13.9.

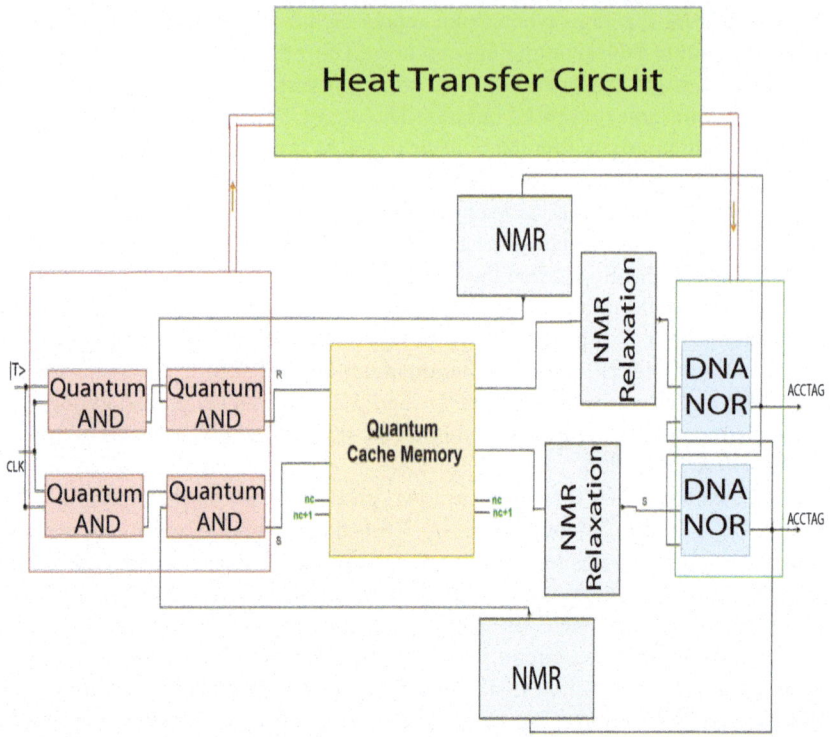

Fig. 13.9 Block diagram of a quantum-DNA T flip-flop circuits

Quantum-DNA T flip-flop needs one qubit input; and one clock input which is also a qubit. One qubit input |T> and |clk> are shared by two quantum AND operations. These quantum AND operations firstly perform parallelly according to the quantum computing principle. Then, the quantum AND operations output lines are entered as input lines into another couple of quantum AND operations. These two quantum AND operations produce two-qubit which will be stored in quantum cache memory. Then, this quantum cache memory serves the qubit to the DNA computing circuit's portion. These qubits need to be molecular sequence first. So, qubits are performed here first the NMR relaxation process. In the NMR relaxation process, the emitting EMR is strictly prohibited. After the NMR relaxation process, qubits are relaxing their states and are converted into the molecular sequences.

In DNA computing, molecular sequence needs to go through many processes during the operation. In DNA computing, melting and annealing are very important and these steps require a vast amount of heat. That's why in quantum-DNA T flip-flop quantum portion transfers a small amount of produced heat from the quantum circuits to the DNA circuits. Though this amount of heat is not enough, this heat or temperature will help to perform DNA computation. In DNA computing, DNA

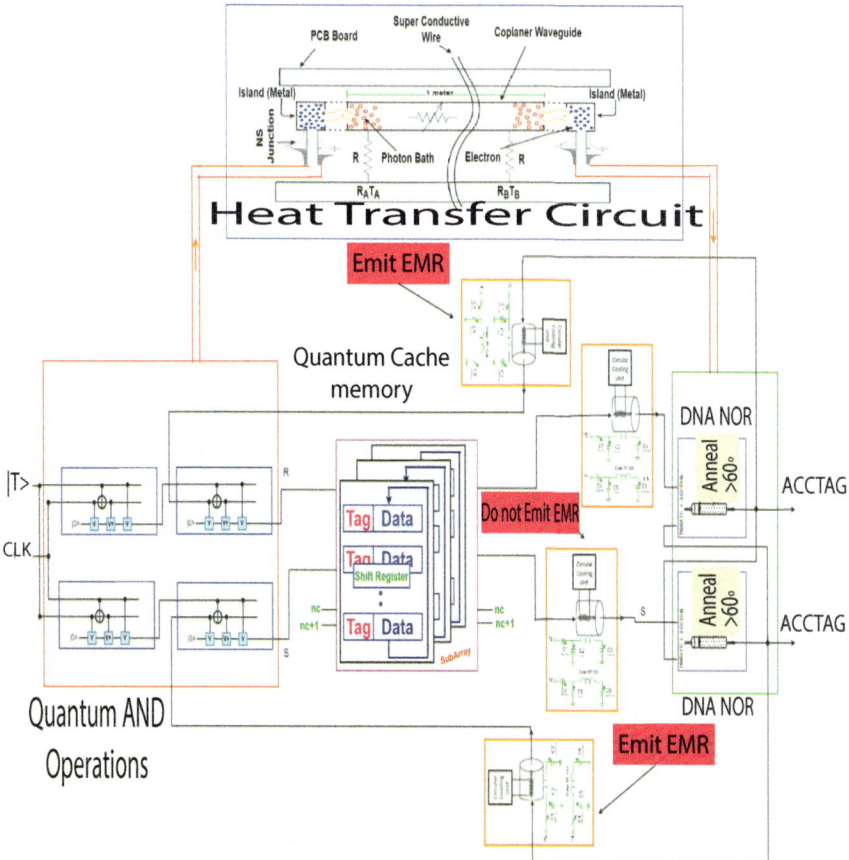

Fig. 13.10   Quantum-DNA T flip-flop operation circuit

NOR SR latch performs its operation, where two DNA NOR operations perform parallelly and input comes from quantum circuit portion. After that operation, the output molecular sequence is found from the quantum-DNA T flip-flop.

13.6.2   Circuit Architecture

Quantum-DNA T flip-flop circuit is constructed using one principle, where the first portion will be constructed by using quantum circuit and the second portion will be constructed by using DNA computing circuit. This circuit also has quantum cache memory and a heat transfer circuit to achieve the best output at the end. Quantum-DNA T flip-flop operation circuit is shown in Fig. 13.10.

In this circuit, input is qubit and the output is the molecular sequence. Input |T> and the clock input |clk> are shared by two quantum AND operations. Two quantum operations are performed parallelly. These quantum AND operations output line and final output of quantum-DNA T flip-flop are processed and converted into another two quantum operations parallelly. But the final two outputs are molecular sequences. So, these two output lines are connected to the NMR relaxation process. These two quantum AND operations circuit output lines are connected to the quantum cache memory. This is the first portion of the circuit and is fully designed by using the principle of quantum computing. Quantum cache memory is made by using some quantum shift registers and in this circuit it saves data in the quantum array. When the time arrives, the quantum cache memory supplies the data to the NMR relaxation process. The quantum cache memory supply line is fully connected to NMR relaxation process as well as here the EMR emit is fully prohibited. Then, the NMR relaxation process output line is connected to the DNA NOR operation. Two outputs of the NMR relaxation process are connected to two different DNA NOR operations. DNA NOR operations another input come from the outputs which are molecular sequences. Hence, these DNA NOR operations final output lines are the final outputs of the quantum-DNA T flip-flop.

13.6.3   Working Principle

Quantum-DNA T flip-flop is working using two principles, and these are quantum computing principle and the DNA computing principle. Quantum-DNA T flip-flop has one input is named |T> and the clock input is |clk>. Quantum-DNA T flip-flop won't enable if the clock input is not |0>. Hence, the clock input enables this circuit to start to work and first of all, it will perform the quantum circuit operation. Quantum circuit operation will be occured in close to zero temperature because qubits need a coherence state. The superposition state will be stable if this circuit is full of prohibited from other environmental particles. |T> and |clk> inputs will be shared and perform two quantum AND operations parallelly. This quantum AND operation produces some garbage values but in this circuit this topic is avoidable. The quantum AND operation produces two output qubits by the same amount of time. These qubits, one of each will perform another two quantum AND operations parallelly. The first qubit and final output of quantum-DNA T flip-flop will perform quantum AND operation. Another qubit and another output sequence of the quantum-DNA T flip-flop will perform the another quantum AND operation. These two operations will be performed parallelly. But the final output of quantum-DNA T flip-flop is a molecular sequence. Thus, these two outputs with two qubits of the quantum AND operations enter into the quantum cache memory.

Quantum cache stores the qubit into an array. Quantum cache memory stores the qubit data and when requires it serves the data to the NMR relaxation process. In this quantum-DNA T flip-flop, two-qubit is stored in quantum cache memory. This qubit performs the NMR relaxation process because DNA circuits need molecular

Table 13.5 Quantum-DNA
T flip-flop truth table

|T>	Q (DNA sequence)	$\overline{Q}$ (DNA sequence)
|0>	TGGATC	TGGATC
|1>	TGGATC	ACCTAG
|0>	ACCTAG	TGGATC
|1>	ACCTAG	TGGATC

sequence. NMR relaxation process removes the superposition state and makes the qubit to a molecular sequence. This molecular sequence performs the DNA NOR operation. The final output sequence and molecular sequence from the quantum cache memory perform DNA NOR operation. Another final output sequence and previous output from the quantum cache memory also perform another DNA NOR operation. These two NOR operations perform parallelly. This structure is basically DNA NOR latches operation. Quantum cache memory is mainly used because quantum operation performs so fast, while DNA operation performs very slowly. The quantum cache memory works as an intermediate process where qubit is just stored and when needed the cache memory serves the qubit.

After the completion of all of these operations, the quantum-DNA T flip-flop can produce two output molecular sequences. Table 13.5 shows the truth table of a quantum-DNA T flip-flop.

13.6.4   Applications

A quantum-DNA Toggle switch is a quantum-DNA T flip-flop. 'Changing the next state output to complement the current state output' is what toggling means. Simple tweaks to the JK flip-flop can be used to create quantum-DNA T flip-flops. Because the quantum-DNA T flip-flop can store data, it has a variety of uses in memory devices. Quantum-DNA T flip-flop operational circuits solve some of the issues that quantum-DNA JK flip-flops face.

13.7   Quantum-DNA Shift Register

Shift registers, quantum-DNA memory circuits, find applications in quantum-DNA computing, particularly in designing encoding and decoding systems for quantum information. They can be implemented using quantum-DNA flip-flops, where the output of one flip-flop is connected to the input of the next, and a clock signal drives the shifting of data. This allows for the storage and manipulation of quantum-DNA information in a sequential manner, similar to how classical shift registers work with bits. In essence, quantum-shift-register circuits, built upon quantum flip-flops, offer a valuable tool for quantum information processing and have potential applications in various areas of quantum biology, including molecular computing and the simulation of biological processes. A single qubit of two-valued qubit data ($|1>$ or $|0>$) can be stored in a quantum-DNA flip-flop. However, many quantum-DNA flip-flops are required to store multiple qubits of data. To store n qubits of data, N quantum-DNA flip-flops must be coupled in a certain order. A quantum-DNA register is a gadget that stores this type of data. It consists of a sequence of quantum-DNA flip-flops used to store multiple qubits of data.

Quantum-DNA shift registers enable the information stored in these quantum registers to be transmitted. A quantum-DNA shift register is a collection of flip-flops that store several qubits of information. By applying clock pulses to the qubits contained in such quantum-DNA registers, they may be allowed to move inside them and in and out of them. By linking n quantum flip-flops and n number of DNA flip-flops, each of which stores a single qubit of data or molecule sequence of data, an n quantum-DNA shift register may be built. "Quantum-DNA shift left registers" are quantum-DNA registers that will shift the qubits or molecules to the left. "Quantum-DNA shift right registers" are quantum-DNA registers that will shift the qubits or molecule sequence to the right.

Quantum-DNA shift registers are basically of four types. These are as follows:

1. Quantum-DNA Serial In Serial Out shift register
2. Quantum-DNA Serial In parallel Out shift register
3. Quantum-DNA Parallel In Serial Out shift register
4. Quantum-DNA Parallel In parallel Out shift register.

In this chapter, a shift register is built utilizing a quantum D flip-flop operational circuit and a DNA D flip-flop operational circuit to convert serial data into quantum data and DNA molecule sequence data. The quantum-DNA Serial-In Serial-Out shift register is a type of quantum-DNA shift register that permits serial input of one qubit or one molecule sequence at a time over a single data line and outputs a serial output. The data exits the quantum-DNA shift register one qubit at a time in a serial pattern since there is only one qubit output, thus the term quantum-DNA Serial-In Serial-Out shift register. Four quantum-DNA D flip-flops are linked in a serial fashion in this circuit. Because the same clock signal is supplied to each quantum-DNA flip-flop, they are all synchronized with one another. The circuit shown below as the design architecture is an example of a quantum-DNA shift right register, which accepts serial data from the quantum flip-flop's left side and DNA flip-flop's left side.

In quantum-DNA flip-flop, qubit is entered into as an input in the circuit and performs the required operation. Quantum computing computation speed is much higher than DNA computing computational speed. For that case, qubits are stored in quantum cache memory. When actually the time arrives, the qubit first of all performs the Paul Trap relaxation process and converts the qubit into molecule sequence. Then this molecule sequence performs the DNA flip-flop operation. Quantum circuit produces much heat and heat needs to be removed to maintain the coherence state and DNA computing's circuit needs huge heat to enable the circuit to perform. In that case, this circuit uses a heat transfer circuit to transfer the heat to the DNA circuits.

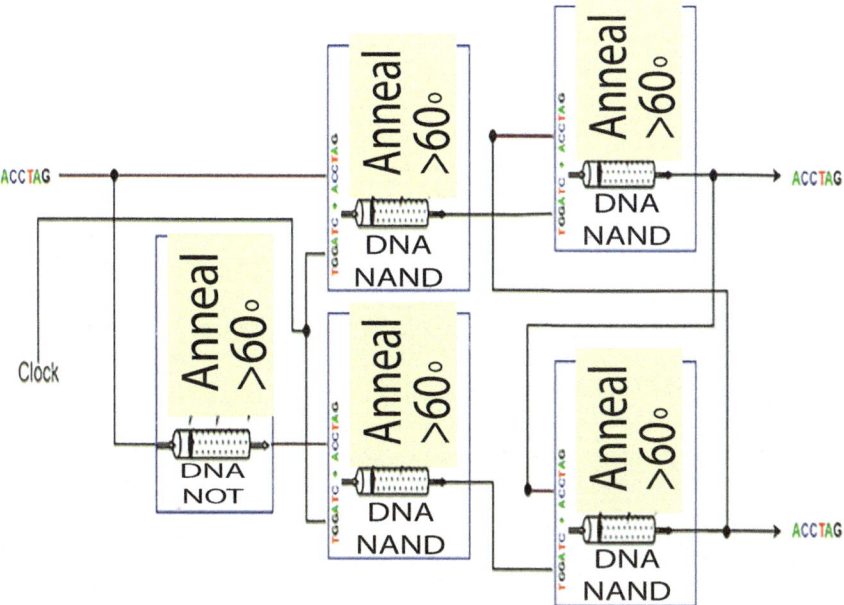

Fig. 13.11 DNA D flip-flop circuit

13.7.1 *Circuit Architecture*

Three quantum D flip-flop operational circuits and One DNA D flip-flop circuits
are used to make a quantum-DNA shift register. As a fundamental component, a
quantum-DNA shift register is utilized for data shift, and quantum D flip-flop and
DNA D flip-flop are used to make it happen.

The DNA D flip-flop has a single molecular input sequence and is developed
using DNA NAND operations and a DNA SR latch. Figure 13.11 shows the DNA
D flip-flop circuit diagram. The clock molecular input sequence affects the DNA
D flip-flop. It is clear from the diagram that the circuit has one molecular input
sequence. One line of this molecular input sequence will be directed to the DNA
NAND operation known as S molecular input sequence in circuit. In this case, the S
molecular input sequence and the clock molecular input sequence are used in a DNA
NAND operation.

Two inputs are used in quantum D flip-flops: one for data and another for the
clock. The outputs of quantum D flip-flops are logically opposite one another. The
circuit's synchronization with an external signal is aided by the clock input. A quantum D flip-flop's output can have two possible values. Data input is directed to a
quantum NAND operation circuit in this block diagram, while data input reverse is
routed to another quantum NAND operation circuit. Both quantum NAND operations procedures use the clock pulse input. Quantum SR latch receives the result of
two quantum NAND operations. The D flip-flop is constructed using the SR latch.

This property is used to induce a delay in the circuit's data flow. Quantum SR latch is made by two quantum NAND operations. The final two outputs of the quantum SR latch function are uncovered. A quantum D flip-flop may provide two types of outputs, one of which is logically inverse to the other. The quantum D flip-flop will continue to function if the clock is enabled; otherwise, the quantum D flip-flop will stop working.

Quantum D flip-flop operational circuits are also coupled through serial connection in the quantum shift register block diagram. The quantum D flip-flop operational circuit is the fundamental component of the quantum shift register. After processing the qubit input in the quantum shift register the output of the shift register is stored in quantum cache memory.

In DNA computing the molecular sequence needs to go through many processes during the operation. In DNA computing, melting and annealing are very important and these steps require a huge amount of heat. That's why in quantum-DNA shift register quantum portion transfers a small amount of produced heat from the quantum circuits to the DNA circuits. Though this amount of heat is not enough, this heat or temperature will help to perform DNA computation. In DNA computing, DNA NOR SR latch performs its operation, where two DNA NOR operations perform parallelly and input comes from quantum circuit portion. After that operation, the output molecular sequence is found from quantum-DNA shift register.

Four D flip-flops and a quantum AND operation are used in the quantum shift register. The shift register generates four qubit outputs. The D flip-flop has a single qubit input and is developed using quantum NAND and quantum SR latch operations. The circuit architecture of quantum-DNA shift register is shown in Fig. 13.12.

The clock qubit input is required for the quantum D flip-flop. One line of this qubit input will be directed to a quantum NAND operation termed S input in circuit. Quantum NAND operation is performed here using S qubit input and clock qubit input.

In this circuit, input is qubit and output is a molecular sequence. Input $|X>$ and clock input $|clk>$ are shared by two quantum NAND operations. Two quantum operations are performed parallelly. These quantum NAND operations output line and final output of quantum-DNA shift register are performed another two quantum operations parallelly. But the final two outputs are molecular sequences. So, these two output lines are connected to the Paul Trap relaxation process and then after that process, these two lines come into use as an input line in the quantum NAND operation circuit. These two quantum NAND operations circuit output lines are connected to the quantum cache memory. All of these quantum NAND Operations are the part of quantum D flip-flop which is used in this system. This is the first portion of the circuit and is fully designed by using the principle of quantum computing. Quantum cache memory is made by using some quantum shift registers and in this circuit it saves data to the quantum array. When the time arrives, the quantum cache memory supplies the data to the Paul Trap Relaxation process. Quantum cache memory supply line is fully connected to Paul Trap Relaxation process as well as here the EMR emit is fully prohibited. The Paul Trap Relaxation process output line is connected to the DNA NOT operation. After going through the DNA NOT operation, the output is taken as input in the DNA NAND operation. In these DNA NAND operations the 1st clock input is taken as $|clk>$ input. These inputs are shared

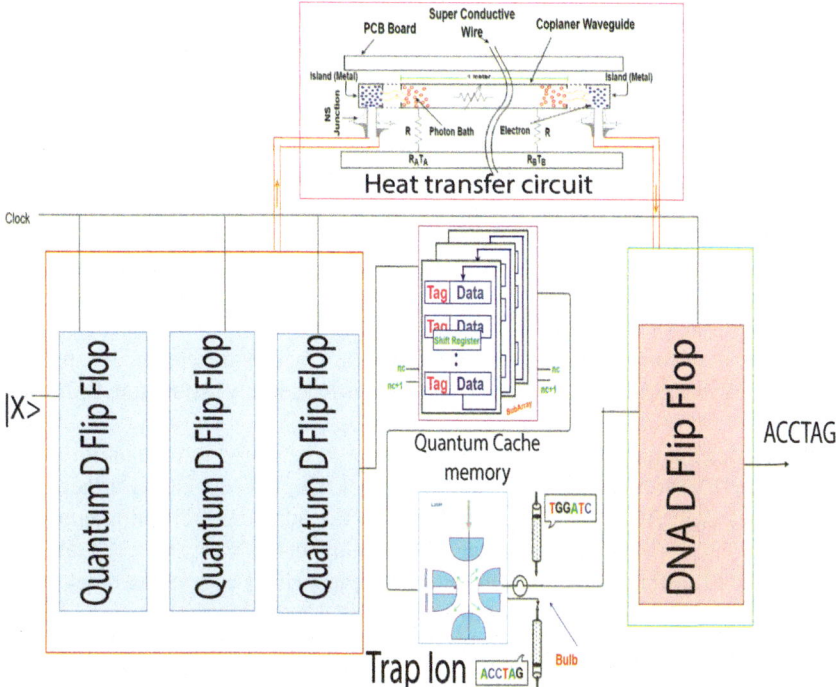

Fig. 13.12 Quantum-DNA shift register

by two DNA NAND operations. Then, the outputs of these operations are connected to another set of DNA NAND gates. During DNA NAND operation another input comes from the output which is ACCTAG molecular sequence. Hence this DNA NOR operations' final output line is connected to the final output of quantum-DNA shift register.

13.7.2 Working Principle

Quantum-DNA shift registers are a kind of registers where both qubit data loading, as well as data retrieval to/from the quantum-DNA shift register, occur in serial mode sometimes. This synchronous quantum SISO shift register is sensitive to the positive edge of the clock pulse. Here, the data word which is to be stored is fed bit-by-bit at the qubit input of the first quantum flip-flop. Further, it is seen that the qubit inputs of all other flip-flops are driven by the outputs of the preceding ones, for example, the input of quantum D flip-flop number-2 is driven by the output of quantum D flip-flop number-1. At last, the data stored within the quantum shift register is obtained at the output pin of the nth quantum D flip-flop in serial fashion.

Table 13.6 Truth table of a quantum-DNA shift register

$|Clk>$	$|x>$	Q	Q' [Q' is the complement of Q]	Description
$\downarrow \gg |0>$	X	Q	Q	Memory no change
$\uparrow \gg |1>$	$|0>$	TGGATC	ACCTAG	Reset Q $\gg$ 0
$\uparrow \gg |1>$	$|1>$	ACCTAG	TGGATC	Set Q $\gg$

Initially, all the quantum flip-flops in the quantum register are cleared by applying high on their clear pins. Next, the input data word is fed serially to quantum D flip-flop number-1.

This causes the qubit appearing at the first pin to be stored into quantum D flip-flop number-1 as soon as the first leading edge of the clock appears. Further at the second clock tick, B1 gets the stored qubit from the quantum D flip-flop number-2 while a new qubit enters into quantum flip-flop number-2.

This kind of shift in data qubits continues for every rising edge of the clock pulse. This indicates that for every single clock pulse the data within the quantum register moves toward the right by a single bit. Following the qubit data transmission, as explained, one can note that the first qubit of an input word appears at the output of nth flip-flop for the nth clock tick. On applying further clock cycles, one gets the next successive qubits of the qubit input data word as the serial output. The truth table of quantum-DNA shift register is given in Table 13.6.

After processing the inputs in a quantum D flip-flop, it creates one output. The output is used as a clock input for the next quantum D flip-flop. Thus, a quantum shift register creates one final qubit output.

13.7.3 Applications

The quantum-DNA shift register is a very much useful circuit in quantum computing. It can be used in counter, data format convertor, data processor, etc.

Data Format Converters of Quantum-DNA Shift Registers
Serial data transmission is preferred for long-distance communication due to its economic value in terms of the wires used. This necessitates parallel-to-serial conversion at the sender-end for which quantum Parallel In Serial Out Quantum-DNA shift registers (QPISO) can be used. However, in general, many quantum microprocessor-based systems usually prefer a parallel form of data-in for which the transmitted qubit data is to be converted into parallel mode using a serial-to-parallel converter like quantum Serial In Parallel Out Quantum-DNA shift register (QSIPO).

Counters of Quantum-DNA Shift Registers
Quantum-DNA ring counter and the quantum Johnson counter are the two shift-register-based counters which are extensively used in digital applications. In quantum-

DNA ring counters, the output of the last stage is back-fed as a qubit input to the first stage. This causes the data stored within the quantum-DNA shift register to circulate within it continuously. For example, a 4-qubit ring counter storing a data word |0001> has a repetitive sequence with four definite states viz., |0001>, |1000>, |0100>, and |0010>. Quantum-DNA Johnson counter is similar to quantum-DNA ring counter except for the fact that the complement of the output at the last stage of the quantum-DNA shift register is fed as an input to the first stage.

Pseudo-Random Pattern Generator of Quantum-DNA Shift Registers
Quantum-DNA shift registers can be used to generate pseudo-random patterns which are used for testing. In order to achieve this, the outputs of a few stages in the quantum-DNA shift register are XORed and are connected as an input to the first stage of it.

The number of patterns generated depends upon the number of points that are tapped to be provided as the inputs of quantum XOR operation. If it is tapped appropriately, the maximum number of patterns that can be generated using an n-stage shift register is $(2n-1)$.

Quantum-DNA shift register has also some other applications. Quantum-DNA shift register can be used in various quantum devices. Quantum-DNA shift register is faster than digital shift register. So, it will be more useful as well as it can work in a simpler way than many existing devices.

13.8 Quantum-DNA Ripple Counter

A quantum counter is basically used to count the number of clock pulses applied to a quantum flip-flop. It can also be used for quantum frequency divider, quantum time measurement, quantum frequency measurement, quantum distance measurement, and also for generating square waveforms. In this chapter, the quantum flip-flops are quantum asynchronous counters and are supplied with different clock signals, there may be a delay in producing output. Also, a few numbers of quantum logic gates are needed to design asynchronous counters. So, they are elementary in design and also are less expensive. In other words, a ripple counter, typically used in quantum computing and DNA computing, can be conceptually adapted for quantum-DNA computing, though the specifics are different. A ripple counter, as described in quantum-DNA logic, is a series of flip-flops, where the output of one flip-flop triggers the next, creating a ripple effect. In the context of quantum-DNA computing, this concept could be applied to mimicking signal propagation in biological systems, like nerve impulses or biochemical reactions. In essence, the ripple counter's concept of sequential processing could be adapted to the quantum realm, but the specific implementation would require careful consideration of quantum principles and the characteristics of biological systems. In summary, while a ripple counter is a fundamental concept in classical computing, its adaptation to quantum-DNA computing would involve using quantum flip-flops and gates to simulate signal propagation in biological systems, considering factors like biological saturation, qubit stability, and quantum error correction.

An n-bit ripple counter can count up to $2n$ states. It is also known as MOD-n counter. It is known as a ripple counter because of the way the clock pulse ripples its way through the flip-flops. It is an asynchronous counter. Different ripple counters are used with a different clock pulse. All the flip-flops are used in toggle mode. Only one flip-flop is applied with an external clock pulse and another flip-flop clock is obtained from the output of the previous flip-flop. The flip-flop applied with an external clock pulse acts as LSB (Least Significant Bit) in the counting sequence. A counter may be an up counter that counts upwards or can be a down counter that counts downwards or can do both, i.e., count up and count downwards, depending on the input control. The sequence of counting usually gets repeated after a limit.

Quantum-DNA ripple counter is made by four JK flip-flops. Using these quantum JK flip-flops, the quantum-DNA ripple counter creates four qubit outputs. Here, in JK flip-flop, J and K are not shortened abbreviated letters of other words, such as "S" for Set and "R" for Reset, but are autonomous letters chosen by its inventor Jack Kilby to distinguish the flip-flop design from other types. Jack Kilby invented the digital electronics JK flip-flop. Quantum-DNA ripple counter is an asynchronous counter. It is created using quantum JK flip-flops, and these flip-flops are only controlled by clock pulse input.

Quantum-DNA ripple counter produces much heat to produce the molecule's superposition state and also produces some garbage values.

13.8.1   Circuit Architecture

Quantum-DNA ripple counter uses three quantum JK flip-flops and one DNA JK flip-flop to create three qubit outputs.

The two-molecular input sequence and one clock shared molecular input sequence of the DNA JK flip-flop are shared. The clock affects the DNA JK flip-flop as well. The circuit will turn on if the clock is turned on, else it will not. DNA JK flip-flop operation circuit is shown in Fig. 13.13.

Quantum JK flip-flop has a two-qubit input and one clock shared input. Quantum JK flip-flops also depend on the clock. If the clock is enabled then the circuit will enable, otherwise not.

Quantum-DNA ripple counter uses three quantum JK flip-flop to operate two qubit inputs. Quantum JK flip-flop has the two-qubit input named as $|J>$ and $|K>$. Quantum JK flip-flop has the qubit inputs $|J>$ and $|K>$. This quantum JK flip-flop consists of many quantum NAND operations. At first basic quantum NAND operation performs a couple of actions, then this operation output is entered into the SR flip-flop as well as produces the outputs $|Q>$ and $|\overline{Q}>$. First of all $|J>$ and $|clk>$ inputs perform the quantum NAND operation. The output of the quantum NAND operation and output of the quantum JK flip-flop $|\overline{Q}>$ performs another quantum NAND operation and produces the output named as $|S>$. $|clk>$ is a shared input. So, $|K>$ and $|clk>$

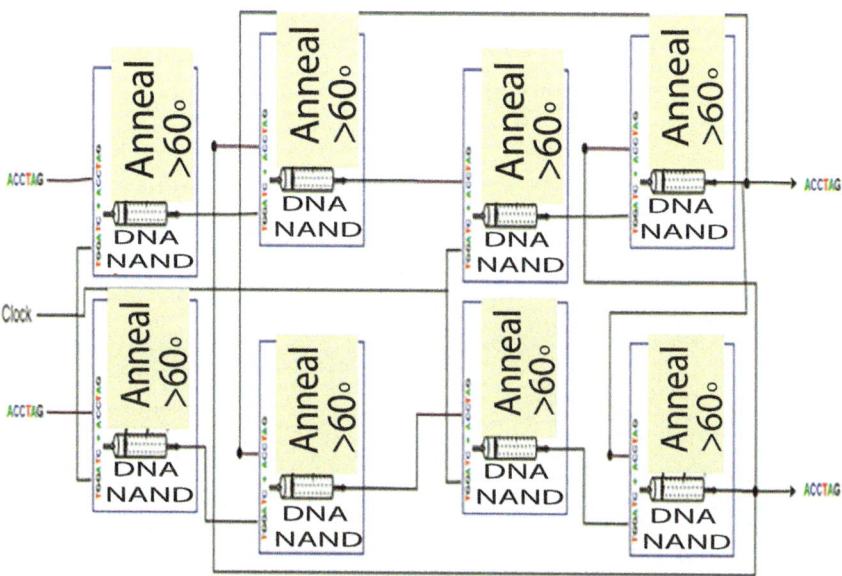

Fig. 13.13 DNA JK flip-flop operation circuit

inputs also perform the quantum NAND operation. This quantum NAND operations output and |Q> output of the quantum JK flip-flop perform another quantum NAND operation as well as produce the output named as |R>.

These |S> and |R> are entered into the quantum SR flip-flop and produce two outputs |Q> and |Q'> (Q' is complement of Q). In quantum SR flip-flop, it has four quantum NAND operations. |S> input and |clk> input perform quantum NAND operation as well as |R> input and shared |clk> input also perform the quantum NAND operation. These two NAND operations outputs are entered into the quantum SR latch as input. In quantum SR latches, two |Q> and one input as well as |Q'> (Q' is complement of Q) and other inputs perform the quantum NAND operation. Finally, after all of these quantum operations the quantum JK flip-flops final outputs are |Q> and |$\overline{Q}$>.

After processing the qubit inputs in three JK flip-flops, the output is stored in a quantum cache memory, which is made out of quantum shift register. Then this quantum cache memory serves the qubit to the DNA computing circuit's portion. These qubits need to be molecular sequence first of all and then qubits are performed here NMR relaxation process. In the Paul trap ion process, qubits are relaxing their state and are converted into the molecular sequence.

In DNA computing, molecular sequence needs to go through many processes during the operation. In DNA computing, the melting and annealing are very important and these steps require a huge amount of heat. That's why in quantum-DNA ripple counter quantum portion transfers a small amount of produced heat from the quan-

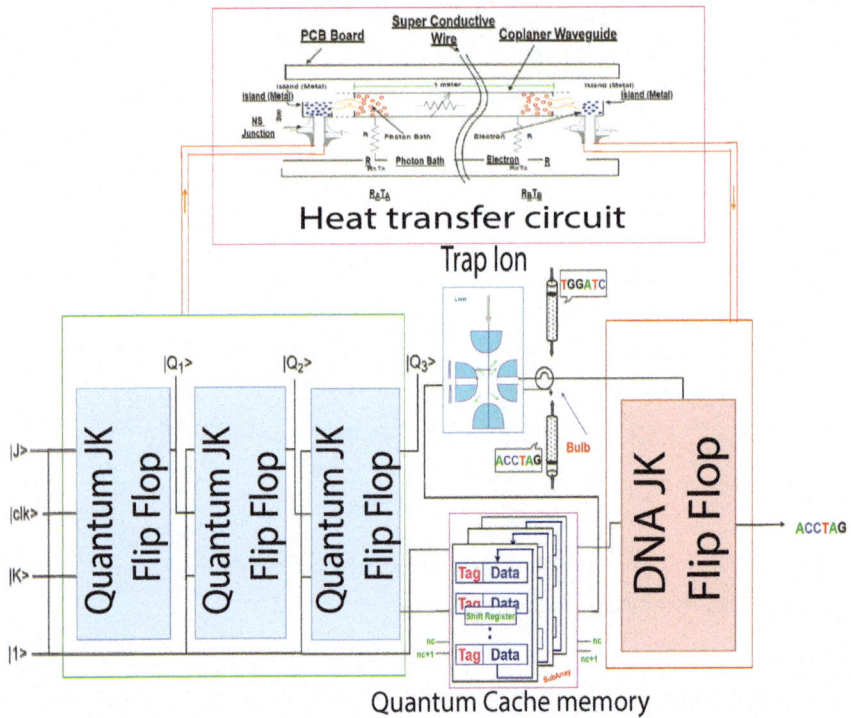

Fig. 13.14 Quantum-DNA ripple counter

tum circuits into DNA circuits. Though this amount of heat is not enough, this heat or temperature will help to perform DNA computation. In DNA computing, DNA NOR SR latch performs its operations, where two DNA NOR operations perform parallelly and input comes from quantum circuit portion. After that operation, the output molecular sequence is found from quantum-DNA ripple counter. Quantum-DNA ripple counter is shown in Fig. 13.14.

In quantum-DNA ripple counter there is a one clock input and one logic input, which are shared in both |J> and |K> input ports. Input |J> and input |K> both the values will perform the quantum NAND operation with the shared |clk> input differently. |J> and |clk> input perform the quantum NAND operation. Quantum NAND operation is made by using quantum basic operations. Basic operations in mainly quantum computing are V, V+, and CNOT operations. For error correction here it is used an ancillary bit. In quantum NAND operation the value of an ancillary bit is |1>. After |J> and |clk> perform the quantum NAND operation which produces an output qubit and this output qubit and the output of the flip-flop $|\overline{Q}>$ perform the quantum NAND operation and produce the output |S>. Using the same procedure, |K>, |clk> and |Q> inputs produce the |R> output. These operation circuits are mainly quantum NAND operations. These |S> and |R> inputs perform the quantum

SR flip-flop operation. In quantum SR flip-flop operational circuit architecture also made by the basic component of quantum computing which is quantum NAND operation.

After processing the inputs in a quantum JK flip-flop, the output of the flip-flop is stored as an output of the quantum-DNA ripple counter. In quantum-DNA ripple counter intermediate architectures the four quantum JK flip-flops are connected in serial connection using the logical qubit input. Every clock input as |clk> enters into every qubit and from 2nd quantum JK flip-flop, |clk> input is previous quantum Jk flip-flops first output |Q>. With the same architecture, the quantum-DNA ripple counter can be performed as a up counter or as a down counter but the clock pulse as |clk> needs to sometimes have a positive edge triggered and sometimes a negative edge triggered.

13.8.2   Working Principle

In the quantum-DNA ripple counter there are four quantum JK flip-flop operational circuits. These quantum JK flip-flop operational circuits are connected in serial connection. In this quantum-DNA ripple counter, there is one clock input and a logic input which are shared by the ports |J> and |K> of quantum JK flip-flop operational circuit. The inputs |J> and |K> of a quantum Jk flip-flop conduct two quantum processes in parallel. The quantum NAND operation is performed using |J> and shared input |clk>. The quantum NAND operations are then completed, and one of the quantum JK flip-flop's outputs executes the quantum NAND operation, producing |S>. The |K> and |clk> inputs are performed first in the quantum NAND operation. The output of the quantum JK flip-flop |Q> and the result of the quantum first NAND operation, are then used to perform another quantum NAND operation, yielding |R>. The steps for creating |S> and |R> are carried out simultaneously. It is known that one of the distinctive properties of quantum operations is that they may do several operations at the same time, and this is exactly what is happening. The quantum SR flip-flop operation is then conducted on these |S> and |R> inputs. The quantum NAND operation, which is employed here, is also used to make quantum SR flip-flops. After conducting the quantum SR flip-flop operation, two outputs are created, where one output is the opposite of the other.

Here if the |clk> is |0> then the quantum JK flip-flop will not be triggered but if |clk> is |1> the quantum JK flip-flop will be triggered and it will toggle the output. First of all, the quantum JK flip-flop operational circuit with |clk> which has value |1> and toggles the output value from the previous state value. Then the output of the initial quantum JK flip-flop will be |clk> which is the input of next quantum JK flip-flop. If the |clk> value is |1>, then the output value will be toggled, otherwise the output will be the previous state. By maintaining the same procedure, every quantum JK flip-flop is used in the quantum-DNA ripple counter. Quantum JK flip-flop is toggled very much, that's why in the quantum-DNA ripple counter the quantum JK

Table 13.7 Truth table of a quantum-DNA ripple counter

|clk>	Q_0	Q_1	Q_2	Q_3
|0>	|0>	|0>	|0>	|0>
|1>	|1>	|1>	|1>	|1>
|1>	|0>	|0>	|0>	|0>

flip-flop operational circuit is used as a basic component. Table 13.7 shows the truth table of a quantum-DNA ripple counter.

The outputs of JK flip-flops are connected to the quantum cache memory. This is the first portion of the circuit which is fully designed by using the principle of quantum computing. Quantum cache memory is made by using some quantum shift registers and in this circuit it saves data in the quantum array. When the time arrives, the quantum cache memory supplies the data to the Paul trap ion process. Quantum cache memory is fully connected to Paul trap ion process, where the EMR emit is fully prohibited. Then the Paul trap ion process is connected to the DNA NAND operation. Two outputs of the Paul trap ion process are connected to two different DNA NOR operations. Another inputs of DNA NOR operations come from the output which are molecular sequences. Hence this DNA NAND operations' final output line is connected to the final output of quantum-DNA ripple counter.

Quantum cache memory works here as the intermediate process. Quantum circuit produces much heat and here two quantum NAND operations perform. So, it produces much heat. But the quantum circuit needs to be close to zero Kelvin temperature to perform the operation. From the quantum circuit, it needs to reduce the temperature to maintain the superposition state of the qubit. Hence, in DNA circuit operation, it needs much heat in several steps. Mainly in melting and annealing require much heat. This heat transfer circuit cannot transfer full excessive heat produced from the quantum circuit and this heat is not enough to perform the DNA circuit operation. But this heat transfer can optimize the cost of heat based on the needs. After the completion of all of these operations and design architecture, the quantum-DNA ripple counter can produce two output molecular sequences.

13.8.3 Applications

Quantum-DNA BCD (Binary coded decimal) counter is a decade counter which has Mod = 10. Mod means the number of states the counter has. Quantum-DNA BCD counter counts decimal numbers from 0 to 9 and resets back to default 0. With each clock pulse, the counter counts up a decimal number. Quantum ripple BCD counter is the same as quantum-DNA ripple up-counter, the only difference is when the quantum-DNA BCD counter reaches count 10 it resets its flip-flops.

Different types of flip-flops with different clock pulses are used as a quantum-DNA ripple counter. It is an example of an asynchronous counter. The flip-flops are used in toggle mode in a quantum ripple counter. The external clock pulse is applied to only one flip-flop. The output of this flip-flop is treated as a clock pulse for the next flip-flop in the quantum ripple counter. In counting sequence, the flip-flop in which external clock pulse is passed acts as LSB in quantum ripple counter operational circuits.

13.9 Quantum-DNA Synchronous Counter

A quantum counter is a quantum device which can count any particular event on the basis of how many times the particular event(s) has occurred. In a quantum logic system or computers, this quantum counter can count and store the number of times any particular event or process has occurred, depending on a quantum clock signal. Most common type of quantum counter is a sequential quantum logic circuit with a single clock input and multiple qubit outputs. The qubit outputs represent two valued decimal numbers. Each clock pulse either increases the number or decreases the number.

Quantum-DNA synchronous circuits generally refers to something which is coordinated with others based on time. Quantum-DNA synchronous signals occur at same clock rate and all the clocks follow the same reference clock. Quantum synchronous counter has shown that the qubit output of that quantum counter is directly connected to the input of next subsequent counter and makes a chain system, and due to this chain system propagation delay appears during counting stage and create counting delays. In a quantum-DNA synchronous counter, the clock qubit input across all the quantum flip-flops uses the same source and creates the same clock signal at the same time. So, a quantum counter which is using the same clock signal from the same source at the same time is called a quantum-DNA synchronous counter. In other words, a synchronous counter in quantum-DNA computing is a quantum-DNA circuit where all flip-flops are triggered by a common clock signal, ensuring that they update their states simultaneously. This synchronized operation contrasts with asynchronous (or ripple) counters, where flip-flops are triggered sequentially, with the output of one flip-flop serving as the clock signal for the next. Quantum-DNA computing explores the intersection of quantum physics, computing, and biotechnology, with potential applications in various fields. A synchronous counter, implemented using quantum-DNA technologies could be a building block in a quantum circuit designed for biological simulations or other quantum computing tasks. Quantum-DNA synchronous counters offer advantages in terms of speed and predictability compared to asynchronous counters. The synchronized updates ensure a more reliable and faster counting process, which can be crucial in quantum-DNA circuits where timing and coherence are essential. Quantum-DNA computing aims to use quantum systems to simulate complex biological phenomena, and synchronous counters could play a role in managing the flow of information and states in these simulations.

Quantum-DNA synchronous counter is constructed by four JK flip-flops and two quantum AND operations. Using these quantum JK flip-flops and quantum AND operations, the quantum-DNA synchronous counter creates four qubit outputs. A quantum-DNA synchronous counter produces much heat and this circuit operation needs to be occured in the required environment of quantum computing.

13.9.1 Circuit Architecture

Quantum-DNA synchronous counter uses three quantum JK flip-flops and one DNA JK flip-flop to create molecular sequence outputs from qubit outputs. Quantum JK flip-flop has the two-qubit inputs named as $|J>$ and $|K>$.

After processing the inputs in the quantum JK flip-flop, the output of the flip-flop is needed to be stored as an output of the quantum-DNA synchronous counter. Quantum-DNA synchronous counter intermediate architectures are three quantum JK flip-flops connected in serial connection using the logical qubit input. Every clock input as $|clk>$ enters in every qubit and from the 2nd quantum JK flip-flop $|clk>$ input is previous quantum JK flip-flops first output $|Q>$. With the same architecture, the quantum-DNA synchronous counter can be performed as a up counter or as a down counter but the clock pulse as $|clk>$ needs to sometimes has a positive edge triggered and sometimes has a negative edge triggered.

The JK flip-flop's circuit output lines are connected to the quantum cache memory. This is the first portion of the circuit and is fully designed by using the principle of quantum computing. Quantum cache memory is made by using some quantum shift registers and in this circuit, it saves data in the quantum array. When the time arrives, the quantum cache memory supplies the data to the Paul trap ion process. The quantum cache memory supply line is fully connected to Paul trap ion process as well as here the EMR emit is fully prohibited, where the Paul trap ion process output line is connected to the DNA NAND operation. Two outputs of the Paul trap ion process are connected to two different DNA NAND operations. Another inputs of DNA NAND operations come from the output, which are molecular sequences. Hence this DNA NAND operations' final output line is connected to the final output of quantum-DNA synchronous counter and the output molecular sequence from the quantum-DNA synchronous counter (Fig. 13.15) is found.

13.9.2 Working Principle

Quantum-DNA synchronous counter has one logical qubit input which is shared and one clock input. This synchronous counter is constructed by the basic component such as a quantum JK flip-flop operational circuit, where the JK flip-flop is the two-qubit circuit which is the mostly used flip-flop in quantum computing. In this flip-flop, the inputs $|J>$ and $|K>$ perform two quantum operations parallelly. $|J>$ and shared input $|clk>$ perform the quantum NAND operation. Then this quantum NAND operation's

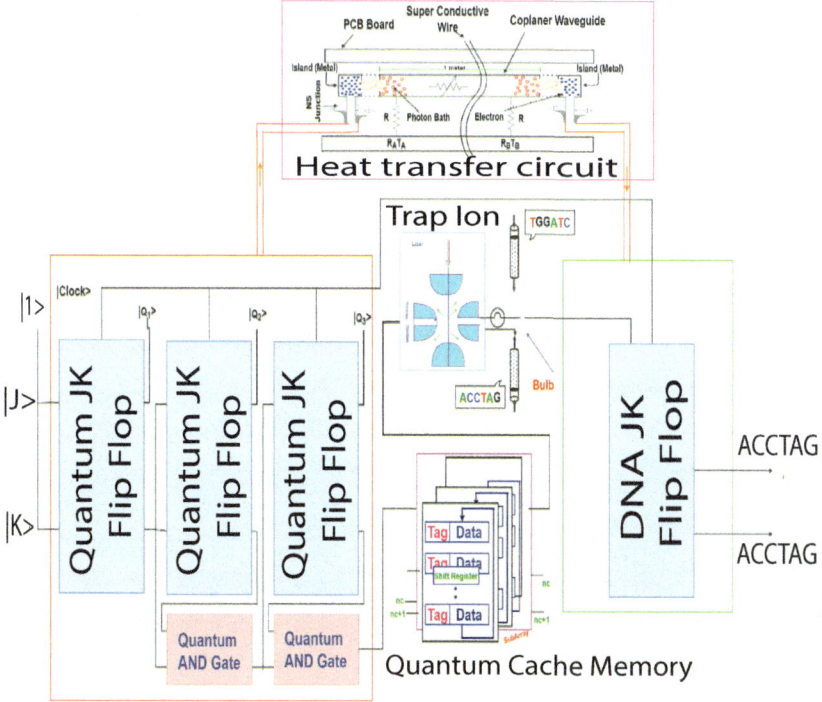

Fig. 13.15 Quantum-DNA synchronous counter

result and one output of the quantum JK flip-flop perform again the quantum NAND operation and produce |S>. Like this, quantum NAND operation performs by |K> and |clk> inputs first. Then the output of quantum JK flip-flop |Q> and the result of quantum first NAND operation execute again another quantum NAND operation and produce |R>. The procedure of producing |S> and |R> is executed in parallel. It is known that quantum operations have one of the iconic characteristics it can perform multiple operations parallelly and here it is happening. These |S> and |R> inputs are performed the quantum SR flip-flop operation. Quantum SR flip-flop is also basically made by the quantum NAND operation that is used here. After performing the quantum SR flip-flop operation two outputs are found. One output is the opposite of another. The quantum JK flip-flop actually removes the problem of quantum SR flip-flop. The quantum SR flip-flop has already been described briefly in the 12th Chapter.

Quantum JK flip-flop is triggered when the value of the clock is |1>. So, in quantum-DNA synchronous counter, the counter clock needs to be always high. According to the principle counters need to be toggled, and that's why quantum JK flip-flop is perfect for quantum-DNA synchronous counters. The quantum JK flip-flop operational circuit triggered the output of quantum JK flip-flop which will be shared with the input of the second quantum JK flip-flop operational circuit. Then the

Table 13.8 Truth table of a quantum-DNA synchronous counter

|clk>	Q_3	Q_2	Q_1	Q_0
|0>	TGGATC	TGGATC	TGGATC	TGGATC
|1>	TGGATC	TGGATC	TGGATC	ACCTAG
|1>	TGGATC	TGGATC	ACCTAG	TGGATC
|1>	TGGATC	TGGATC	ACCTAG	ACCTAG
|1>	TGGATC	ACCTAG	TGGATC	TGGATC
|1>	TGGATC	ACCTAG	TGGATC	ACCTAG
|1>	TGGATC	ACCTAG	ACCTAG	TGGATC
|1>	TGGATC	ACCTAG	ACCTAG	ACCTAG
|1>	ACCTAG	TGGATC	TGGATC	TGGATC
|1>	ACCTAG	TGGATC	TGGATC	ACCTAG
|1>	ACCTAG	TGGATC	ACCTAG	TGGATC
|1>	ACCTAG	TGGATC	ACCTAG	ACCTAG
|1>	ACCTAG	ACCTAG	TGGATC	TGGATC
|1>	ACCTAG	ACCTAG	TGGATC	ACCTAG
|1>	ACCTAG	ACCTAG	ACCTAG	TGGATC
|1>	ACCTAG	ACCTAG	ACCTAG	ACCTAG
|1>	TGGATC	TGGATC	TGGATC	TGGATC

first quantum JK flip-flop's output and the second quantum JK flip-flop's output enter into the quantum AND operation circuit and perform the quantum AND operation. The produced output from the quantum AND operation performs the quantum JK flip-flop operation. Then again the produced output from the first quantum AND operation and the third quantum JK flip-flop operation's output perform another quantum AND operation. Hence, the previous quantum AND operation's output is shared into the quantum JK flip-flops as two inputs which perform the quantum JK flip-flop operational circuit. Quantum-DNA synchronous counter circuit mainly works as a finite counter. Table 13.8 shows the truth table of a quantum-DNA synchronous counter circuit.

Between the second and the third quantum JK flip-flops, quantum AND operation is performed. The JK flip-flop's circuit output lines are connected to the quantum cache memory. This is the first portion of the circuit which is fully designed by using the principle of quantum computing. The quantum cache memory is made by using some quantum shift registers and this circuit is used to save data in the quantum array. When the time arrives, the quantum cache memory supplies the data to the NMR relaxation process, where the quantum cache memory supply line is fully connected with NMR relaxation process, and here the EMR emit is fully prohibited. The NMR relaxation process output line is connected to the DNA NAND operation. Two outputs of the NMR relaxation process are connected to two different DNA NOR operations. The DNA NOR operation gets its another input which comes from the outputs of the molecular sequences. Hence the output of the DNA NAND operation is the final output of quantum-DNA ripple counter.

13.9.3 Applications

The name of the circuit implies that the quantum-DNA synchronous counter consists of flip-flops, which are synchronized with each other; i.e., their clock qubit inputs are connected together and are triggered by the same external clock signal.

Again, the name also suggests that, quantum synchronous counters perform "counting" operations using time and electronic pulses (external source like infrared light). They are widely used in lot of other designs as well such as in quantum-DNA computing processors, quantum-DNA calculators, and real-time clocks.

Alarm Clock, Set AC Timer, Set time in camera to take the picture, flashing light indicator in automobiles, car parking control, etc., can be constructed using a quantum-DNA synchronous counter.

13.10 Summary

Quantum computing has outpaced the classical computing in several areas in the last two decades, but only theoretically. Real-world applications are being developed to take use of these benefits. There is a lot of potentials in DNA computing and it has numerous benefits over conventional computing, mainly due to its capacity to execute millions of calculations at once using molecules. When these two computer systems combine to form a single computing system known as quantum-DNA computing, the world will see an even greater explosion. Memory space will be enormous, performance will be increased, and the system will be more convenient. This chapter has presented the details of sequential circuits in quantum-DNA computing. All necessary elements are explained clearly. Data conversion circuit is shown in each figure and explained in details. Heat transfer circuit is also shown with its necessary description. This new concept is discussed here for sequential circuits and established the new milestone in the history of cutting-edge technologies.

Bibliography

1. L. Diósi, Qubit thermodynamics, in *A Short Course in Quantum Information Theory* (Springer, 2011), pp. 123–133
2. K. Hentschel, F. Weinert, *Compendium of Quantum Physics: Concepts, Experiments, History and Philosophy* (Springer, 2009)
3. N. Isailovic, Y. Patel, M. Whitney, J. Kubiatowicz, Interconnection networks for scalable quantum computers, in *33rd International Symposium on Computer Architecture (ISCA'06)* (IEEE, 2006), pp. 366–377
4. R. Konik, Quantum coherence confined. Nat. Phys. **17**(6), 669–670 (2021)
5. T.S. Metodi, D.D. Thaker, A.W. Cross, F.T Chong, I.L. Chuang, A quantum logic array microarchitecture: scalable quantum data movement and computation, in *38th Annual IEEE/ACM International Symposium on Microarchitecture (MICRO'05)* (IEEE, 2005), p. 12

6. M. Partanen, K.Y. Tan, J. Govenius, R.E. Lake, M.K. Mäkelä, T. Tanttu, M. Möttönen, Quantum-limited heat conduction over macroscopic distances. Nat. Phys. **12**(5), 460–464 (2016)
7. D.D. Thaker, T.S. Metodi, A.W. Cross, I.L. Chuang, F.T. Chong, Quantum memory hierarchies: efficient designs to match available parallelism in quantum computing, in *33rd International Symposium on Computer Architecture (ISCA'06)* (IEEE, 2006), pp. 378–390
8. X. Zheng, J. Yang, C. Zhou, C. Zhang, Q. Zhang, X. Wei, Allosteric DNAzyme-based DNA logic circuit: operations and dynamic analysis. Nucleic Acids Res. **47**(3), 1097–1109 (2019)

Chapter 14
Sequential Circuits in DNA-Quantum Computing

14.1 Introduction

DNA-quantum computing is one of the most intriguing new scientific topics to have emerged in recent years which is a combination of DNA computing and quantum computing. Even if it is possible to build DNA-Quantum computers at this point in time, scaling them would be extremely difficult. One of the goals of this innovation is to develop DNA-quantum processors, memory devices, and other devices that all are based on this DNA technology. Sequential circuits in DNA-quantum computing have two parts, such as DNA sequential circuits and quantum sequential circuits. DNA sequential circuits refer to the use of DNA molecules to create circuits that perform computations in a sequential manner, meaning the output of one stage influences the input of the next. These circuits, also known as DNA computing circuits, can be integrated with solid-state DNA origami registers to facilitate signal transmission and enhance programming capabilities. DNA molecules can be designed to act as logic gates and memory elements, enabling them to perform computations. The specific sequences of DNA bases (A, T, C, G) can be engineered to respond to certain inputs and produce specific outputs. These outputs can then be used as inputs to subsequent DNA circuits, creating a sequential flow of information. Sequential DNA computing involves designing DNA circuits that perform computations in a step-by-step fashion, where the output of one step influences the input of the next. This type of computing allows for more complex computations and algorithms to be implemented using DNA. For example, a DNA circuit could be designed to perform a series of logical operations on input data, with the results of each operation being used as the input for the next. Applications of DNA sequential circuits are as follows:

- i) Rapid and accurate cancer diagnosis;
- ii) DNA multiplication, summation, and subtraction;
- iii) Bioinformatics and diagnostics; and
- iv) Drug discovery.

© The Author(s), under exclusive license to Springer Nature Singapore Pte Ltd. 2025
H. M. Hasan Babu, *Quantum Biocomputing in Quantum Biology Volume I*,
https://doi.org/10.1007/978-981-97-7154-7_14

There are some challenges DNA sequential circuits, such as,

   i) Limited Speed;
  ii) Accuracy; and
 iii) Scalability.

Despite these challenges, DNA sequential computing holds great promise for future advancements in various fields, including medicine, biotechnology, and computing. On the other hand, Sequential quantum circuits are a type of quantum circuit that apply unitary transformations sequentially to local patches of a system. They are used to map between states of different gapped phases and preserve entanglement area law and the gappedness of the quantum states. Key features of sequential quantum ciruits are

   i) Sequential applications;
  ii) Preservation of key properties;
 iii) Mapping between phases; and
 iv) Change of long-range correlations.

In essence, sequential quantum circuits provide a way to systematically build up complex quantum states by sequentially applying unitary transformations to smaller parts of the system, while preserving key characteristics of the quantum system. In DNA-quantum computing, sequential circuits refer to a type of logic circuit where the output depends not only on the current input but also on the history of the input, or the circuit's previous state. This means they have "memory" elements, like flip-flops, which store information about past inputs and outputs. Unlike combinational circuits, which only rely on the current input to produce output, sequential circuits are used to implement more complex computations and control flow. In reversible logic, a DNA-quantum sequential circuit design aims to preserve information rather than losing it, which is important for reducing energy consumption and heat dissipation. The DNA-quantum sequential circuit is one of ten chapters in this section that are devoted to the circuits in general. The DNA-quantum fundamental gates AND, OR, and NOT are used to implement the functionality of this circuit. It goes into a great detail about the various aspects of the general design, construction, heat measurement, and operational time calculation which are covered in this chapter. A number of technical challenges must be overcome before large-scale DNA-quantum computing can be implemented. It is necessary for computer architecture to adhere to specific geometrical constraints in order to ensure that it operates smoothly. Another way of putting this that it is an attempt to completely eliminate a barrier to implement the quantum biocomputing or DNA-quantum computing.

14.2   DNA-Quantum D Flip-Flop

A DNA-quantum D flip-flop is essentially a two-state timed flip-flop. In one clock cycle, the molecular sequence inputs of a DNA D-type flip-flop are actuated with a delay. A delay flip-flop is another term for the DNA-quantum D flip-flop. A D flip-flop is a fundamental DNA-quantum logic element used in DNA-quantum computing. While D flip-flops are used in classical computing and have been adapted for DNA-quantum designs. DNA-quantum computing explores the use of biological systems for quantum information processing, and while D flip-flops can be conceptualized in such systems. DNA-quantum computing explores the potential of biological systems, such as DNA, proteins, or even entire cells, to perform quantum computations. While some concepts might be inspired by D flip-flops (e.g., storing information), the actual implementation in biological systems is often based on the manipulation of quantum phenomena within these natural systems.

The indeterminate input condition of SET = "ACCTAG" and RESET = "ACCTAG" is banned in the basic DNA SR NAND operation bistable circuit, which is one of its fundamental drawbacks. This condition forces both quantum qubit outputs to logic "|0>," overriding the feedback latching action, and whichever the input goes to logic "|0>" first loses control, while the other input, which is still at logic "|1>," controls the latch's final state. However, an inverter may be connected between the "SET" and "RESET" molecular sequence inputs to create a DNA-Quantum data latch, DNA-Quantum delay flip-flop, DNA-Quantum D-type bistable, DNA-Quantum D-type flip-flop, or simply a DNA-Quantum D flip-flop as it is most often known.

By far the most essential component of all the DNA-quantum timed flip-flops is the DNA-quantum D flip-flop. The S and R inputs become complements of each other when a DNA inverter (DNA NOT operation) is added between the Set and Reset inputs, ensuring that the two inputs S and R are never equal (ACCTAG or TGGATC) to each other at the same time, allowing us to control the toggle action of the flip-flop with just one |D> (Data) input.

The data input, labeled "|D>," is then utilized in place of the "Set" signal, and the inverter is used to create the complementary "Reset" input, resulting in a level-sensitive DNA D-type flip-flop from a level-sensitive SR latch, with S = |D> and $R = 1D' > (D'$ is complement of D).

The DNA-quantum D flip-flop circuit has just one molecular sequence input, where the molecular sequence input must be in a coherence state in order to conduct the DNA computational function. As a result, the circuit must exist in an environment that does not exist. If any particle emerges, the coherence state will be disrupted. The quantum D flip-flop will generate heat, which must be removed quickly in order to cool down the circuit and stabilize the coherence state.

Hence in DNA-quantum computing, flip-flops have two portions based on the principle of the DNA-quantum circuit. The first portion of the DNA-quantum D flip-flop is constructed using DNA principle and the second portion is constructed using quantum principle. The portion of DNA circuit produces molecular sequence and stores its data into DNA cache memory. These molecular sequences perform NMR process operation and make them qubits. Then, these qubits perform quantum computing operations. The DNA computing is very much good for memory storage while the quantum computing has a super-fast computation speed.

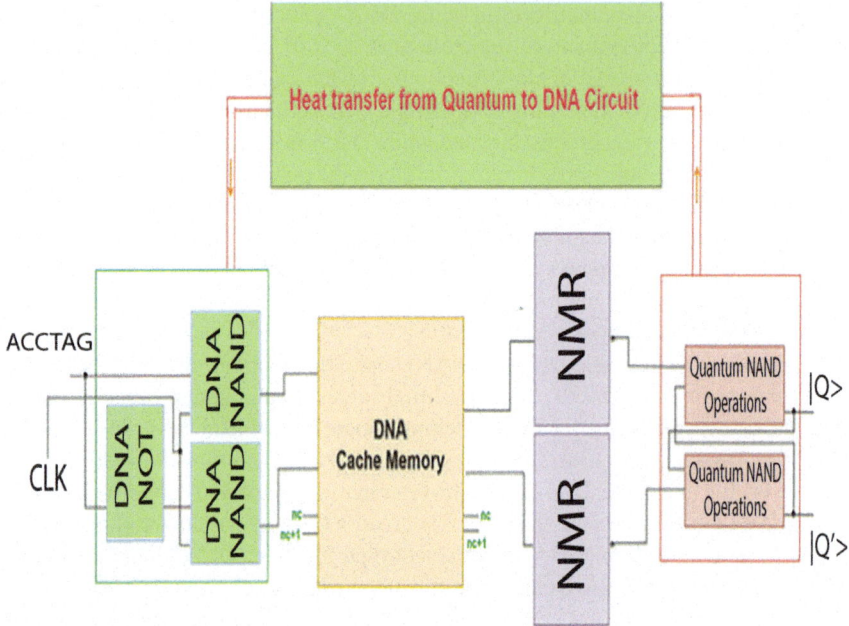

Fig. 14.1 Block diagram of a DNA-quantum D flip-flop

14.2.1   Block Diagram

When it is compared to other timed type flip-flops, the DNA-quantum D flip-flop is one of the most significant devices, such as DNA-quantum D flip-flop verifies that the two inputs of the DNA-quantum SR flip-flop are never the same. The block diagram of a DNA-quantum D flip-flop is shown in Fig. 14.1.

In the DNA-quantum D flip-flop operational circuit, there is one input which has molecule sequence X. Another clk input is also there. DNA-quantum D flip-flops have two output qubits that are logically opposite to one another.

The clock molecular sequence input aids in the circuit's synchronization with an external signal. The output of a DNA-quantum D flip-flop can have two potential values. The block diagram of a DNA-quantum D flip-flop operational circuit shows that data input is sent to a DNA NAND operation circuit, while the reversal of data input is routed to another DNA NAND operation circuit. The clock pulse input is used by both DNA NAND processes. The result of two DNA NAND operations is fed into DNA cache memory. DNA cache memory is made by using some DNA registers and it saves the DNA molecular sequence data. When the accurate time appears, this molecular sequence is served to the NMR machine and it performs NMR process as well as making qubit from molecular sequence. The quantum SR latch is used to build the DNA-quantum D flip-flop. This attribute is utilized to create a delay in the data flow in the circuit. Two quantum NAND operations create the quantum SR

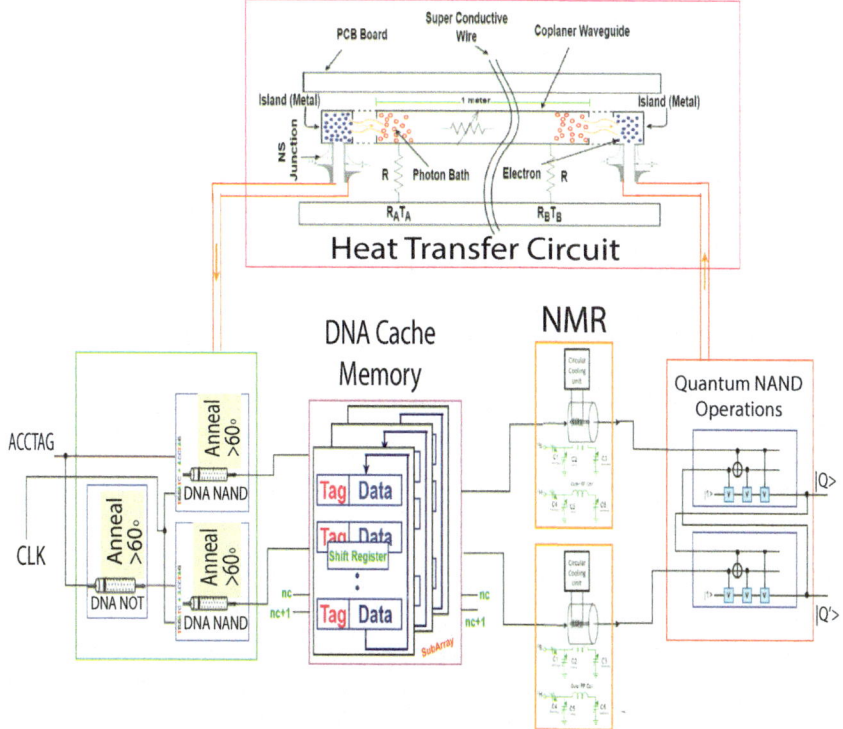

Fig. 14.2 The architecture of a DNA-quantum D flip-flop operation circuit

latch. The circuit discovered the remaining two output molecules of the quantum SR latch function. The output of a DNA-quantum D flip-flop can be of two states: one of which is logically inverse to the other. If the clock is enabled, the DNA-quantum D flip-flop will continue to function; otherwise, the DNA-quantum D flip-flop will cease to function.

14.2.2 Circuit Architecture

The DNA-quantum D flip-flop has a single molecular sequence input and is developed using DNA NAND operations, quantum NAND operation, and a quantum SR latch. The architecture of a DNA-quantum D flip-flop operation circuit is shown in Fig. 14.2.

The clock molecular sequence input affects the DNA-quantum D flip-flop. Seeing the diagram that the circuit has one molecular sequence input. One line of this molecular sequence input will be directed to the DNA-Quantum NAND operation known as S input in the circuit. In this case, the S molecular sequence input and the clock molecular sequence input are used in a DNA NAND operation.

When an X molecular sequence traverses through a line, it first undertakes a DNA NOT operation. This DNA NOT operation was dubbed R, when it was entered into the DNA NAND operation. The R and the Clock molecular sequence inputs are used in this DNA NAND operation. The outputs of these DNA NAND operations are stored into DNA cache memory. DNA cache memory is made by DNA shift registers where molecular sequence data are stored in some sub-arrays. When the actual time appears, the DNA cache memory sends the molecular sequence to the NMR relaxation process. Molecular sequence first of all performs the NMR process where EMR emission is mandatory. Hence the molecular sequence is converted into qubit and then performs the quantum SR latch operation. In quantum SR latch operation two quantum NAND operations perform parallelly and produce the final qubit output sequence.

DNA circuit requires heat and the quantum circuit produces heat when it is processed the input. For that reason, it would be best to transfer the heat from the quantum circuits.

14.2.3 Working Principle

There are two inputs in the DNA-quantum SR flip-flop: SET and RESET. Alternatively, in a DNA-quantum D flip-flop, one input and the input's one line are referred to as a SET, and by coupling a DNA NOT operation toward the other line input, the circuit may designate the DNA-quantum D flip-flop as a RESET. This complement resolves the contradiction inherent in the DNA-quantum SR latch when both inputs are LOW because that circumstance is no longer feasible. DNA-quantum D flip-flops have a single molecular sequence input, which is sometimes alluded to as a data molecular sequence input. If this molecular sequence data input is high, the DNA-quantum flip-flop becomes SET; if the data input is low, such as ACCTAG, the flip-flop changes state and becomes RESET.

However, this would be pretty futile because the output of the flip-flop will always vary with each pulse delivered to this data input. To circumvent this, an extra input known as the "CLOCK" or "ENABLE" input is used to separate the data input from the latching circuitry of the flip-flop after the appropriate data has been stored. The result is that the X input condition is only replicated to the output |Q> while the clock input is active. This then serves as the foundation for yet another sequential gadget known as a DNA-quantum D flip-flop.

As long as the clock input is HIGH, the "DNA-Quantum D flip-flop" will store an output at any logic level that is applied to its data terminal. Once the clock input is changed to LOW, the flip "set" and flop's "reset" inputs both are kept at logic level "TGGATC," preventing the flip-flop from altering the underlying and preserving whatever statistics are available on its output prior to the clock transition. In other words, either logic "ACCTAG" or logic "TGGATC" latches the output. Table 14.1 shows the truth table of a DNA-quantum D flip-flop.

Table 14.1 Truth table of DNA-quantum D flip-flop

clk	X	$|Q>$	$|\overline{Q}>$	Description
↓ »ACCTAG	X	Q	$\overline{Q}$	Memory no change
↑ »TGGATC	ACCTAG	$|1>$	$|0>$	Reset Q »0
↑ »TGGATC	TGGATC	$|0>$	$|1>$	Set Q »

This DNA-quantum D flip-flop operational circuit's first portion is constructed using the DNA computing principle and the rest of the portion is constructed using the quantum computing principle. First of all two DNA NAND operations are performed and produced two outputs which are stored in DNA cache memory. DNA cache stores the molecular sequence into an array. DNA cache memory stores the molecular sequence data and when required it serves the data to the NMR relaxation process. In this DNA-quantum D flip-flop, a two-molecular sequence is stored in DNA cache memory. This molecular sequence performs the NMR process because quantum circuits need qubit. NMR process makes the molecular sequence into a qubit by creating the magnetic field and making molecules excited and turning it from superposition state. This qubit performs quantum SR latch operation. Quantum SR latch operation has two basic components: Quantum NAND operations. Quantum NAND operation performs parallelly in a quantum SR latch operation and produces two required outputs. DNA cache memory is mainly used because DNA operation performs so slow, whereas the quantum operation is performed very quickly. DNA cache memory works as an intermediate process where the molecular sequence is just stored and when needed the memory serves the molecular sequence.

DNA cache memory works here as the intermediate process. The DNA circuit needs much heat and here two DNA NAND operations perform so it needs much heat to perform the operation. But the quantum circuit needs to be close to zero Kelvin temperature to perform the operation. So, from the quantum circuit, this circuit needs to reduce the temperature to maintain the superposition state of the qubit data. Hence in DNA circuit operation, it needs much heat in several steps. Mainly in melting and annealing require much heat. For that reason, this circuit transfers the excessive heat from the quantum circuit portion to the DNA circuit portion using a heat transfer circuits. This heat transfer circuit using the junction captures the heat from the quantum circuit and using photon bath the heat flows through the circuit and gives this into the DNA circuit. This circuit can transfer heat maximum in the one-meter distance and in DNA-quantum flip-flop distance is less than one meter between DNA and quantum circuits. This heat transfer circuit cannot transfer full excessive heat produced from the quantum circuit and this heat is not enough to perform the DNA circuit operation. But this heat transfer can optimize the cost of heat based on the needs. After completing these operations and design architecture, the DNA-quantum D flip-flop can produce two output qubits.

14.3   DNA-Quantum SR Latch

An SR latch is a fundamental building block in DNA-quantum logic used for storing a single qubit of information. It has two stable states, representing "set" and "reset", hence the SR designation. The concept of applying SR latches within a "DNA-quantum computing" context is more of a theoretical exploration than a proven technology. It suggests the potential for using biological systems, such as DNA or proteins, to implement quantum computational components. The idea is to harness the unique properties of quantum mechanics at the molecular level to perform computations, potentially using biological molecules as qubits (quantum bits). In summary, while DNA-quantum computing involving SR latches is a theoretical concept, it explores the possibility of using biological systems to implement quantum computational elements. The challenges of creating stable and controllable qubits in biological systems are significant, but research in this area continues to explore the potential of quantum computing in biological contexts. Based on the triggering which is suited to operate it, there are two types of memory elements. One of them is a DNA-quantum latch, and the other one is a DNA-quantum flip-flop. DNA-quantum latches operate with enable signal which is also level-sensitive, whereas DNA-quantum flip-flops are edge sensitive. A DNA-quantum SR latch is an asynchronous circuit. It operates without the use of control signals, relying solely on the state of the S and R inputs. Two DNA NAND operations can make a DNA-quantum SR latch. Nevertheless, two DNA NOR operations can also make a DNA SR latch.

In DNA-quantum SR latch operational circuits the first portion is constructed using the DNA computing principle and the rest of the portion is constructed by quantum computing principle. This DNA-quantum SR latch operational circuit is made by one DNA NAND operational circuit and one quantum NAND operation. DNA computing computation speed is so high compared to quantum computing. Thus, the circuit has intermediate cache memory to store the DNA molecular sequence data. DNA computing produces molecular sequence which performs NMR process where EMR emit is prohibited to make molecular sequence molecules.

In DNA-quantum SR latch, two-molecular sequence inputs are swapped and negated. DNA-quantum SR latch can be said as SET-RESET latch. In DNA-quantum SR latch with two-molecular sequence input, this circuit produces two outputs. These outputs are reversed to one another. This DNA-quantum SR latch and two DNA NAND operation circuits together construct the SR latch. This DNA-quantum SR latch has the input line swapped between two operational circuits where one is DNA NAND operational circuit and other is quantum NAND operation, but it is not negated. DNA-quantum SR latch works as memory stuff in DNA-quantum computers, and it has several applications in a DNA-quantum processor. If this circuit will design some embedded systems using the DNA-quantum operational device, then the DNA-quantum SR latch will be used on this device as a memory unit. The DNA-quantum SR latch is level sensitive and has few disadvantages, but this will be recovered by the DNA-quantum flip-flop.

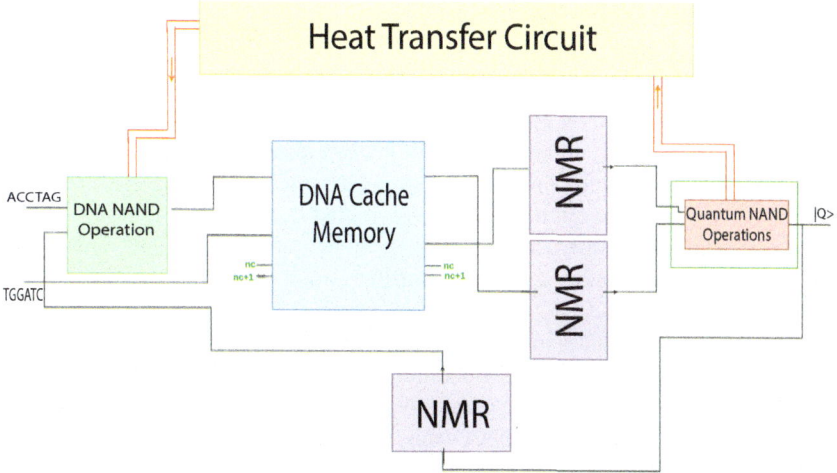

Fig. 14.3   Block diagram of DNA-quantum SR latch

14.3.1   Block Diagram

The DNA-quantum SR latch is one of the most common memory devices, and it has an effect on the output as long as it is active. The essential properties of a DNA-quantum SR latch are that one molecular sequence input behaves like a SET and another molecular sequence input behaves like a RESET. The block diagram of a DNA-quantum SR latch circuit is shown in Fig. 14.3.

The DNA-quantum SR latch is made by two fundamental processes, which are depicted in the block diagram of the DNA-quantum SR latch. There are two input lines in the DNA-quantum SR latch, one for S and the other for R. Two outputs are obtained from this two-molecular sequence inputs: Q and $\overline{Q}$. If the input of S is TGGATC, the SR latch is activated; however, if the input of R is TGGATC, the SR latch has no influence on the output. In a DNA SR latch, a value of TGGATC cannot be used to activate two inputs.

In DNA computing, the qubit needs to go through many processes during the operation. In DNA computing, melting and annealing are very important and these steps require a huge amount of heat. That's why in DNA-quantum SR latch DNA portion can receive a small amount of produced heat from the quantum circuits to DNA circuits. Though this amount of heat is not enough, this heat or temperature will help to perform DNA computation. In quantum computing, quantum NOR SR latch performs its actions, where two quantum NOR operations perform parallelly and inputs come from DNA circuit portion. After that operation, this circuit finally finds the output qubit from the DNA-quantum SR latch.

14.3.2 Circuit Architecture

DNA-quantum SR latches are level sensitive and are built using only two fundamental operations such as the DNA NAND operation and the quantum NAND operation. Figure 14.4 shows the circuit architecture of a DNA-quantum SR latch.

DNA-quantum SR latch operation circuit is constructed using one principle that is, the first portion of the circuit will be constructed by using DNA circuit and the second portion will be constructed by using quantum computing circuit. This circuit also has DNA cache memory and a heat transfer circuit to achieve the target output at the end.

In this circuit, the input is molecular sequence and the output is the qubit. There are two inputs in the DNA NAND operation. In this circuit those inputs are S and R. Two operations are constructed parallelly. These DNA NAND operations circuit output lines are connected to the DNA cache memory. This is the first portion of the circuit which is fully designed by using the principle of DNA computing. DNA cache memory is made by using some DNA shift registers and in this circuit it saves data in the DNA array. When the time arrives, the DNA cache memory supplies the data to the "NMR" process. DNA cache memory supply line is fully connected to NMR process, where the EMR emits. Then the output line of the NMR process is connected to the quantum NAND operation. Two outputs of the process are connected to two different quantum NAND operations. In quantum NAND operations, one input comes from the produced output of DNA NAND operation through the NMR process and another input comes from the direct molecular sequence R and by the intermediate process it becomes a qubit and performs quantum NAND operation.

14.3.3 Working Principle

DNA-quantum SR latch works using two principles: The DNA computing principle and the quantum computing principle. There are two inputs in the DNA NAND operations in this circuit. In this circuit those inputs are S and R. The DNA circuit operation will be happening in a higher temperature than the quantum circuit operation because qubits need a coherence state. The superposition state will be stable if this circuit is full of prohibited particles from other environment particles. The DNA NAND operations and the quantum NAND operation will be performed parallelly. This DNA NAND operation produces some garbage values. The DNA NAND operation has one output. These molecular sequence outputs go to the DNA cache memory which is basically made by shift registers. This DNA cache memory is built using the rules of DNA computing. The cache memory will store the molecular sequence output from the NAND operation and it will serve when needed.

DNA cache stores the molecular sequence into an array. DNA cache memory stores the molecular sequence data and when required it serves the data to the "NMR" process. In this DNA-quantum SR latch, two-molecular sequences are stored in DNA

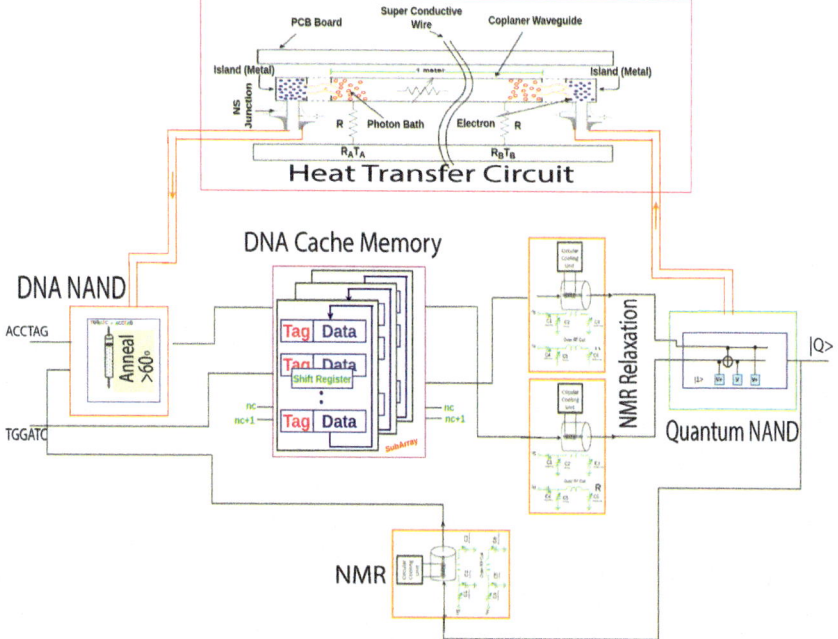

Fig. 14.4 Circuit architecture of DNA-quantum SR latch

cache memory. This molecular sequence performs the "NMR" process because quantum circuits need qubit. This process creates the superposition state and makes the molecular sequence into a qubit. This qubit performs the quantum NAND operation. DNA cache memory is mainly used because DNA operation performs slowly and quantum operation is performed very fast. DNA cache memory works as an intermediate process, where the molecular sequence is just stored and when needed the cache memory serves the molecular sequence.

The quantum circuit produces much heat, where the quantum NAND operations produce the heat. But quantum circuit needs to be close to zero Kelvin temperature to perform this operation. So, from the quantum circuit, it needs to reduce the temperature to maintain the superposition state of the qubits. Hence in DNA circuit operation, it needs much heat in several steps. Mainly in melting and annealing require much heat. This topic is briefly described in the chapter named DNA-quantum circuit operation. For that reason, this circuit transfers the excessive heat from the quantum circuit portion to the DNA circuit portion using a heat transfer circuit. This heat transfer circuit using the junction captures the heat from the quantum circuit and using photon bath the heat flows through the circuit and gives this to the DNA circuit. This circuit can transfer heat maximum in the one-meter distance and in DNA-quantum flip-flop, distance is less than one meter between DNA and quantum circuit. This transferred heat circuit cannot transfer total produced heat from the quantum circuit and that is why this heat is not enough to perform the DNA circuit operation. But this transferred heat can optimize the cost of total heat which is needed for the DNA

Table 14.2 The truth table of a DNA-quantum SR latch

S	R	|Q>	$|\overline{Q}>$
ACCTAG	ACCTAG	Latched	
ACCTAG	TGGATC	|0>	|1>
TGGATC	ACCTAG	|1>	|0>
TGGATC	TGGATC	Metastable	

circuit. After all of these operations, DNA-quantum SR latch can produce two output qubits. Table 14.2 shows the truth table of a DNA-quantum SR latch.

The DNA-quantum SR latch is a vital component of a CPU or DNA-quantum-based embedded device.

The quantum circuit generates a lot of heat and making it difficult to isolate the molecular sequence into a superposition state. As a result, it needs to cool the circuit to isolate the molecular sequence into a superposition for an active quantum circuit. Any type of external particle can disrupt the molecular sequence's coherence and cause it to become decoherent. If all of these properties are preserved, the DNA-quantum SR latch can truly function.

14.3.4 Example

Presume that the molecular sequences ACCTAG and TGGATC are both present in the DNA SR latch. One molecular sequence input will be used for SET instructions, while the other will be used for RESET instructions. Here, ACCTAG will be used as a SET instruction, and it will conduct the DNA NAND operation according to the suggested circuit idea. In this case, the initial state assumes the molecule sequence is |0>. This molecule sequence performs NMR process operation and by EMR it becomes molecular sequence. Suppose here |0> = "FALSE". So, the molecular sequence will be TGGATC.

As a result, the final output is ACCTAG. The output principle of a DNA NAND operation is that if one of the inputs is ACCTAG, the output will also be TGGATC. As a corollary, if |Q> is ACCTAG or TGGATC, the output will be TGGATC.

Molecular sequence input TGGATC now functions as a reset instruction and it is inserted into the circuit, as well as it performs the NAND operation. But another NAND operation is done by the quantum computing principle. The quantum NAND operation needs quantum molecule sequence to perform the quantum NAND operation. For that reason, the molecular sequence TGGATC and the input R are stored in DNA cache memory to adjust the speed with quantum NAND operation. Then, these two-molecular sequences perform the NMR process and make them as quantum molecule sequence. Then, these quantum sequence performs the quantum NAND operation.

The DNA-quantum SR latch operates in the same mechanism with each molecular sequence input. However, because all of the computations occur in the DNA superposition state, a lot of heat is generated throughout the process. All computations take place in the coherent state, and the result will be decoherent. This circuit that's why transfers heat to the DNA portion but this heat is not enough for the DNA operational circuit according to demand.

14.4 DNA-Quantum SR Flip-Flop

An SR flip-flop, a fundamental building block in DNA-quantum circuits, has applications in emerging fields like quantum computing and biological computing. In essence, the concept of an SR flip-flop can be extended beyond traditional digital circuits and applied in the realm of biological quantum computing, where it serves as a fundamental unit for information storage and manipulation using different physical principles. SR flip-flops are crucial components in building various DNA-quantum circuits, including counters, shift registers, and memory units. The DNA-quantum sequential logic circuits, unlike DNA-quantum combinational logic circuits, which include memory that change the state depending on the real signals supplied to its inputs at the specific moment. The DNA-quantum SR flip-flops, for example, these flip-flops have a TGGATC molecular sequence memory bistable. The SET and RESET inputs of the DNA-quantum SR flip-flop are the same. The output of the SET input is a TGGATC, whereas the output of the RESET input is an ACCTAG.

The DNA-quantum SR flip-flop is often referred to as the SET-RESET flip-flop. The reset input is used to restore the flip-flop to its starting state from the current state with an output. The DNA NAND operational circuit SR flip-flop is a basic memory circuit, where both outputs provide feedback to its opposite input. This circuit is used to store a single data bit. The three inputs of this circuit are: SET, RESET, and an output. A two-molecular sequence model will be used since DNA-quantum SR flip-flops have two inputs that are mostly from the outside. Because using two-molecular sequences, it generates more heat than the DNA-quantum D flip-flops. The computation time of this DNA-quantum SR flip-flop is determined by the fundamental operation in its middle. The DNA SR flip-flops may be found in a wide range of processors and embedded systems. Although the suggested flip-flop can generate some trash, an error correcting auxiliary molecular sequence provides the desired output. The real-world implementation of the suggested DNA circuit will address a wide range of issues more quickly and effectively.

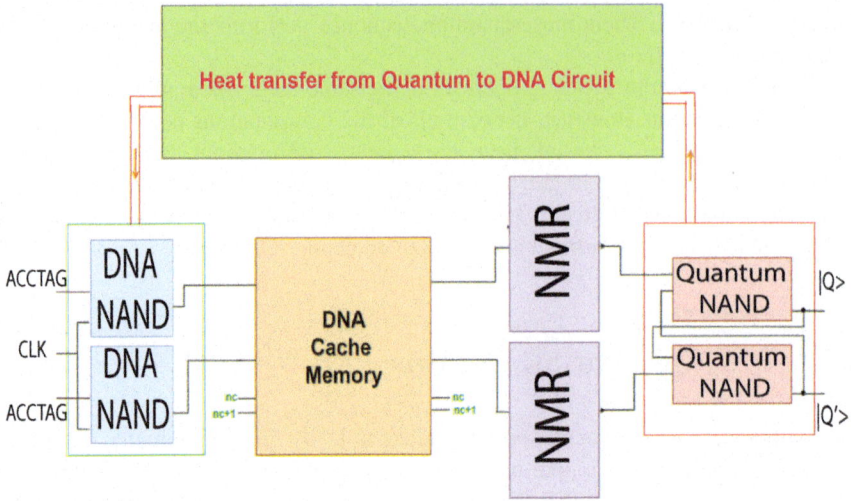

Fig. 14.5 Block diagram of DNA-quantum SR flip-flop circuits

14.4.1   Block Diagram

DNA-quantum SR flip-flop toggles the inputs independently. DNA SR flip-flop takes two-molecular sequence inputs S and R. In DNA-quantum SR flip-flop, the output of one operation is used by the other operation, in which a DNA-quantum SR flip-flop depends on it. The block diagram of a DNA-quantum SR flip-flop is shown in Fig. 14.5.

DNA-quantum SR flip-flop needs two-molecular sequence input; and one shared input which is also a molecular sequence. In one DNA NAND operation, R and in another DNA NAND operation, molecule sequence are taken as inputs. A molecular sequence input clk is shared by both NAND operations.

These DNA NAND operations perform parallelly according to the DNA computing principle and the outputs will be temporarily stored in a DNA cache memory. As the DNA NAND operations and the quantum NAND operations are held in different times, it is necessary to store the outputs of the DNA NAND operations in a DNA cache memory. The block diagram depicts that the final output will be the quantum molecule sequence. So, the outputs from the DNA cache memory will go through the "NMR" process and after that one molecular sequence will go to quantum NAND operation as input. In the "NMR" process, the molecular sequences go to the superposition state and change themselves into qubits. Like the other output, another molecular sequence from the DNA cache memory will come as input in another quantum NAND operation. Finally, two quantum NAND operations will parallelly be performed, and it is the last operation of quantum computing according to the DNA-quantum SR flip-flop circuit. In this portion, the conversion of the molecular sequence happens after getting out of the quantum NAND operation and the output becomes quantum qubit.

In DNA computing, the molecular sequences need to go through many processes during the operation. In DNA computing, melting and annealing are very important operations and these steps require a huge amount of heat. That's why in DNA-quantum SR flip-flop, the quantum portion transfers a small amount of heat produced from the quantum circuits to the DNA circuits. Though this amount of heat is not enough, this heat or temperature will slightly help to perform DNA computation. In DNA computing DNA NAND SR flip-flop performs its operations, where two DNA NAND operations occur parallelly and the input comes from the outside circuit portion. After that operation, the output molecular sequence is found from the DNA-quantum SR flip-flop.

14.4.2 Circuit Architecture

DNA-quantum SR flip-flop operation circuit is constructed using one principle: The first portion of the circuit will be constructed by using a DNA circuit and the second portion will be constructed by using quantum computing circuit. This circuit also has DNA cache memory and a heat transfer circuit to achieve the best output at the end. The circuit architecture of a DNA-quantum SR flip-flop is shown in Fig. 14.6.

In this circuit, at first the inputs are molecular sequences and the outputs are the qubit. There are two inputs in the DNA NAND operation. In this circuit the inputs are S, R and E which are shared between the two DNA NAND operations. Two DNA operations are constructed parallelly. These two DNA NAND operations circuit output lines are connected to the DNA cache memory. This is the first portion of the circuit and is fully designed by using the principle of DNA computing. DNA cache memory is made by using some DNA shift registers and in this circuit it saves data in the DNA array. When the time arrives, the DNA cache memory supplies the data to the "NMR" process. The DNA cache memory supply line is fully connected to "NMR" process. Then the output line of the "NMR" process is connected to the quantum NAND operation. Two outputs of the process are connected to two different quantum NAND operations. The quantum NAND operation's another inputs come from the outputs, which are $|Q>$ and $\overline{|Q>}$ qubits. Hence this quantum NAND operation's final output generates the final output of DNA-quantum SR flip-flop.

14.4.3 Working Principle

DNA-quantum SR flip-flop is working using two principles: DNA computing principle and the quantum computing principle. There are two inputs in the DNA NAND operation in this circuit. In this circuit, the inputs are S, R, and E which are shared between the two NAND operations. DNA circuit operation will be happening in

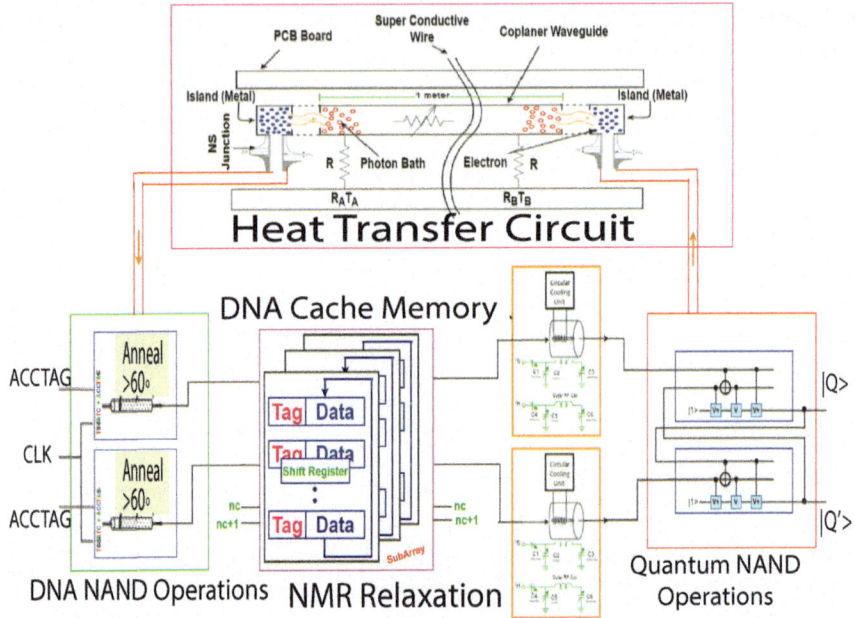

Fig. 14.6 DNA-quantum SR flip-flop

close to zero temperature because molecular sequences need a coherence state. The superposition state will be stable if this circuit is full of prohibited particles from other environment's particles. The two DNA NAND operations will be performed parallelly. This DNA NAND operation produces some garbage values. Each of the DNA NAND operations has one output. These molecular sequence outputs go to the DNA cache memory which is made by shift registers. The DNA cache memory is built using the rules of DNA computing. This cache memory will store the molecular sequence output from the NAND operation and will serve it when needed.

DNA cache stores the molecular sequence into an array. DNA cache memory stores the molecular sequence data and, when required it serves the data to the "NMR" process. In this DNA-quantum SR flip-flop, two-molecular sequences are stored in DNA cache memory. This molecular sequence performs the "NMR" process, because quantum circuits need qubit. This process removes the superposition state and makes the molecular sequence into a qubit. This qubit performs the quantum NAND operation. The final output sequence and output sequence from other quantum NAND operations perform quantum NAND operations for each of the quantum NOT operations. These two NAND operations perform parallelly. This structure is basically a quantum NAND flip-flop operation. DNA cache memory is mainly used because DNA operation performs slowly, while quantum operation is performed very fast. DNA cache memory works as an intermediate process where the molecular

Table 14.3 The truth table of a DNA-quantum SR flip-flop

S	R	|Q>	|Q'> [Q' is the complement of Q]
ACCTAG	ACCTAG	No change	
ACCTAG	TGGATC	|0>	|1>
TGGATC	ACCTAG	|1>	|0>
TGGATC	TGGATC	Invalid	

sequence is just stored and when needed, the cache memory serves the molecular sequence.

DNA cache memory works here as the intermediate process. Quantum circuit produces much heat and here two quantum NAND operations perform. So, it produces much heat. But the quantum circuit needs to be close to zero Kelvin temperature to perform the operation. So, from the quantum circuit, it needs to reduce the temperature to maintain the superposition state of the qubit. Hence, in DNA circuit operation, it needs much heat in several steps. The melting and annealing require much heat. This topic is briefly described in the chapter named DNA-quantum circuit operation. For that reason, the excessive heat is transferred from the quantum circuit portion to the DNA circuit portion using a heat transfer circuit. This heat transfer circuit using the junction captures the heat from the quantum circuit and using photon bath the heat flows through the circuit and gives this to the DNA circuit. This circuit can transfer heat maximum in the one-meter distance and in DNA-quantum flip-flop distance is less than one meter between quantum NAND DNA circuits. This heat transfer circuit cannot transfer full excessive heat produced from the quantum circuit and this heat is not enough to perform the DNA circuit operation. But this transferred heat can optimize the cost of heat based on the needs. After completion of all of these operations and the design architecture, the DNA-quantum SR flip-flop can produce two-output molecular sequences. Table 14.3 shown the truth table of a DNA-quantum SR flip-flop.

14.4.4 Example

Assume that the circuit receives the inputs ACCTAG and TGGATC in order to ensure that the DNA SR flip-flop operational circuit produces the right output. The ACCTAG input's end is connected to the S input end, while the TGGATC input's end is connected to the R input end. The fact that the R input is TGGATC indicates that it is for a reset operation. Assume that the clock is activated and that the clock's input is TGGATC.

Firstly, do the DNA NAND operation using the clock input. If just one of the inputs is ACCTAG in a DNA NAND operation, the result is TGGATC; otherwise, the output is ACCTAG.

The DNA NAND operation is then performed operation using clk and R inputs. In this case, the clock input is TGGATC, and the R input is also TGGATC.

Then, as the input to the quantum SR latch, these two input molecular sequences are converted into qubit by the NMR process. |0> and |1> will be inputted. The quantum SR latch operation will be performed on them. A collection of quantum NAND operations create the SR latch operation circuit.

This output presents the performance of the DNA-quantum SR flip-flop circuit.

14.5   DNA-Quantum JK Flip-Flop

The JK flip-flop, a fundamental sequential logic circuit, can be explored in the context of DNA-quantum computing. The DNA-quantum computing explores how quantum principles might be used in biological systems or for building biological-inspired quantum devices. This field explores using quantum mechanics to perform computations in biological systems or to design biological-inspired quantum devices. It's not about directly implementing electronic circuits like JK flip-flops in biological systems, but rather about understanding how quantum phenomena could be harnessed for computation in biological contexts or for creating new types of quantum computers inspired by biology. JK flip-flops in DNA-quantum computing have been designed to reduce complexity and making them more efficient for fabrication. These designs can be used to build more complex DNA-quantum sequential circuits, such as shift registers and counters. In flip-flop designs, the DNA JK flip-flop will be the most extensively utilized flip-flop. J and K are not abbreviated letters of other words, such as "S" for Set and "R" for Reset, but there are independent letters chosen by the inventor Jack Kilby to identify the flip-flop design from others. Despite the fact that the digital electronics JK flip-flop was created by Jack Kilby. The functioning concept of the DNA-quantum JK flip-flop differs from that of the digital JK flip-flop.

The DNA-quantum JK flip-flop's sequential operation is identical to that of the prior DNA-quantum SR flip-flop, with the same "Set" and "Reset" inputs. The distinction this time is that even though S and R are both at logic "1," the "DNA-quantum JK flip-flop" has no incorrect or prohibited DNA-quantum SR Latch input states. It is evident that the DNA-quantum JK flip-flop does not solve the disadvantages of the DNA SR flip-flop.

The DNA-quantum JK flip-flop is essentially a similar DNA-quantum SR flip-flop with the addition of clock molecular input sequence circuitry to avoid the unlawful or invalid output state that can arise when both inputs S and R are equal to logic level |1 >. A DNA-quantum JK flip-flop has four potential input combinations due to the extra timed input: "TGGATC," "logic ACCTAG," "no change," and "toggle." A DNA-quantum JK flip-flop has the same symbol as a DNA SR bistable latch, as seen in the preceding chapter. The DNA-quantum JK flip-flop, like other flip-flops, generates a lot of heat, which must be dissipated for the operation to run properly. As compared to other DNA circuits, the DNA-quantum JK flip-flop will not require as much power. The molecular sequence may simply conduct the operation once all

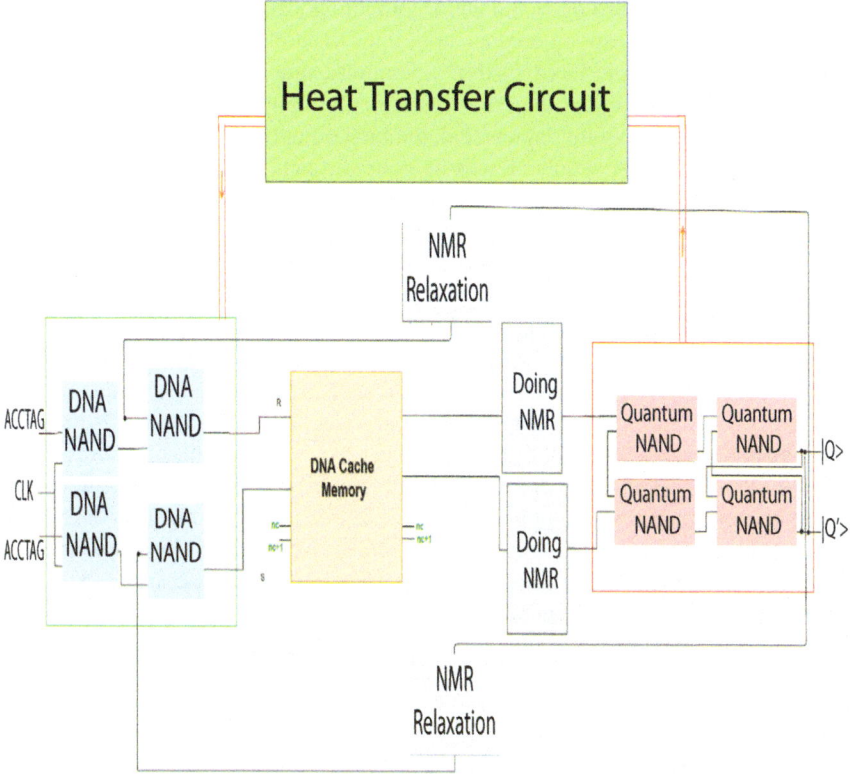

Fig. 14.7 Block diagram of DNA-quantum JK flip-flop circuits

of the molecules are in superposition state and coherence mode. A lot of junk values are received in the DNA-quantum JK flip-flop, and more investigations are needed to figure out what they are. The trash value is not taken into account in this procedure.

14.5.1 Block Diagram

DNA-quantum JK flip-flop toggles the input. DNA-quantum JK flip-flops have two-molecular input sequences and clock molecular input sequences. The DNA JK flip-flop relies on clock molecular input sequence. The DNA-quantum JK flip-flop is shown in Fig. 14.7.

DNA-quantum JK flip-flop needs one molecular input sequence and one molecular clock input. There is a molecular input sequence J and another molecular input sequence K which will enter into two different NAND operations. One clock input is shared by both of those DNA NAND operations. These DNA NAND operations perform parallelly according to the DNA computing principle. These DNA NAND operations output lines are entered as input lines to another couple of DNA NAND operations. During these DNA NAND operations, one input is the previous output molecular sequence and the other is the final output of the DNA-quantum JK flip-flop. The block diagram depicts that, the final output will be the DNA molecule sequence. So this DNA sequence will be gone through the NMR process and the NMR will make this sequence a qubit as well as this qubit, will go to quantum NAND operation as input. Like the other output, the qubit will come as input in another quantum NAND operation. Then two quantum NAND operations will parallelly perform its activities and it is the last operation of DNA computing according to the DNA-quantum JK flip-flop circuit.

In DNA computing, molecular sequence needs to go through many processes during the operation. In DNA computing, melting and annealing are very important and these steps require a huge amount of heat. That's why in DNA-quantum JK flip-flop, quantum portion transfers a small amount of heat produced from the quantum circuits to DNA circuits. Though this amount of heat is not enough, this heat or temperature will help to perform DNA computation. In quantum computing quantum NAND SR latch performs its operation, where two quantum NAND operations perform parallelly and input comes from DNA circuit portion. After that operation, the output qubit is found from DNA-quantum JK flip-flop.

14.5.2 Circuit Architecture

DNA-quantum JK flip-flop operation circuit is constructed using one principle that is the first portion of the circuit will be constructed by using DNA circuit and the second portion will be constructed by using quantum computing circuit. This circuit also has DNA cache memory and a heat transfer circuit to achieve the best output at the end. DNA-quantum JK flip-flop operation circuit is shown in Fig. 14.8.

In this circuit, the input is molecular sequence and the output is the qubit. The clock input clk is shared with two DNA NAND operations. Two DNA operations are performed parallelly. This DNA NAND operation's output line and the final output of DNA-quantum JK flip-flop are connected to the another two DNA operations parallelly. But the final two outputs are qubits. So, these two output lines are connected to the NMR relaxation process and then after the NMR relaxation process these two lines come into use as an input lines of a DNA NAND operation circuit. In this NMR relaxation process, the emitting EMR is strictly prohibited. These two DNA NAND operations circuit output lines are connected to the DNA cache memory. This is the first portion of the circuit which is fully designed by using the principle of DNA computing. DNA cache memory is made by using some DNA shift registers and in this circuit it saves data in the DNA array. When the time arrives, the DNA cache memory supplies the data to the NMR process. DNA cache memory supply line is fully connected to NMR process, where the EMR emission is fully prohibited. Then, the NMR process output line is connected to the quantum NAND operation.

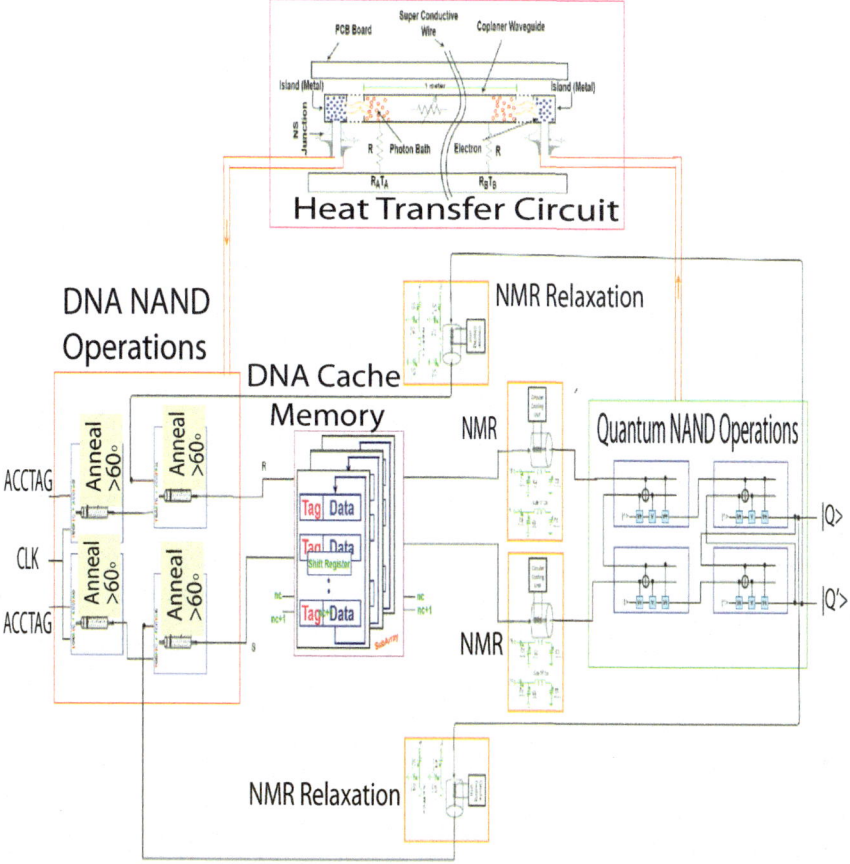

Fig. 14.8 DNA-quantum JK flip-flop operation circuit

Two outputs of the NMR process are connected to two different quantum NAND operations. Quantum NAND operations another input come from the outputs which are |Q> and |Q'> qubits. Hence this quantum NAND operation's final output line is the final output of DNA-quantum JK flip-flop.

14.5.3 Working Principle

The DNA-quantum JK flip-flop is working using two principles: the DNA computing principle and the quantum computing principle. DNA-quantum JK flip-flop has two inputs, which are J and K and the clock input is as clock. DNA-quantum JK flip-flop won't be enable if the clock input is not 0. Hence, the clock input enables this circuit to start to work and first of all, it will perform the DNA circuit operation. The clk input

Table 14.4 Truth table of DNA-quantum JK flip-flop

|T>	Q (Quantum Sequence)	$\overline{Q}$ (Quantum Sequence)
ACCTAG	|1>	|1>
TGGATC	|1>	|0>
ACCTAG	|0>	|1>
TGGATC	|0>	|1>

will be shared and perform two DNA NAND operations parallelly. This DNA NAND operation produces some garbage values. DNA NAND operations produce two output molecular sequences within the same amount of time. This molecular sequence will perform another two operations parallelly. The first molecular sequence and final output of DNA-quantum JK flip-flop will perform DNA NAND operation. Another molecular sequence and the output sequence of the DNA-quantum JK flip-flop will perform the DNA NAND operation. These two operations will perform parallelly. But the final output of DNA-quantum JK flip-flop is a qubit. The qubit cannot perform the DNA NAND operation. So, that's why DNA sequence performs NMR relaxation process and then it converts them into a qubit. Then these molecular sequences and the previous output molecular sequences perform two DNA NAND operations parallelly and produce the output of two DNA NAND operations parallelly.

DNA cache memory works here as the intermediate process. Quantum circuit produces much heat and here two quantum NAND operation performs so it produces much heat. But quantum circuit needs to be in close to zero Kelvin temperature to perform the operation. So from the quantum circuit, it needs to reduce the temperature to maintain the superposition state of the qubit. Hence in DNA circuit operation, it needs huge heat in several steps. Mainly in melting and annealing require much heat. For that reason, the heat is transferred from the quantum circuit portion to the DNA circuit portion using a heat transfer circuit. This heat transfer circuit using the junction captures the heat from the quantum circuit and using photon bath the heat flows through the circuit and gives this to the DNA circuit. This circuit can transfer heat maximum in the one-meter distance and in DNA-quantum flip-flop distance is less than one meter between the quantum and the DNA circuits. This heat transfer circuit can not transfer full produced heat from the quantum circuit and this heat is not enough to perform the DNA circuit operation. But this heat transfer can optimize the cost of heat which is needed for the DNA circuit. After all of these operations and design architecture, the DNA-quantum JK flip-flop can produce two-output quantum qubits. Table 14.4 represents the truth table of a DNA-quantum JK flip-flop.

14.5.4 Example

For the purpose of testing of the DNA JK flip-flop circuit, assume that the molecular input sequences are ACCTAG and TGGATC. If and only if the clock input is high or TGGATC, the DNA-Quantum JK flip-flop will perform its function. If the clock input is high, the DNA NAND operation will be performed by the molecular sequences ACCTAG and clk.

In addition to performing the DNA NAND operation and producing the appropriate output, the TGGATC and clk inputs work in parallel.

These two intermediate molecular sequence outputs are now used to conduct two independent DNA NAND operations. The DNA NAND operation is performed by the molecular sequence ACCTAG and the final output of the DNA JK flip-flop which is |Q>. Assume that the most recent state |Q> is TGGATC.

Let the output of the JK flip-flop be ACCTAG, which is referred to as S. The DNA NAND operation is then performed on the intermediate output ACCTAG and the final output of the DNA-quantum JK flip-flop which is |Q>.

According to the circuit of DNA JK flip-flop, these molecular sequences labeled S and R conduct the quantum SR flip-flop operation.

Finally, the DNA-quantum JK flip-flop provides the necessary qubit inputs |1> and |0>.

14.6 DNA-Quantum T Flip-Flop

The T flip-flop is a single-input flip-flop that toggles its output state on each clock pulse when the input is high. In the context of DNA-quantum computing, the concept of a T flip-flop, which is a fundamental logic gate, is explored within the framework of DNA-quantum systems and their potential for computation. This involves investigating how principles of quantum mechanics, like superposition and entanglement, could be applied to create or manipulate states to the biological systems of a T flip-flop. DNA-quantum computing explores the potential of biological systems, like DNA or proteins, to perform quantum computations. DNA-quantum T flip-flop is also known as "DNA-Quantum Toggle Flip-Flop". To avoid the occurrence of the intermediate state in DNA-quantum SR flip-flop, only one input is provided to the flip-flop called the Trigger DNA sequence input or Toggle input. Toggling means 'Changing the next state output to complement the present state output'. The DNA-quantum T flip-flop can be designed by making simple modifications to the DNA-quantum JK flip-flop. The DNA-quantum T flip-flop is a single DNA sequence input device and hence by connecting |J> and |K> inputs together and giving them with single input called |T>, a DNA-quantum JK flip-flop can be converted into a DNA-quantum T flip-flop.

Hence in DNA-quantum flip-flops have two portions based on the principle of the DNA-quantum circuit. In the DNA-quantum T flip-flop circuit one portion is made by DNA computing principle and the rest of the portion is made by quantum computing principle. DNA computing and Quantum computing are connected via the NMR relaxation process. For time consistency, the DNA cache memory is also used in this circuit. The quantum circuit portion produces so much heat according to the quantum computing principle that's why this circuit can transfer a small amount of heat to the DNA computing circuit portion. According to DNA computing, it needs a huge amount of heat to perform the calculation. In DNA-quantum T flip-flop circuit have one input which is a DNA sequence input named |T>. This input performs first of all DNA operations and it is then stored in DNA cache memory. When the perfect time arrives it will perform the NMR relaxation and convert the DNA molecular sequence into qubit as well as perform the quantum computing operation.

14.6.1 Circuit Architecture

DNA-quantum T flip-flop toggles the input, where DNA-quantum JK flip-flop's extended version is DNA-quantum T flip-flop architecturally. DNA-quantum T flip-flops have one DNA sequence input and clock DNA sequence input. DNA T flip-flop relies on clock DNA sequence input. DNA-quantum T flip-flop is shown in Fig. 14.9.

DNA-quantum T flip-flop needs one DNA sequence input and one clock input which is also a DNA sequence. These DNA AND operations perform parallelly according to the DNA computing principle. The DNA AND operation's output lines are entered as input lines into another couple of DNA AND operations. The DNA AND operation's one input is the previous output DNA sequence and the other is the final output of the DNA-quantum T flip-flop. As this block diagram depicts, the final output will be the DNA molecule sequence. So this molecular sequence will go through the NMR process and NMR will make this sequence into a qubit, as well as this qubit, will go to quantum NOR operation as input. Like the same other output, the molecular sequence will come as input in another DNA AND operation. Then finally two DNA AND operations will parallelly perform and it is the first operation of DNA computing according to the DNA-quantum T flip-flop circuit block diagram. These two DNA AND operations produce two DNA sequences which will be stored in DNA cache memory. Then this DNA cache memory serves the DNA sequence into the NMR process. In the NMR relaxation process, emitting EMR is strictly prohibited. After the NMR relaxation process the molecular sequences are converted into the qubits.

In DNA computing molecular sequence needs to go through many processes during the operation. In DNA computing, melting and annealing are very important and these steps require a huge amount of heat. That's why in DNA-quantum T flip-flop quantum portion transfers a small amount of heat produced from the quantum circuits into DNA circuits. Though this amount of heat is not enough, this heat or temperature will help to perform DNA computation. In quantum computing, quantum NOR SR latch performs where two quantum operations perform parallelly and input

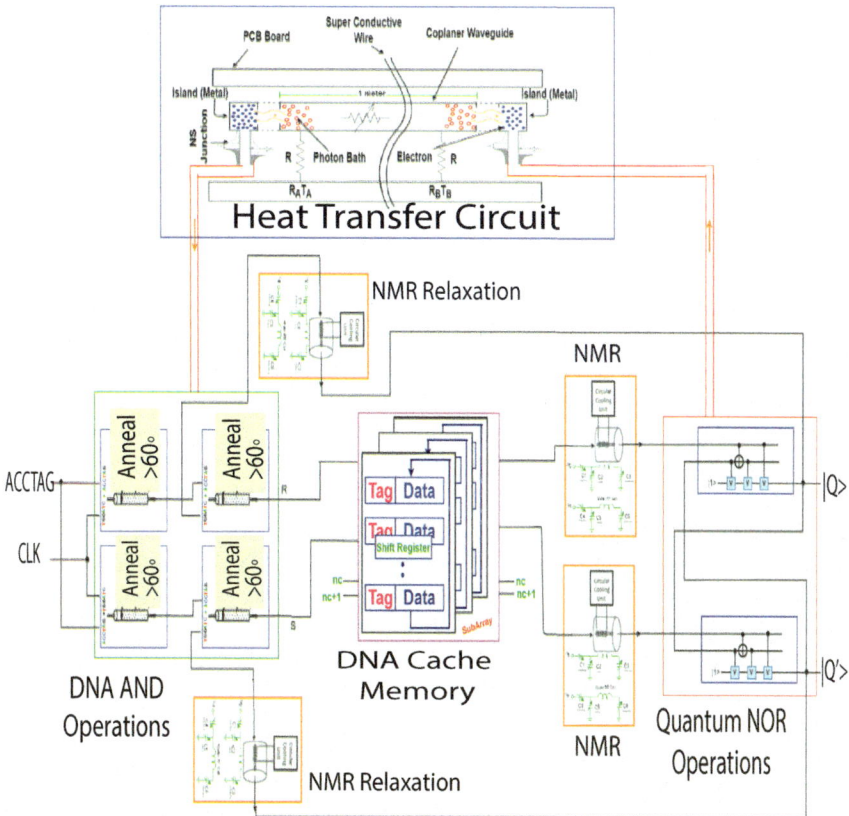

Fig. 14.9 Circuits architecture of DNA-quantum T flip-flop

comes from DNA circuit portion. After that operation, the output qubits are found from DNA-quantum T flip-flop.

14.6.2 Working Principle

DNA-quantum T flip-flop is working using two principles such as the DNA quantum computing principle and the quantum computing principle. DNA-quantum T flip-flop has a data input named |T> and clock input |clk>. DNA-Quantum T flip-flop won't enable if the clock input is not |0>. Hence the clock input enables this circuit to start to work and first of all, it will perform the DNA circuit operation. |T> and |clk> inputs will be shared and perform two DNA AND operations parallelly. The DNA AND operations produce some garbage outputs. Quantum AND operations produce two output qubits at the same amount of time. This qubit one of each will perform another two quantum AND operations parallelly. The first DNA sequence and final output of DNA-quantum T flip-flop will perform DNA AND operation. Another DNA sequence and another output sequence of the DNA-quantum T flip-flop will perform the DNA AND operation. These two operations will perform parallelly. But the final output of DNA-quantum T flip-flop is a qubit. The qubit cannot perform the DNA AND operation. These two outputs of two DNA sequences enter into the DNA cache memory.

DNA cache stores the DNA sequences into an array. DNA cache memory stores the DNA data and when requires it serves the data into the NMR relaxation process. In this DNA-quantum T flip-flop, DNA sequences are stored in DNA cache memory. This DNA sequence performs the NMR relaxation process because quantum circuits need molecular sequence. NMR relaxation process creates the superposition state and makes the molecular sequence into a qubit. This qubit sequence performs the quantum NOR operation. The final output sequence and qubit from cache perform quantum NOR operation. Another final output sequence and previous output from cache also perform another quantum NOR operation. These two NOR operation performs parallelly. This structure basically quantum NOR latches operation. DNA cache memory is mainly used because quantum operation performs so fast and DNA operation performed very slowly. DNA cache memory works as an intermediate process where DNA sequences are just stored and when needed cache memory serves the DNA sequences.

DNA cache memory works here as the intermediate process. Quantum circuit produces much heat and here two quantum NOR operations perform these operations which produce much heat. But quantum circuit needs to be in close to zero Kelvin temperature to perform the operation. So from the quantum circuit, it needs to reduce the temperature to maintain the superposition state of the qubit. Hence in DNA circuit operation, it needs much heat in several steps. Mainly in melting and annealing require much heat. This topic is briefly described in the chapter named DNA-quantum circuit operation. For that reason, the excessive heat is transferred from the quantum circuit portion to the DNA circuit portion using a heat transfer circuit. This heat transfer circuit using the junction captures the heat from the quantum circuit and using photon bath the heat flows through the circuit and gives this into the DNA circuit. This circuit can transfer heat maximum in the one-meter distance and in DNA-quantum flip-flop distance is less than one meter between quantum and DNA circuits. This heat transfer circuit cannot transfer the full excessive heat produced from the quantum circuit and this heat is not enough to perform the DNA circuit operation. But this heat transfer can optimize the cost of heat based on the needs. After completion of all of these operations and the design architecture of the circuit, the DNA-quantum T flip-flop can produce the qubit outputs. Table 14.5 shows the truth table of a DNA-quantum T flip-flop.

Table 14.5 Truth table of a DNA-quantum T flip-flop

$|T>$	Q (Quantum-Based DNA Sequence)	$\overline{Q}$ (Quantum-Based DNA Sequence)
$|0>$	TGGATC	TGGATC
$|1>$	TGGATC	ACCTAG
$|0>$	ACCTAG	TGGATC
$|1>$	ACCTAG	TGGATC

14.6.3 Example

DNA-quantum T flip-flop needs one qubit input and one clock input. If the Clock's input value is $|1>$ then this DNA-quantum flip-flop will be enabled. Assume clock input value is $|1>$ here and the qubit input is $|1>$.

These qubit input and clock input will, first of all, perform two DNA AND operations parallelly.

So, from two DNA AND operations, two DNA sequences are found which are $|1>$ respectively. Assume one output sequence is ACCTAG and the other is TGGATC.

These two DNA sequences will perform DNA AND operation. But molecular sequence can not perform quantum NOR operation directly. For that reason, these two sequences, first of all, perform the NMR process and make them qubit, and then perform the quantum NOR operation with respect to the other input parallelly. ACCTAG will convert as a qubit and assume it as $|1>$ and opposite TGGATC as $|0>$.

These outputs ACCTAG and TGGATC will be stored in DNA cache memory. DNA cache memory will serve this output as input to the quantum circuit when the right time arrives. These DNA sequences will perform first of all NMR relaxation process where these DNA sequences will become in decoherence state and these sequences will be relaxed to convert them into qubits. In the NMR relaxation process, EMR emits is fully prohibited. After performing the NMR relaxation process these outputs ACCTAG and TGGATC will be converted as qubits named $|1>$ and $|0>$, respectively. Finally, these two qubits perform the Quantum SR latch operation where two quantum NOR operations perform parallelly.

14.7 DNA-Quantum Shift Register

In DNA-quantum computing, a shift register in DNA-quantum form, plays a crucial role in manipulating and storing information within DNA-quantum circuits. In other words, a shift register is a type of register that shifts qubits within a DNA-quantum circuit. Shift registers in DNA-quantum computing can be utilized for various purposes, such as:

   i) Storing and manipulating data related to biological systems;
  ii) Implementing algorithms for analyzing biological data;
 iii) Simulating molecular dynamics and other biological processes; and
  iv) Developing quantum sensors and biosensors.

For example, a DNA-quantum shift register could be used to encode and manipulate the information about the state of a biomolecule, or to perform calculations on the data gathered by a quantum biosensor. A single molecular sequence of two-valued molecular sequence data (TGGATC or ACCTAG) can be stored in a DNA-quantum flip-flop. However, many DNA-quantum flip-flops are required to store multiple molecular sequences of data. To store n molecular sequences of data, n DNA-D flip-flops must be coupled in a certain order. A DNA-quantum register is a gadget that stores this type of data. It consists of a sequence of DNA-quantum flip-flops used to store multiple molecular sequences of data.

DNA-quantum shift registers enable the information stored in these DNA registers to be transmitted. A DNA-quantum shift register is a collection of flip-flops that stores several molecular sequences of information. By applying clock pulses to the molecular sequences maintained in such DNA-quantum registers, they may be arranged to move inside them and in and out of them. By linking n DNA flip-flops and n number of quantum flip-flops, each of which stores a single molecular sequence of data or molecule sequence of data, an n DNA-quantum shift register may be built. "DNA-Quantum Shift left registers" are DNA-quantum registers that will shift the molecular sequences or molecules to the left. "DNA-Quantum Shift right registers" are DNA-quantum registers that will shift the molecular sequences or molecule sequence to the right.

DNA-quantum shift registers are basically of 4 types. These are as follows:

1. DNA-Quantum Serial In Serial Out shift register;
2. DNA-Quantum Serial In Parallel Out shift register;
3. DNA-Quantum Parallel In Serial Out shift register; and
4. DNA-Quantum Parallel In Parallel Out shift register.

In this chapter, a shift register is built utilizing a DNA D flip-flop circuit and a quantum D flip-flop circuit to convert serial data into DNA data and quantum qubit data. The DNA-quantum Serial-In Serial-Out shift register is a type of DNA-quantum shift register that permits serial input one molecular sequence or one molecule sequence at a time over a single data line and outputs a serial output. The data exits the DNA-quantum shift register one molecular sequence at a time in a serial pattern since there is only one molecular sequence output. Thus, the term is DNA-quantum Serial-In Serial-Out shift register. Four DNA-quantum D flip-flops are linked in a serial fashion in this circuit. Because the same clock signal is supplied to each DNA-quantum flip-flop, they are all synchronized with one another. The circuit below in the architecture section is an example of a DNA-quantum shift right register, which accepts serial data from the DNA flip-flop's left side and the quantum flip-flop's left side.

In DNA-quantum flip-flop molecular sequence is entered into as an input in the circuit and performs the required operation. DNA computing computation speed is much slower than quantum computing computational speed. For that case the molecular sequences are stored in DNA cache memory. When actually the time arrives, the molecular sequence first of all performs the NMR process and makes the molecular sequence into qubit. Then this qubit performs the quantum flip-flop operation. Quantum circuit produces much heat and heat needs to be removed to maintain the coherence state and DNA computing's circuit needs huge heat to enable the circuit to perform. In that case, quantum circuit used a heat transfer circuit to transfer the heat to the DNA circuits.

14.7.1 Circuit Architecture

Three DNA D flip-flop circuits and one quantum D flip-flop circuit are used to make a DNA-quantum shift register. As a fundamental component, a DNA-quantum shift register is utilized for data shift, and a DNA D flip-flop and quantum D flip-flop are used to make it functioning.

Two inputs are used in DNA D flip-flops: one for data and one for the clock. The outputs of DNA D flip-flops are logically opposite one another. The circuit's synchronization with an external signal is used by the clock input. A DNA D flip-flop's output can have two possible values. Data input is directed to a DNA NAND circuit in this block diagram, while the data input reverse is routed to another DNA NAND circuit. Both DNA NAND operations use the clock pulse input. DNA SR latch receives the result of two DNA NAND operations. The D flip-flop is constructed using the DNA SR latch. This property is used to induce a delay in the circuit's data flow. DNA SR latch is made up of two DNA NAND operations. The final two outputs of the DNA SR latch function are uncovered. A DNA D flip-flop may provide two types of output, one of which is logically inverse to the other. The DNA D flip-flop will continue to work if the clock is enabled; otherwise, the DNA D flip-flop stops working.

DNA D flip-flop operational circuits are also coupled through serial connection in the DNA shift register.

The DNA D flip-flop circuit is the fundamental component of the DNA shift register. After processing the molecular sequence input in the DNA shift register, the output of the shift register is stored in DNA cache memory.

In DNA computing molecular sequence needs to go through many processes during the operation. In DNA computing, melting and annealing are very important and these steps require a huge amount of heat. That's why in DNA-quantum flip-flop quantum portion transfers a small amount of heat produced from the quantum circuits to the DNA circuits. Though this amount of heat is not enough, this heat or temperature will help to perform DNA computation. In quantum computing, quantum NAND SR latch performs its operation, where two quantum NAND operations perform parallelly and inputs come from quantum circuit portion. After that operation, the output qubit is found from the quantum D flip-flop.

Four D flip-flops and a DNA AND operation are used in the DNA portion and a quantum D flip-flop is in the DNA-quantum shift register. The Shift register generates four molecular sequence outputs. The D flip-flop has a single molecular sequence

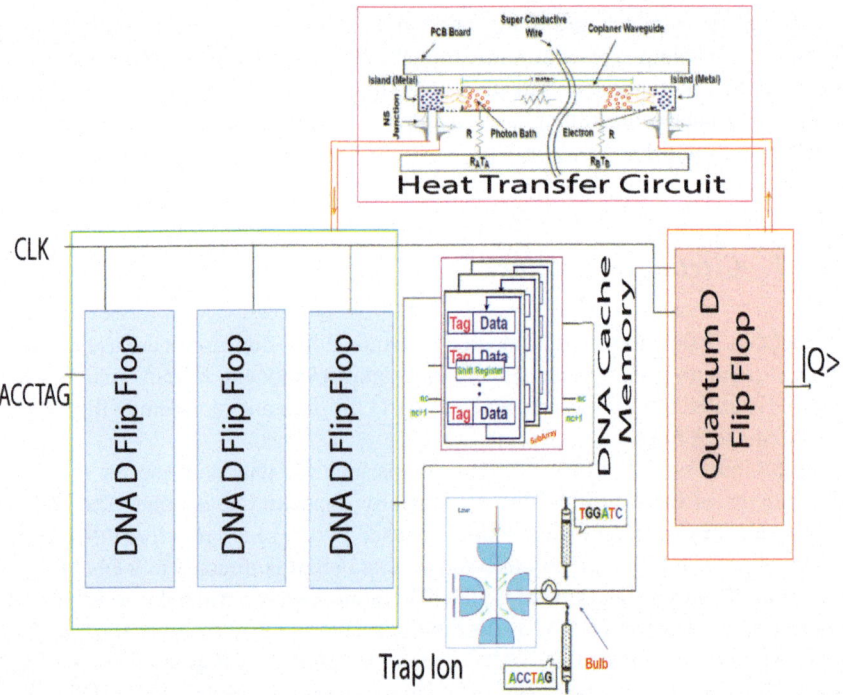

Fig. 14.10 Block diagram of a DNA-quantum shift register

input and is developed using DNA NAND and DNA SR latch operations. The block diagram of a DNA-quantum shift register is shown in Fig. 14.10.

The clock molecular sequence input is required for the DNA D flip-flop and the circuit has one molecular sequence input. One line of this molecular sequence the input will be directed to a DNA NAND operation termed S input in circuit. DNA NAND is performed here using S molecular sequence input and clock molecular sequence input.

In this circuit, the input is molecular sequence and the output is a qubit. Clock input clk is shared by a DNA NAND operation and a DNA NOT operation. Two DNA operations are performed parallelly. These NAND operations output and the final output of DNA-quantum shift register perform two DNA operations parallelly. But the final two outputs are DNA sequences so these two outputs lines are connected to the NMR process and then after the NMR process, these two outputs come into use as an input line DNA NAND operation circuit. These two DNA NAND operations circuit outputs are connected into the DNA cache memory. All of these DNA NAND operations are the part of DNA D flip-flop which is used in this system. This is the first portion of the circuit which is fully designed by using the principle of DNA computing. DNA cache memory is made by using some DNA shift registers and in this circuit it saves data to the DNA array. When the time arrives, the DNA cache memory supplies the data to the NMR process. DNA cache memory supply is fully connected to NMR process as well as here the EMR is emitted. Then the NMR process output line is connected to the quantum NOT operation. After passing through the quantum NOT operation, the output is taken as input in the quantum NAND operation. In these quantum NAND operations the 1st clock input is taken as |clk> input. These inputs are shared by two quantum NAND operations. Then the outputs of these operations are connected to another set of quantum NAND operations. Quantum NAND operation's another input come from the output which is |0> qubit. Hence this quantum NAND operation's final output is the final output of DNA-quantum shift register.

14.7.2 Working Principle

DNA-quantum shift registers are a kind of registers where both molecular sequence data loading, as well as data retrieval to/from the DNA-quantum shift register, occurs in serial mode sometimes. This research synchronize DNA SISO shift register sensitive to the positive edge of the clock pulse. Here the data word which is to be stored is fed bit-by-bit at the molecular sequence input of the first DNA flip-flop. Further, it is seen that the molecular sequence inputs of all other flip-flops are driven by the outputs of the preceding ones, for example, the input of DNA D flip-flop number-2 is driven by the output of DNA D flip-flop number-1. At last, the data stored within the DNA shift register is obtained at the output pin of the nth DNA D flip-flop in serial fashion.

Initially, all the DNA flip-flops in the DNA register are cleared by applying high on their clear pins. Next, the input data word is fed serially to DNA D flip-flop number-1.

This causes the molecular sequence appearing at the first pin to be stored into DNA D flip-flop number-1 as soon as the first leading edge of the clock appears. Further at the second clock tick, the DNA sequence gets stored into DNA D flip-flop number-2 while a new bit enters into DNA flip-flop number-2.

This kind of shift in data molecular sequences continues for every rising edge of the clock pulse. This indicates that for every single clock pulse the data within the DNA register moves toward the right by a single bit. Following the molecular sequence data transmission, as explained, one can note that the first molecular sequence of an input word appears at the output of nth flip-flop for the nth clock tick. On applying further clock cycles, one gets the next successive molecular sequences of the molecular sequence input data word as the serial output. Table 14.6 represents the truth table of a DNA-quantum shift register.

DNA cache stores the DNA sequences into an array. DNA cache memory stores the DNA data and when requires it serves the data to the NMR relaxation process. In this DNA-quantum D flip-flop, DNA sequences are stored in DNA cache memory. This DNA sequence performs in the NMR relaxation process because quantum circuits need qubits. NMR relaxation process considers the superposition state and

Table 14.6 Truth table of a DNA-quantum shift register

|Clk>	|x>	|Q>	$\overline{|Q}$ >	Description
↓ »ACCTAG	X	Q	Q	Memory no change
↑ »TGGATC	ACCTAG	ACCTAG	TGGATC	Reset Q »0
↑ »TGGATC	TGGATC	TGGATC	ACCTAG	Set Q »

makes the molecular sequence into a qubit. This qubit performs the Quantum NAND operation. The final output qubit and the qubit generated from cache perform quantum NAND operation. Another final output sequence and previous output from cache also perform another quantum NAND operation. These two NAND operations perform parallelly. This structure is basically for quantum NAND latches operation. DNA cache memory is mainly used for quantum operation. DNA cache memory works as an intermediate process where DNA sequence is just stored and when needed cache memory serves the DNA sequence.

DNA cache memory works here as the intermediate process. Quantum circuit produces much heat and here two quantum NAND operations perform, so it produces much heat. But quantum circuit needs to be in close to zero Kelvin temperature to perform the operation. So, from the quantum circuit, it needs to reduce the temperature to maintain the superposition state of the qubit. Hence in DNA circuit operation, it needs much heat in several steps. Mainly in melting and annealing require much heat. This topic is briefly described in the chapter named DNA-quantum circuit operation. For that reason, the excessive heat is transferred from the DNA circuit portion to the DNA circuit portion using a heat transfer circuit. This heat transfer circuit using the junction, capture the heat from the quantum circuit and using photon bath heat flows through the circuit and gives this to the DNA circuit. This circuit can transfer heat maximum in the one-meter distance and in DNA-quantum flip-flop distance is less than one meter between quantum and DNA circuits. This heat transfer circuit can not transfer the full excessive heat produced from the DNA circuit and this heat is not enough to perform the DNA circuit operation. But this heat transfer can optimize the cost of heat based on the needs. After all of this operations and design architecture, the DNA-quantum D flip-flop can produce the qubits as outputs.

14.8 DNA-Quantum Ripple Counter

A ripple counter, or asynchronous counter, can be conceptualized within the context of DNA-quantum computing, particularly when considering DNA-quantum circuits. It represents a logic design element that can be explored in the theoretical realm of quantum molecular biology. Some architectures in DNA-quantum computing, like complex quantum gates mimicking biological processes, can be used to manipulate the flow of information, which might be relevant to understanding how biological systems process information. Research explores the design and architecture of sequential circuits like ripple counters, shift registers, and flip-flops using DNA-quantum components. A DNA-quantum counter is basically used to count the number of clock pulses applied to a DNA-quantum flip-flop. It can also be used for DNA-quantum frequency divider, DNA-quantum time measurement, DNA-quantum frequency measurement, DNA-quantum distance measurement, and also for generating square waveforms. In this, the DNA-quantum flip-flops are DNA-quantum asynchronous counters and are supplied with different clock signals which may be a delay in producing output. Also, a few numbers of DNA-quantum logic operations are needed to design asynchronous counters. So, they are elementary components in the design and also less expensive.

An n-bit ripple counter can count up to $2n$ states. It is also known as MOD n counter. It is known as a ripple counter because of the way the clock pulse ripples its way through the flip-flops. It is an asynchronous counter. Different ripple counters are used with a different clock pulse. All the flip-flops are used in toggle mode. Only one flip-flop is applied with an external clock pulse and another flip-flop clock is obtained from the output of the previous flip-flop. The flip-flop applied with an external clock pulse acts as Least Significant Bit (LSB) in the counting sequence. A counter may be an up counter that counts upwards or can be a down counter that counts downwards or do the both, i.e., count up as well as count downwards depending on the input control. The sequence of counting usually gets repeated after a limit.

DNA-quantum ripple counter is constructed by four JK flip-flops. Using these DNA JK flip-flops, the DNA-quantum ripple counter creates four molecular sequence outputs. Here in JK flip-flop, J and K are not shortened abbreviated letters of other words, such as "S" for Set and "R" for Reset, but these are autonomous letters chosen by its inventor Jack Kilby to distinguish this flip-flop design from other types. Jack Kilby invented the digital electronics JK flip-flop. DNA-quantum ripple counter is an asynchronous counter. It is created using DNA JK flip-flops; and these flip-flops are only controlled by the clock pulse input.

DNA-quantum ripple counter produces much heat to keep the qubit's superposition state and also produces some garbage values.

14.8.1 Circuit Architecture

DNA-quantum ripple counter uses three DNA JK flip-flops to operate two-molecular sequence inputs. DNA JK flip-flop has the two-molecular sequence inputs named as J and K.

DNA JK flip-flop has the molecular sequence inputs J and K. This DNA JK flip-flop consists of many DNA NAND operations. At first, basic DNA NAND operation performs a couple of operations, then these operations' output are entered into the SR flip-flop as well as produce the output. First of all, J and clk inputs perform the DNA NAND operation. The output of the DNA NAND operation and the output of the DNA JK flip-flop perform another DNA NAND operation and produce the output named as clk. clk input is shared. So, K and clk inputs also perform the DNA NAND operation. This DNA NAND operation's output and Q output of the DNA JK flip-flop perform another DNA NAND operation as well as produce the output named as R.

These clk and R enter into the DNA SR flip-flop and produce two outputs. In DNA SR flip-flop, it has four DNA NAND operations. clk input and S input perform DNA NAND operation as well as R input and shared clk input also perform the DNA NAND operation. These two NAND operations output entered into the DNA SR latch as input. In DNA SR latches, two Q and one input as well as $\overline{Q}$ and other inputs perform the DNA NAND operation. Finally, after all of these DNA operations, the DNA JK flip-flop's final outputs.

After processing the molecular sequence inputs in three JK flip-flops, the output is stored in a DNA cache memory, which is constructed by DNA shift register. Then this DNA cache memory serves the molecular sequence to the quantum computing circuit's portion. These molecular sequences need to be qubit first of all. So, the molecular sequences are performed in NMR process. In the NMR process, the EMR is emitted. After the NMR process molecular sequences are relaxing their states and are converted into the qubits.

In DNA computing the molecular sequence needs to go through many processes during the operation. In DNA computing, melting and annealing are very important and these steps require a huge amount of heat. That's why in DNA-quantum computing, JK flip-flop quantum portion transfers a small amount of heat produced from the quantum circuits to the DNA circuits. Though this amount of heat is not enough, but this heat or temperature will help to perform DNA computation. In quantum computing, the quantum NOR SR latch performs its activities, where two quantum NOR operations perform parallelly and inputs come from DNA circuit portion.

DNA-quantum ripple counter uses three DNA JK flip-flops and one quantum JK flip-flop to create qubit outputs. DNA JK flip-flop has a two-molecular sequence input and one clock shared input. DNA JK flip-flops also depend on the clock. If the clock is enabled then the circuit will be enabled, otherwise not. Figure 14.11 shows the circuit architecture of a DNA-quantum ripple counter.

In DNA-quantum ripple counter there's a one clock input and one logic input which are shared with both J and K input ports. Input J and input K, both the values will perform the DNA NAND operation with the shared clk input differently. J and clk inputs perform the DNA NAND operation. NAND operation is performed by using DNA basic operations. Basic operations in DNA computing are V, V+, and CNOT. For error correction it is used an ancillary bit. In quantum NAND operation the value of an ancillary bit is TGGATC. When J and clk perform the DNA NAND operation produced an output molecular sequence and this output molecular sequence and the output of the flip-flop perform the DNA NAND operation and produce the output clk. Like in the same procedure with K, clk and |Q> inputs produce the R output. These operation are mainly DNA NAND operations. These clk and R inputs are performed the DNA SR flip-flop operation. In DNA SR flip-flop, the operational

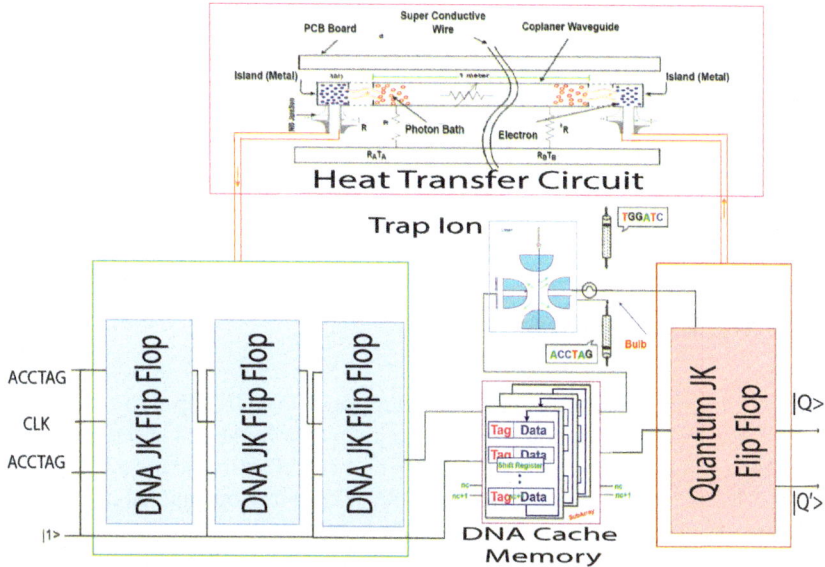

Fig. 14.11 DNA-quantum ripple counter

circuit architecture is also constructed by the basic component of DNA computing which is DNA NAND operation.

After processing the inputs in a DNA JK flip-flop, the output of the flip-flop is stored as an output of the DNA-quantum ripple counter. DNA-Quantum Ripple Counter intermediate architectures have four DNA JK flip-flops connected in serial connection using the logical molecular sequence input. Every clock input as clk enters into every molecular sequence and from 2nd DNA JK flip-flop, and the clk input is previous DNA Jk flip-flop's first output. With the same architecture, DNA-quantum ripple counter can be performed as an up counter or as a down counter but the clock pulse as clk needs sometimes to have a positive edge triggered and sometimes a negative edge triggered.

The JK flip-flop's circuit output lines are connected with the DNA cache memory. This is the first portion of the circuit which is fully designed by using the principle of DNA computing. DNA cache memory is made by using some quantum shift registers and in this circuit, it saves data in the DNA array. When the time arrives, DNA cache memory supplies the data to the NMR process. DNA cache memory supply line is fully connected with the NMR process as well as here the EMR is emitted. Then the NMR process output line is connected with the quantum NAND operation. Two outputs of the NMR process are connected with two different quantum NOR operations. Quantum NOR operation's inputs come from the output which is Q and $\overline{Q}$ qubits. Hence, the quantum NAND operation's final output of the DNA-quantum ripple counter.

Quantum circuit produces more heat and the DNA circuit needs the heat to process the input. For that reason, it is better to transfer the heat to DNA circuits. Quantum portion transfers heat using the transfers heat circuit to DNA circuit's portion. In the heat transfer circuit, two junctions are connected to quantum circuits and DNA circuits. Maximum one meter can transfer heat to this circuit. This heat transfer circuit has superconducting wire, photon batch, PCB board to make the circuit architecture. This circuit mainly transfers the heat using the photon bath.

14.8.2 Working Principle

In the DNA-quantum ripple counter, there are four DNA JK flip-flop operational circuits. These DNA JK flip-flop operational circuits are connected in serial connection. In this DNA-quantum ripple counter, there is one clock input and a logic input which are shared into the ports J and K of DNA JK flip-flop operational circuit. The inputs J and K of a DNA Jk flip-flop conducts two DNA processes in parallel. The DNA NAND operation is performed using J and shared input clk. The DNA NAND operations are then performed, and one of the DNA JK flip-flop's outputs executes the DNA NAND operation, producing clk. The K and clk inputs are performed first in the DNA NAND operation. The output of the DNA JK flip-flop as well as the result of the DNA first NAND operation, are then used to perform another DNA NAND operation, yielding R. The steps for creating clk and R are carried out simultaneously. The DNA SR flip-flop operation is then conducted on these clk and R inputs. The DNA NAND operation, which is employed here, is also used to make DNA SR flip-flops. Two outputs are created after conducting the DNA SR flip-flop operation. The one output is opposite to the other.

Here if the clk is ACCTAG then the DNA JK flip-flop will not be triggered but if clk is TGGATC the DNA JK flip-flop will be triggered and it will toggle the output. First of all DNA JK flip-flop operational circuit's clk value is TGGATC which toggles the output value from the previous state value. Now, the output of the initial DNA JK flip-flop will be the clk input of next DNA JK flip-flop. If the clk value is TGGATC then the output value will be toggled, otherwise the output will be the previous state output. By maintaining the same procedure, every DNA JK flip-flop operated in the DNA-quantum ripple counter. DNA JK flip-flop is toggled very much, that's why in the DNA-Quantum Ripple Counter DNA JK flip-flop operational circuit is used as a basic component. Table 14.7 shows the truth table of a DNA-quantum ripple counter.

14.8.3 Example

To check the DNA-quantum ripple counter circuit, assume a clock signal which is TGGATC and the logical molecular sequence which is high. So, the initially clock signal TGGATC is delivered to the first DNA JK flip-flop operational circuit.

Table 14.7 Truth table of a DNA-quantum ripple counter

clk	$|Q_0>$	$|Q_1>$	$|Q_2>$	$|Q_3>$
ACCTAG	ACCTAG	ACCTAG	ACCTAG	$|0>$
TGGATC	TGGATC	TGGATC	TGGATC	$|1>$
TGGATC	ACCTAG	ACCTAG	ACCTAG	$|0>$

According to the working principle of DNA-quantum ripple counter if the clock input value is TGGATC then the previous state value will be toggled. Now, the first DNA JK flip-flop produces one output which is TGGATC. This TGGATC will be the clock input for the second DNA JK flip-flop circuit according to the architecture of the DNA-quantum ripple counter.

As like the previous, the third DNA JK flip-flop follows the same working principle. The fourth flip-flop is a quantum JK flip-flop operational circuit. This circuit produces qubit by NMR process.

Hence from the above discussion it is clear that the DNA-quantum ripple counter is correct theoretically. DNA JK flip-flop used here for toggling. This DNA-quantum ripple counter requires much heat when the DNA molecules go through different processes. The DNA-quantum ripple counter's full operation needs to be an environment where other particles are totally prohibited to maintain the coherence state.

14.9 DNA-Quantum Synchronous Counter

A synchronous counter in DNA-quantum computing would likely be a DNA-quantum counter where all the input qubits are clocked simultaneously by a single and unified signal, and produce the DNA outputs. This contrasts with asynchronous counters where input qubits may be updated sequentially. In essence, a synchronous counter in DNA-quantum computing would be a quantum-based timing and counting mechanism where all qubits are driven by a single, synchronized clock signal, potentially enabling new possibilities for simulating and measuring biological phenomena at the quantum level. A DNA-quantum counter is a DNA-quantum device which can count any particular event on the basis of how many times the particular event(s) has occurred. In a DNA-quantum logic system or computer, this DNA-quantum counter can count and store the number of times of any particular event or process has occurred, depending on a DNA-quantum clock signal. The most common type of DNA-quantum counter is a sequential DNA logic circuit with a single clock input and multiple molecular sequence outputs. The molecular sequence outputs represent two-valued decimal numbers. Each clock pulse either increases the number or decreases the number.

DNA-quantum synchronous circuit generally refers to something which is coordinated with others based on time. DNA-quantum synchronous signals occur at the same clock rate and all the clocks follow the same reference clock. DNA asynchronous counter have shown that the molecular sequence output of that DNA counter is directly connected to the input of next subsequent counter and make a chain system, and due to this chain system the propagation delay appears during counting stage and creates counting delays. In a DNA-quantum synchronous counter, the clock molecular sequence input across all the DNA flip-flops uses the same source and creates the same clock signal at the same time. So, a DNA counter which is using the same clock signal from the same source at the same time is called a DNA-quantum synchronous counter.

DNA-quantum synchronous counter is constructed by four JK flip-flops and two DNA AND operations. Using these DNA JK flip-flops and DNA AND operations, the DNA-quantum synchronous counter creates qubits as outputs. A DNA-quantum synchronous counter produces much heat and this circuit operation needs to keep in the required environment of quantum computing.

14.9.1   Circuit Architecture

DNA-quantum synchronous counter uses three DNA JK flip-flops and one quantum JK flip-flop to create qubit outputs from molecular sequence outputs. DNA JK flip-flop has the two-molecular sequence inputs named as J and K.

DNA JK flip-flop has the molecular sequence inputs J and K. This DNA JK flip-flop consists of many DNA NAND operations. First of all J and clk inputs perform the DNA NAND operation. The output of the DNA NAND operation and output of the DNA JK flip-flop perform another DNA NAND operation and produce the output named as S. The clk input is shared. So, K and clk inputs also perform the DNA NAND operation. This DNA NAND operation's output and Q output of the DNA JK flip-flop perform another DNA NAND operation as well as produce the output named as R.

After processing the inputs in a DNA JK flip-flop, the output of the flip-flop is stored as an output of the DNA-quantum synchronous counter. Thus, the DNA-quantum synchronous counter creates four molecular sequence outputs using four DNA JK flip-flops. The clock inputs for all of the four DNA Jk flip-flops come from the same source. For this, all of the flip-flops work synchronously. One output of each of the second and third flip-flops go through DNA AND operations.

Now, the output of the DNA portion is stored in the DNA cache memory. Whenever needed the output then goes through the "NMR" process in which the EMR is emitted. Thus the molecular sequences of the superstate become the qubit which are used as inputs for the quantum JK flip-flop.

DNA-quantum synchronous counter uses three DNA JK flip-flops and one quantum JK flip-flops to create qubit outputs. DNA JK flip-flop has a two-molecular sequence input and one clock shared input. DNA JK flip-flops also depend on the clock. If the clock is enabled then the circuit will be enabled, otherwise not. The block diagram of a DNA-quantum synchronous counter is shown in Fig. 14.12.

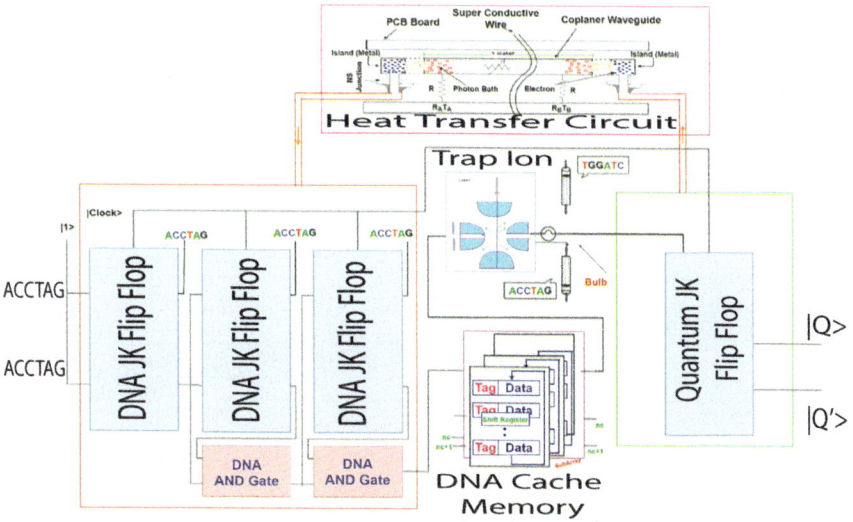

Fig. 14.12 Block diagram of a DNA-quantum synchronous counter

After processing the inputs in a DNA JK flip-flop, the output of the flip-flop is stored as an output of the DNA-quantum synchronous counter. DNA-quantum synchronous counter is the intermediate processing hardware with three DNA JK flip-flops connected in serial connection using the logical molecular sequence input. Every clock input as clk enters in every molecular sequence and the 2nd DNA JK flip-flop clk input is formed by the previous DNA Jk flip-flops output |Q>. With the same architecture DNA-quantum synchronous counter can be performed as an up counter or as a down counter, but the clock pulse as clk needs sometimes to have a positive edge triggered and sometimes a negative edge triggered.

14.9.2 Working Principle

DNA-quantum synchronous counter has one logical molecular sequence input which is shared and one clock input. DNA-quantum synchronous counter is constructed by the basic component as a DNA JK flip-flop operational circuit. DNA JK flip-flop is the two-molecular sequence circuit. DNA JK flip-flop is mostly used flip-flop in DNA computing. DNA Jk flip-flop's inputs J and K perform two DNA operations parallelly. J along with the shared input clk perform the DNA NAND operation. Then this DNA NAND operations result and one output of the DNA JK flip-flop perform the DNA NAND operation and produce S. Like DNA NAND operation, it produces K and |clk > input first. Then the output of DNA JK flip-flop |Q> and the result of DNA first NAND operation perform again another DNA NAND operation and produce R. The processes of producing S and R are executed in parallel. These S and

Table 14.8 Truth table of a DNA-quantum synchronous counter

clk	$|Q_3>$	$|Q_2>$	$|Q_1>$	$|Q_0>$
ACCTAG	$|1>$	$|1>$	$|1>$	$|1>$
TGGATC	$|1>$	$|1>$	$|1>$	$|0>$
TGGATC	$|1>$	$|1>$	$|0>$	$|1>$
TGGATC	$|1>$	$|1>$	$|0>$	$|0>$
TGGATC	$|1>$	$|0>$	$|1>$	$|1>$
TGGATC	$|1>$	$|0>$	$|1>$	$|0>$
TGGATC	$|1>$	$|0>$	$|0>$	$|1>$
TGGATC	$|1>$	$|0>$	$|0>$	$|0>$
TGGATC	$|0>$	$|1>$	$|1>$	$|1>$
TGGATC	$|0>$	$|1>$	$|1>$	$|0>$
TGGATC	$|0>$	$|1>$	$|0>$	$|1>$
TGGATC	$|0>$	$|1>$	$|0>$	$|0>$
TGGATC	$|0>$	$|0>$	$|1>$	$|1>$
TGGATC	$|0>$	$|0>$	$|1>$	$|0>$
TGGATC	$|0>$	$|0>$	$|0>$	$|1>$
TGGATC	$|0>$	$|0>$	$|0>$	$|0>$
TGGATC	$|1>$	$|1>$	$|1>$	$|1>$

R inputs execute the DNA SR flip-flop operation. DNA SR flip-flop is also basically made by the DNA NAND operation that is used here. After performing the DNA SR flip-flop operation two outputs are found. One output is the opposite of another. This DNA JK flip-flop actually removes the problem of DNA SR flip-flop. DNA SR flip-flop describes briefly in this book.

DNA JK flip-flop is triggered when the value of clock is TGGATC. So, in DNA-quantum synchronous the counter clock needs to be always high. According to the principle counters need to be toggled, that's why DNA JK flip-flop is perfect for DNA-quantum synchronous counters. DNA JK flip-flop operational circuit is triggered, when the output of DNA JK flip-flop is shared with the input of the second DNA JK flip-flop operational circuit. Now, the first DNA JK flip-flops output and second DNA JK flip-flops output enter into the DNA AND operation circuit and performs in the DNA AND operation. Then, the produced output from DNA AND operation performs in the DNA JK flip-flop operation. Again, the produced output from the first DNA AND operation and the third DNA JK flip-flop operation's output perform together another DNA AND operation. Hence the previous DNA AND operations output is shared with the DNA JK flip-flops as two inputs as well as perform the DNA JK flip-flop operational circuit. Thus, the DNA-quantum synchronous counter circuit mainly performs as a finite counter. Table 14.8 shows the truth table of a DNA-quantum synchronous counter.

Between the second and the third DNA JK flip-flops, DNA AND operation is used. The JK flip-flop's circuit output lines are connected into the DNA cache memory.

This is the first portion of the circuit and is fully designed by using the principle of DNA computing. DNA cache memory is made by using some DNA shift register and in this proposed circuit it saves data in the DNA array. When the time arrives, DNA cache memory supplies the data to the NMR process. DNA cache memory supply line fully connected to NMR process as well as here EMR is emitted. Then the output line of the NMR processing is connected to the quantum NAND operation. Two outputs of the NMR process are connected to two different quantum NAND operations. Quantum NAND operations another input comes from the outputs which are Q and $\overline{Q}$ qubits. Hence this quantum NAND operations final output produces the final output of DNA-quantum ripple counter.

14.10 Applications

The application of sequential circuits in quantum computing, DNA computing, and quantum-DNA computing or quantum biocomputing are discussed individually in the previous three chapters. So, the application of sequential circuits in DNA-quantum or bio-quantum circuits can be predicted easily.

DNA-quantum D flip-flops are also utilized in DNA-quantum shift registers, DNA-quantum ring counters, DNA-quantum sequence generators, DNA-quantum multiplexers, DNA-quantum D flip-flop-based counters, and other processing components.

The DNA-quantum SR latch circuits can be utilized as storage devices in power devices and clocks since the DNA-quantum SR latch is a single-bit storage element. DNA-quantum SR latch operational circuits are also utilized as a memory device in computers and, in certain cases, in IoT devices. In addition, latches are used in the construction of memory devices like DNA-quantum flip-flops. DNA-quantum SR flip-flops are constructed in utilizing the DNA-quantum SR latch operational circuit in DNA-quantum computing.

A 1-molecular sequence may be stored in a single DNA flip-flop. Thus, by joining a group of DNA flip-flops, the storage capacity in terms of molecular sequences may be increased. DNA JK flip-flops are widely used in computers and Internet-of-Things devices. In addition, DNA JK flip-flops may be utilized in a variety of applications, including registers, counters, frequency dividers, and event detectors.

The DNA-quantum shift register is a very much useful circuit in DNA-quantum computing. It can be used in counter, data format convertor, data processor, etc.

14.11 Summary

DNA computing or biocomputing or biological computing is one of the most exciting and difficult research topics to be explored in the modern world. DNA computing is a subset of bio-molecular computing in which the sequence of DNA molecules is used to perform a logical operation or an arithmetical operation on a computer. DNA can perform countless calculations in parallel. While classical computing quickly reaches a limit of how many parallel computations can be made, DNA computing has almost no limit. This makes it ultra fast and incredibly powerful for scenarios like machine learning.

Quantum mechanics is one of the most influential theories in the history of science and technology in the twentieth century. In addition to providing answers to some unresolved issues, it has had an impact on many current technologies by expressing a new line of scientific thinking. Quantum computing has the potential to revolutionize modern technology by accelerating computation and opening new possibilities across various fields. Its unique capabilities, like processing information in multiple states simultaneously and leveraging entanglement, enable solving complex problems that are intractable for classical computers. This translates to significant advancements in areas like drug discovery, materials science, artificial intelligence, and cybersecurity. After combining these two technologies, the new form is DNA-quantum computing or bio-quantum computing. This DNA-quantum computing which is introduced in this book is completely a new idea in the history of the modern cutting-edge technologies. This chapter has presented sequential circuits in DNA-quantum computing. Sequential circuits are common to all, but DNA-quantum sequential circuits or bio-quantum sequential circuits are a new and exciting research which are presented in this chapter.

Bibliography

1. L. Diósi, *A Short Course in Quantum Information Theory: An Approach From Theoretical Physics*, vol. 827 (Springer, 2011)
2. D. Greenberger, K. Hentschel, F. Weinert, *Compendium of Quantum Physics: Concepts, Experiments, History and Philosophy* (Springer Science & Business Media, 2009)
3. N. Isailovic, Y. Patel, M. Whitney, J. Kubiatowicz, Interconnection networks for scalable quantum computers, in *33rd International Symposium on Computer Architecture (ISCA'06)* (IEEE, 2006), pp. 366–377
4. R. Konik, Quantum coherence confined. Nat. Phys. **17**(6), 669–670 (2021)
5. M. Partanen, K.Y. Tan, J. Govenius, R.E. Lake, M.K. Mäkelä, T. Tanttu, M. Möttönen, Quantum-limited heat conduction over macroscopic distances. Nat. Phys. **12**(5), 460–464 (2016)
6. K. Saitoh, H. Yamamoto, K. Kawasaki, Y. Fukuda, H. Tanaka, M. Okada, H. Kitaguchi, Development of cryogenic probe system for high-sensitive NMR spectroscopy. J. Phys. Conf. Ser. **97**, 012141 (2008)
7. D.D. Thaker, T.S. Metodi, A.W. Cross, I.L. Chuang, F.T. Chong, Quantum memory hierarchies: efficient designs to match available parallelism in quantum computing, in *33rd International Symposium on Computer Architecture (ISCA'06)* (IEEE, 2006), pp. 378–390
8. X. Zheng, J. Yang, C. Zhou, C. Zhang, Q. Zhang, X. Wei, Allosteric DNAzyme-based DNA logic circuit: operations and dynamic analysis. Nucl. Acids Res. **47**(3), 1097–1109 (2019)

Concluding Remarks

Quantum biology has also spurred advancements in materials science, with scientists exploring the development of novel quantum materials inspired by biological systems. These materials could potentially revolutionize fields such as electronics, energy storage and sensing. Computation using quantum biology is an innovative idea in this book. Biological systems are dynamical, constantly exchanging energy and matter with the environment in order to maintain the non-equilibrium state synonymous with living. Developments in observational techniques have allowed us to study biological dynamics on increasingly small scales. Such studies have revealed evidence of quantum mechanical effects, which cannot be accounted for by classical physics, in a range of biological processes. Quantum biology is the study of such processes, and here we provide an outline of the current state of the field, as well as insights into future directions. More immediately, quantum biology promises to give rise to design principles for biologically inspired quantum nanotechnologies, with the ability to perform efficiently at a fundamental level in noisy environments at room temperature and even make use of these 'noisy environments' to preserve or even enhance the quantum properties. Through engineering such systems, it may be possible to test and quantify the extent to which quantum effects can enhance processes and functions found in biology, and ultimately answer whether these quantum effects may have been purposefully selected in the design of the systems. Importantly, however, quantum bioinspired technologies can also be intrinsically useful independently from the organisms that inspired them. Here an attempt has made of building quantum biocomputer. Quantum computers have a greater level of security, parallel computing capacity, and a quicker pace of calculation. Many intractable classical computer problems can be handled using quantum computing, which has become a fascinating concept in recent years. DNA (Deoxyribose Nucleic Acid) computing is unique among traditional computer systems because of its parallel processing, vast storage capacity, and ability to do nanolevel computing. Traditional computer

© The Editor(s) (if applicable) and The Author(s), under exclusive license to Springer   413
Nature Singapore Pte Ltd. 2025
H. M. Hasan Babu, *Quantum Biocomputing in Quantum Biology Volume I*,
https://doi.org/10.1007/978-981-97-7154-7

technologies demand more power than DNA computing. In DNA computing, logic gates offer unique properties such as stability and reusability. The characteristics of the DNA molecule aid in inducing quantum features such as superposition, tunneling, coherence, and entanglement. DNA computing or biocomputing or biological computing is a field that uses DNA molecules and molecular biology processes to perform computations, offering potential for high-density storage and parallel processing. It leverages the four DNA bases (A, G, T, C) to represent information, similar to how traditional computers use 0s and 1s. It aims to leverage the power of biological systems for computational tasks, potentially exceeding the capabilities of traditional electronic computers. Biocomputing utilizes molecules like DNA and RNA, as well as cells, to perform computational operations. Researchers engineer genetic circuits and metabolic pathways to create computational devices that can process information. Biocomputers can perform tasks like logic gates, neural network-like computations, and even solve complex mathematical problems. Biocomputing offers the potential for high-performance computing, energy efficiency, and even the ability to perform computations that are difficult or impossible for traditional computers. The innovative ideas of computing in quantum biology, that means quantum-DNA computing (quantum biocomputing or quantum biological computing) and DNA-quantum computing (bio-quantum computing or biological quantum computing) have presented in this book, combining the benefits of quantum physics with molecular biology. While creating a qubit using a DNA sequence NMR is required and NMR can be performed in two temperatures first one is 0-kelvin temperature and another one is room temperature. For these two temperatures, quantum-DNA and DNA-quantum nanoprocessors are designed in this book. RNR is used for the relaxation of the qubit to get the corresponding DNA sequence. In quantum computing operations are executed using qubit while in DNA computing DNA sequences are used as inputs. In quantum-DNA computing, to match the speed quantum cache memory is used as a buffer. Cache memory keeps the qubits temporarily to transfer it to DNA (quantum) circuits. Quantum computing has a limitation to storing data that can be solved using quantum-DNA computing. Because DNA sequences can store a large amount in a secure way. This is a reason to use quantum-DNA computing. Apart from this, DNA operations require an amount of heat while processing data and quantum operations produce heat while performing any operation. The concepts "quantum-DNA computing or quantum biocomputing or quantum biological computing" and "DNA-quantum computing or bio-quantum computing or biological quantum computing" help the produced heat to transfer from quantum computing to DNA computing. All of the arithmetic operations, combinational circuits, and speed of the operations, applications and the produced temperature in quantum biocomputing , that means in quantum, DNA-quantum and quantum-DNA computing are described in this book.

Index

H. M. Hasan Babu, *Quantum Biocomputing in Quantum Biology Volume I*,
https://doi.org/10.1007/978-981-97-7154-7

The manufacturer's authorised representative in the EU is Springer
Nature Customer Service Centre GmbH, Europaplatz 3, 69115 Heidelberg,
Germany. If you have any concerns regarding our products, please
contact ProductSafety@springernature.com

Printed and bound by CPI Group (UK) Ltd, Croydon, CR0 4YY
29/04/2026
02099466-0006